S0-GHR-630

Pesticides and Neurological Diseases

Donald J. Ecobichon

Department of Pharmacology and Therapeutics
McGill University
Montreal, Quebec, Canada

Robert M. Joy

Department of Physiological Sciences
School of Veterinary Medicine
and
Department of Pharmacology
School of Medicine
University of California, Davis
Davis, California

CRC Press, Inc.
Boca Raton, Florida

Library of Congress Cataloging in Publication Data

Ecobichon, Donald J.
  Pesticides and neurological diseases.

  Includes bibliographical references and
index.
  1. Nervous system—Diseases. 2. Pesticides
—Toxicology. I. Joy, Robert M. II. Title.
[DNLM: 1. Pesticides—Poisoning. 2. Nervous
system diseases—Chemically induced. WA 240
E19p]
RC346.E28   616.8'0471   81-15488
ISBN 0-8493-5571-0   AACR2

This book represents information obtained from authentic and highly regarded sources. Reprinted material is quoted with permission, and sources are indicated. A wide variety of references are listed. Every reasonable effort has been made to give reliable data and information, but the author and the publisher cannot assume responsibility for the validity of all materials or for the consequences of their use.

All rights reserved. This book, or any parts thereof, may not be reproduced in any form without written consent from the publisher.

Direct all inquiries to CRC Press, Inc., 2000 Corporate Blvd., N.W., Boca Raton, Florida, 33431.

© 1982 by CRC Press, Inc.

International Standard Book Number 0-8493-5571-0

Library of Congress Card Number 81-15488
Printed in the United States

PREFACE

To capture the interest of a reader and to encourage him/her to delve beyond the first few pages, it is, in my opinion, essential to explain why a book is written. In 1974, one of us (D.J.E.) was involved in the treatment and monitoring of a patient accidentally exposed to an unknown but rather substantial amount of an organophosphorus insecticide. The duration of the subtle central and neuromuscular effects of this particular agent was a suprise, "toxicity" being observed long after the patient was considered to have recovered on the basis of normal values for erythrocytic and plasma cholinesterase analysis. Psychiatric sequelae were observed for at least three months after the so-called "recovery". Episodes of fatigue and muscular weakness persisted for several months, becoming less frequent and latterly occurring only after excess exertion. Continual observation of the patient along with discussions with her physicians were never conclusive but did suggest that something was still amiss. Being intrigued that such effects could occur following exposure to a widely used, seemingly "safe" insecticide, a search of the literature was begun only to find that, in general, such signs were either unobserved, overlooked, or dismissed as having no relationship to the poisoning. In many cases, once the inhibited enzymatic activities approached normal values, observation of the patients ceased and no extended follow-up of the individuals' well being occurred. There was, however, sufficient published literature mentioning long-term physical and behavioral effects to whet the curiosity.

An examination of the literature revealed that such signs and symptoms were not restricted to acute and chronic poisoning by organophosphorus agents but were observed following exposure to chlorinated hydrocarbon insecticides, carbamate esters, fungicides, etc. This book, then, is our attempt to bring together the literature dealing with the covert toxicity of pesticides, particularly that involving the nervous systems. The aim of the monograph is to provide the reader with a relatively complete, critical, yet readable survey of what is known. We have considered animal toxicity studies in relation to human poisonings as a means of determining mechanisms of action of the various agents. We have attempted to sift good evidence from bad or inconclusive data and hope that it is clear to the reader when we are stating facts, conjecture and/or opinion.

AUTHORS

Donald Ecobichon received his undergraduate training in Pharmacy at the University of Toronto, obtaining his B. Sc. in 1960 and proceeding directly to graduate studies in Pharmacology at Toronto where he worked under Dr. Werner Kalow, using organophosphorus esters to characterize and identify electrophoretically separated human tissue esterases thereby beginning a long association with pesticides. Following his doctoral thesis in 1964 and a year as a postdoctoral fellow in Ottawa at the National Research Council of Canada, he joined the faculty of the Ontario Veterinary College, University of Guelph, and initiated studies on mechanisms of action of chlorinated hydrocarbon insecticides, then current problems in the field of veterinary medicine. In 1969, he joined the Faculty of Medicine, Dalhousie University, Halifax, Canada and, with his graduate students, initiated a number of toxicological studies involving the hepatotoxicity and structure-activity relationships of polychlorinated and polybrominated biphenyls, the biotransformation of organophosphorus ester insecticides, and the transplacental and milk transfer of chemicals in pregnant and lactating animals. In 1977, he moved to the Faculty of Medicine, McGill University, Montreal, Canada where he has continued research on the acquisition of chemicals by perinatal animals transplacentally and via milk and has become deeply involved in toxicological aspects associated with the aerial spraying of insecticides in eastern Canadian forests for spruce budworm abatement.

Robert M. Joy received his B.S. degree in Pharmacy from Oregon State University and his Ph.D. degree in Pharmacology from Stanford University. After postdoctoral training in Neuropharmacology at the University of California, Davis, Dr. Joy joined the faculty in the School of Veterinary Medicine in 1970. He is presently the co-director of the Health Sciences Neurotoxicity Unit, Associate Professor of Pharmacology and Toxicology, School of Veterinary Medicine, and Associate Professor of Pharmacology, School of Medicine at that Institution.

Dr. Joy has been active in research into mechanisms of neurotoxicity, particularly for the chlorinated hydrocarbon insecticides. He has been the recipient of numerous grants for research relating to DDT, dieldrin, and related compounds. He has authored over 30 papers in this area.

Dr. Joy remains active in the area of research and education in neuropharmacology and neurotoxicology. He is a member of many scientific societies, including the American Society of Pharmacology and Experimental Therapeutics, the Society of Toxicology, and the Society for Neurosciences. Dr. Joy is also a Diplomate of the American Board of Toxicology.

TABLE OF CONTENTS

Chapter 1
Introduction 1
Donald J. Ecobichon

I. Introduction 1

II. Historical Development 2
    A. Chlorinated Hydrocarbon Insecticides 2
    B. Organophosphorus Esters 5
    C. Carbamic Acid Esters 7

III. Toxicology 10

IV. Purpose and Scope 11

References 12

Chapter 2
Environmental Dynamics and Toxicokinetics of Pesticides 15
Donald J. Ecobichon

I. Introduction 15

II. Pesticide Economics 15

III. Environmental Distribution 17

IV. Environmental Kinetics 27

References 49

Chapter 3
The Nervous System 53
Robert M. Joy

I. Introduction 53

II. Structural Organization of the Nervous System 54
    A. The Basic Structural Elements 54
    B. The Neuron 54
    C. The Synapse 58

III. Comparative Anatomy of the Nervous System 59
    A. Phylogenetic Development 59
    B. The Insect Nervous System 59
    C. The Mammalian Nervous System 61

IV.	Electrochemical Properties of Neurons		63
	A.	Background Concepts	63
		1. Unicellular Measurements	63
		2. Ion	64
		3. Coulomb	64
		4. Ampere	64
		5. Volt	64
		6. Ohm	64
		7. Permeability	64
	B.	The Generation of a Membrane Potential	65
	C.	The Resting Potential	67
	D.	The Action Potential	67
	E.	Electrotonic Potentials	69
	F.	Synaptic Potentials	71
V.	Neurotransmission and Neurotransmitters		72
	A.	Acetylcholine	74
	B.	Norephinephrine	75
	C.	Dopamine	76
	D.	Serotonin	76
	E.	Gamma-Aminobutyric Acid (GABA)	76
	F.	Glycine	77
	G.	Glutamine	77
VI.	Functional Organization of the Nervous System		77
	A.	Sensory Systems	77
		1. Receptors	77
		2. Cutaneous, Deep and Visceral Sensation	77
		3. Visual	78
		4. Auditory	78
		5. Vestibular	79
		6. Olfactory	79
		7. Gustatory	79
	B.	Motor Systems	79
		1. Spinal Cord	79
		2. Brainstem	81
		3. Basal Ganglia	81
		4. Cerebellum	82
		5. Motor Cortex	82
	C.	The Autonomic Nervous System	82
		1. Afferent Components	82
		2. Efferent Components	83
		a. Parasympathetic Division	83
		b. Sympathetic Division	83
		3. Integrative Components	83
		4. Comparative Roles of Sympathetic and Parasympathetic Divisions	86
VII.	The Higher Functions of the Brain		87
	A.	Consciousness	88
	B.	Speech	89
	C.	Learning and Memory	89
	D.	Esthetics, Ethics, and Related Behaviors	90

References..90

Chapter 4
The Chlorinated Hydrocarbon Insecticides.....................................91
Robert M. Joy

I. Introduction..91

II. Classification and Chemistry..............................93
    A. Dichlorodiphenylethane Derivatives....................93
    B. Hexachlorocyclohexane Derivatives.....................95
    C. Cyclodiene Derivatives................................96

III. Pharmacokinetics in Biological Systems..................98

IV. Comparative Toxicity of Chlorinated Hydrocarbon Insecticides...........98

V. Neurotoxicity of the Dichlorodiphenylethane Derivatives...99
    A. Signs and Symptoms of Poisoning in Animals...........99
        1. Acute Exposure...................................99
            a. Insects....................................99
            b. Vertebrates...............................100
        2. Chronic Exposure................................101
    B. Sites of Action....................................101
        1. Insects......................................102
        2. Vertebrates..................................104
        3. The "Basic Neural Unit"......................106
    C. Effects on Integrated Central Systems..............106
        1. Cerebellum...................................106
        2. Brain Electrical Activity....................107
        3. Behavior.....................................108
    D. Mechanisms of Action...............................110
    E. Synopsis of Mechanisms, Effects, and Symptom Development......115
    F. Neurotoxicity of DDT and its Analogues to Man.....116
        1. Acute Exposure...............................116
        2. Chronic Exposure.............................118
        3. Persisting Neurological Sequelae.............118

VI. Neurotoxicity of the Cyclodiene and Hexachlorocyclohexane
    Derivatives...121
    A. Signs and Symptoms of Poisoning....................121
        1. Acute Exposure...............................121
            a. Insects..................................121
            b. Vertebrates.............................122
        2. Chronic Exposure.............................123
    B. Sites of Action....................................123
    C. Effects on Integrated Central Systems..............125
        1. Brain Electrical Activity....................125
        2. Central Sensory — Motor Reflex Status.........126
        3. Behavior.....................................129
    D. Mechanisms of Action...............................130
    E. Neurotoxicity of the Cyclodiene and Hexachlorocyclohexane
        Derivatives to Man.................................132

	1.	Acute Exposure 132
	2.	Chronic Exposure 134
	3.	Persisting Neurological Sequelae 135

VII. Neurotoxicity of Chlordecone (Kepone®) 137
    A. Toxicity to Animals 138
    B. Toxicity to Humans 139
    C. Neurotoxicity 141
    D. Treatment 141

References 142

Chapter 5
Organophosphorus Ester Insecticides 151
Donald J. Ecobichon

I. Introduction 151

II. Chemistry 151
    A. Nomenclature 151
    B. Physiochemical Properties 154
    C. Biotransformation 157
        1. Phase I Detoxifications 158
            a. Oxidative Biotransformation 158
                1. Oxidative Desulfuration 158
                2. Oxidative Dealkylation 159
                3. Thioether Oxidation 159
                4. Oxidative Dearylation 160
                5. Minor Oxidative Reactions 160
            b. Glutathione Transferases 160
            c. Hydrolases 161
        2. Phase II Detoxifications 163

III. Toxicity 164
    A. Acute Toxicity 164
        1. Animal 164
        2. Human 165
    B. Chronic Toxicity 167
        1. Human 167
        2. Psychopathological Effects 167

IV. Neurological Lesions 172
    A. Historical 172
    B. Signs and Symptoms 174
    C. Mechanisms 176
        1. Psychopharmacological 178
        2. Demyelination 180

V. Structure — Activity Relationships and Neurotoxicity 182
    A. Neurotoxic Esterase 189

VI. Skeletal Muscle Necrosis 191

VII. Treatment 192
References 193

Chapter 6
Carbamate Ester Insecticides 205
Donald J. Ecobichon

I. Introduction 205

II. Chemistry 205
    A. Nomenclature 205
    B. Physiochemical Properties 206
    C. Biotransformation 208
        1. Hydrolysis 209
        2. Oxidation 211
        3. Conjugation 212

III. Toxicity 213
    A. Mechanism of Action 214
    B. Acute Toxicity — Animal 216
    C. Subacute and Chronic Toxicity — Animal 218
    D. Acute Toxicity — Human 219
    E. Subacute Toxicity — Human 223

IV. Neurological Lesions 224

V. Treatment 228

References 229

Chapter 7
The Mercurial Fungicides 235
Donald J. Ecobichon

I. Introduction 235

II. Chemistry 235

III. Toxicity 240
    A. Acute Mercurialism 240
    B. Chronic Mercurialism 242
    C. Fungicide-Induced Toxicity in Humans 243
    D. Perinatal Poisoning 248

IV. Neurological Lesions 249

V. Treatment 253

References 255

Index 263

Chapter 1

INTRODUCTION

Donald J. Ecobichon

I. INTRODUCTION

The agricultural chemicals commonly labeled as pesticides are perhaps the largest group of poisonous substances being disseminated throughout our environment. The term "pesticide" denotes any agent intended for preventing, destroying, repelling, or mitigating any pest and this classification may be subdivided into groups such as insecticides, acaricides, nematocides, herbicides, avicides, rodenticides, molluscicides, etc. depending upon the species of pest. Historically, pesticides were developed by the age-old empirical method of trial-and-error screening of innumerable inorganic and organic chemicals which might selectively kill a species of pest. In the last 50 years, there has been a change in the approach to pesticide development, considerable attention being focussed on the physical and chemical properties of the agent and, following experimentation, selection of the most potent analogs of a particular chemical structure for further development. From such an approach, we have achieved chemical mastery over a wide range of pests, but only at a cost which even now has not been fully appreciated. It would be reasonable to state that neither could we have achieved nor can we maintain the standard of living which we enjoy today without the use of these chemicals. There is no question of stopping the annual application of these agents on our gardens and crops, our forests and fields, or on our animals and ourselves. The pesticide dilemma is not whether to use or not use — but the choice of agent, when it must be used, and how much should be applied. As has been stated recently, there is little evidence that we have really developed a philosophy on pesticides which will ensure the kind of discretionary use that the nature of these materials requires.[1]

Grain treated with organomercurial or chlorinated hydrocarbon fungicides has been responsible for some spectacular poisonings of epidemic-sized proportions. In 1956, 1960, and again in 1971 to 1972, ethylmercuric and methylmercuric chloride-treated seed grain was consumed by rural people in Iraq, resulting in severe poisoning. In the last and most serious episode, the government of Iraq acknowledged that some 6530 individuals were hospitalized and 459 died.[2,3] An outbreak of poisoning of 100 people occurred in West Pakistan in 1961 following the ingestion of seed wheat with a mixture of phenylmercuric acetate and ethylmercuric chloride.[4] The ingestion of hexachlorobenzene, a chlorinated aromatic hydrocarbon fungicide, in treated grain was responsible for the induction of several thousand cases of acquired toxic cutaneous porphyria in southeastern Turkey between 1955 and 1959.[5,6] Large scale poisonings have been caused by the insecticides DDT,[7] endrin,[8,9] parathion,[10] and malathion.[11] Parathion was also implicated as the causative agent associated with the poisoning of orchard workers in California.[12] More recently, Kepone® (1,1a,3,3a,4,4a,5b,6-decachloro-octahydro 1,3,4-methano-2H-cyclobuta [c,d] pentalen-2-one) has been implicated in the serious poisoning of chemical workers involved in its manufacture.[13]

These are only a few of the many incidents in which agricultural chemicals have been involved in widespread poisonings, but they serve to at least highlight the toxicological problems arising from these agents. It should be noted that many of these incidents were a consequence of abuse, misuse, or ignorance of the chemicals involved. The U.S. Environmental Protection Agency has estimated that each year in the U.S., some 45,000 individuals are poisoned by pesticides, some 3000 serious cases which require hospitalization, and an estimated 200 deaths occur. The fatalities primarily occur among people handling pesticides: farmers, crop-dusters, and factory workers involved in the manufacture of these chemicals.[14]

II. HISTORICAL DEVELOPMENT

As long as man has been combating pests ravaging his crops, he has been interested in anything which would selectively eliminate them. In early times, sulfur was used as a fumigant and insecticide. Inorganic salts including copper sulfate, lime sulfur (mixture of sulfur and lime), copper arsenite (Paris Green), lead arsenate, and copper sulfate mixed with lime (Bordeaux Mixture) all were found to be useful insecticides and fungicides. Extracts from tobacco leaves which contained nicotine were used as the first naturally occurring insecticides. A brief visit to any "garden" shop will confirm that many of these old preparations are still in use today. In the mid-1800s, two economically important and effective natural insecticides were introduced. Derris, the most active component of which is the structurally complex compound, rotenone (Figure 1, Structure I), was isolated from the powdered roots of *Derris ellipticus* (Malaya and East Indies), and from a species of South American plant, *Lonchocarpus*. When diluted with an adsorbant clay and dusted on plants, cattle, and sheep, the rotenoids have proven to be useful insecticides for exoparasites as well as being environmentally safe since they are readily degraded by sunlight and air to inactive oxidized products. The second natural insecticide is pyrethrum, obtained from the flowers of *Chrysanthemum cinerarinefolium* and grown commercially in several regions of the world (Kenya, Ecuador, New Guinea, Iran, Japan). The complex mixtures of esters collectively known as pyrethrins, the structure of pyrethrin I (Figure 1, Structure II) being one example, are extracted with ethylene dichloride or kerosene from dried and ground chrysanthemum flowers. While a major disadvantage of pyrethrum is a lack of persistence associated with instability in light and air, it has been and still is used successfully, alone or in combination with other insecticides or synergists, in aerosolized dispensers for the control of insects in the home. In contrast to the rotenoids which were uneconomical to synthesize, considerable success has been achieved recently in the synthesis of pyrethroid esters with subsequent modification of the chemical structure and improvement of insecticidal potency.[15]

A. Chlorinated Hydrocarbon Insecticides

The first chlorinated hydrocarbon insecticide, hexachlorocyclohexane (HCH) (Figure 1, Structure III), incorrectly called benzene hexachloride (BHC), was synthesized by Michael Faraday in 1825, though the significant insecticidal properties were not recognized until over 100 years later. Prepared by the chlorination of benzene, the product loses the aromatic characteristics of the benzene ring. While its chemical properties were studied extensively because of the interesting isomeric forms ($\alpha, \beta, \gamma, \delta$, and ε isomers) depending upon the positions of the chlorines and the "chair" (*trans*) and "boat" (*cis*) forms of the cyclohexane ring, it was not until World War I (1914 to 1918) that the insecticidal properties were considered. While it was never used at that time, related compounds such dichlorobenzene were used as fumigants.[16] While protective patents were issued in the 1930s, no real attempt was made to develop HCH until World War II when sources of commonly used derris and pyrethrum were shut off. HCH was rediscovered by both France and England between 1940 and 1942. The gamma isomer, γ-HCH (γ—1,2,3,4,5,6—hexachlorocyclohexane), commonly known as lindane, was recognized as a highly potent insecticide and was marketed in France by Solvay et Cie as "Isogam" and in England by Imperial Chemical Industries Ltd. as "Gammexane®".[17] An interesting narrative on the development and testing is pre-

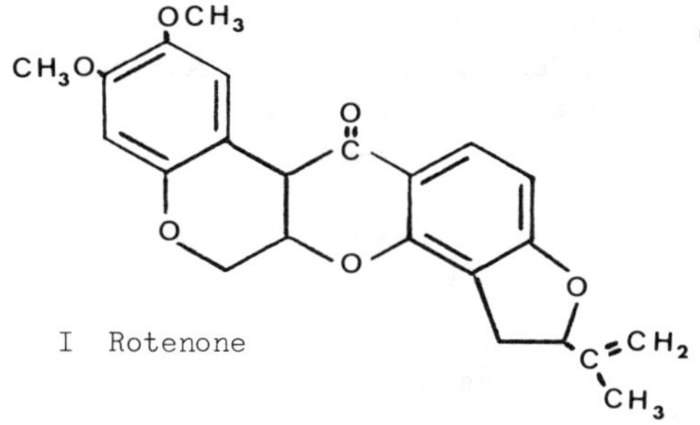

I Rotenone

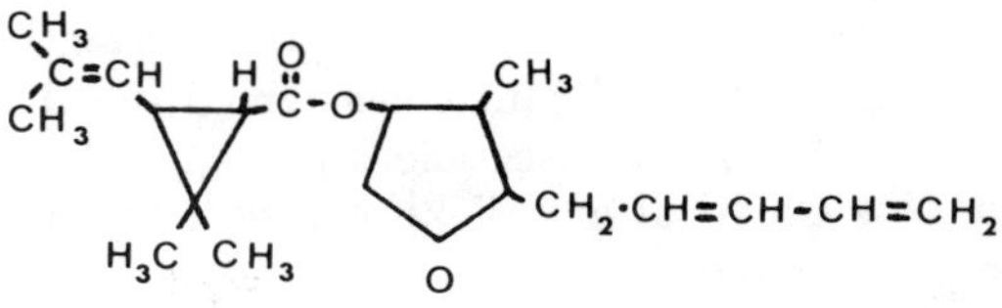

II Pyrethrin I

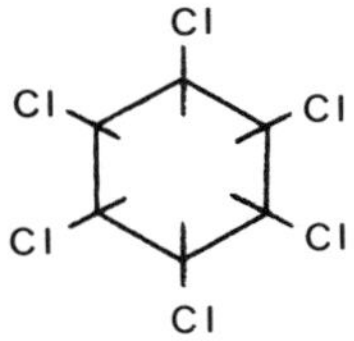

III Hexachloro-
cyclohexane

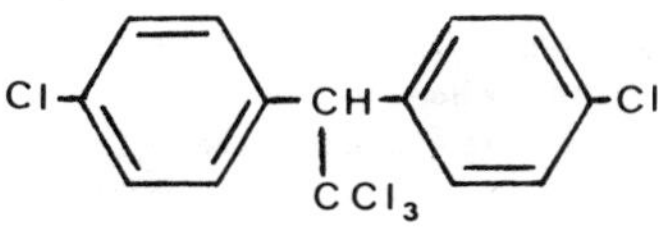

IV p,p - DDT

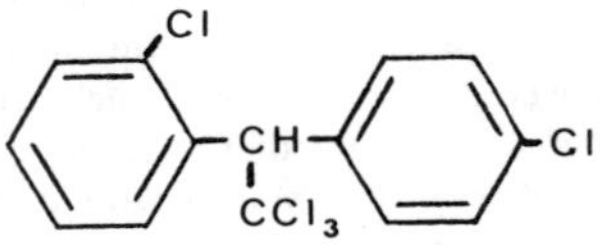

V o,p - DDT

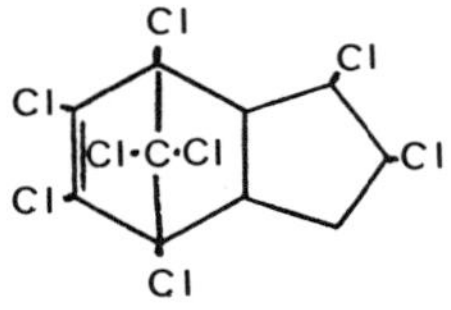

VI Chlordane

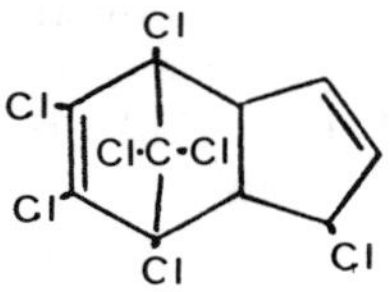

VII Heptachlor

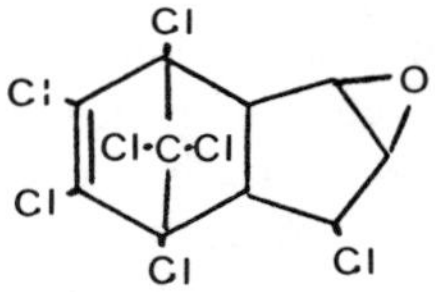

VIII Heptachlor
Epoxide

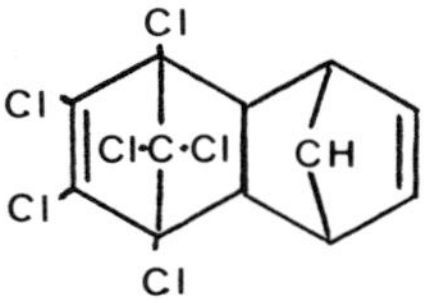

IX Isodrin (endo-endo isomer)

XI Aldrin (endo-exo isomer)

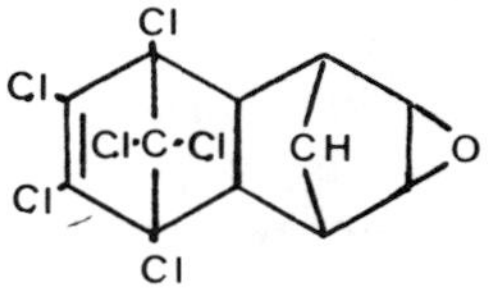

X Endrin (endo-endo isomer)

XII Dieldrin (endo-exo isomer)

FIGURE 1. Structures of commonly-used "natural" and chlorinated hydrocarbon insecticides.

sented by Brooks.[17] This chemical has been used extensively in controlling insects of importance to public health, having greater acute toxicity toward a wide variety of insects than DDT while having comparable acute toxicity to that of DDT in most mammalian species.

Dichlorodiphenyltrichloroethane or DDT (o,p′- and p,p′-isomers, Figure 1, Structures IV and V) was first synthesized in 1874 by Ziedler but the insecticidal potency was not appreciated until 1939 when Dr. Paul H. Muller of the Geigy Company of Switzerland began experimenting with this molecule and some analogs.[18] His discoveries earned him a Nobel Prize in Medicine in 1948 and his work ushered in a new era in pesticide chemistry, one in which attention was drawn to the chemical structure of the agent and to the influence which substituent groups had on lipid solubility, chemical stability, and insecticidal potency of the molecule. The story of the development of this chemical is well known but, for the uninitiated, post-DDT generation, bears repeating in part.

The search which led to the discovery of DDT began in 1932 when P. Lauger of the Geigy Company, searching for new moth-proofing agents, was experimenting with triphenylmethane dyes. The synthesis of some chlorinated benzene carbinols led P. H. Muller to synthesize 1,1,1-trichloro-2,2-(p-chlorophenyl) ethane (p,p'-DDT) and a number of analogs, testing their potency on a number of insect species. While his first paper on DDT was published in 1946, it is obvious from various records that Dr. Muller knew that he had discovered an important insecticide since patent applications were made in Switzerland in March 1940. The agent, marketed as Gesarol®, Gesaron®, and Neocid®, was tested against houseflies, cockroaches, mosquitoes and agricultural pests (potato beetle, European corn borer, alfalfa weevils, codling moth, fruit moth, etc.). Field trials for the control of typhus carried by body lice were conducted in the Balkans in 1942 in collaboration with the international Red Cross. The toxicological assessment was well in hand when news of DDT came to the attention of the Allied governments in 1942, then faced with the problems of maintaining the health of military personnel in regions where insect-carried diseases were endemic. By mid-1943, supplies of DDT were available to soldiers but the first extensive test of its effectiveness came in Naples in December 1943 where typhus threatened to become epidemic among the civilian population. Some 1,300,000 people were "dusted" at two "delousing" stations during January 1944 and medical history was made since this was the first time that a typhus epidemic was halted in mid-winter before it even started. The negligible human toxicity observed in individuals exposed to what now seem to be massive amounts of this agent greatly aided in heralding DDT as the boon of mankind. As can be observed from relevant production data, the world-wide use of this agent in the post-war years expanded apace until the mid 1960s when concerns about its environmental impact began to be expressed.[17]

With all of the fanfare which accompanied the introduction of DDT into widespread agricultural and human health use, the cyclodiene-type chlorinated hydrocarbons emerged on the scene with little attention being paid to them. This unique group of highly stable insecticides is prepared from hexachlorocyclopentadiene with cyclopentadiene by the Diels-Alder reaction giving rise to chlordene which, after further chlorination, gives rise to the potent insecticides chlordane and heptachlor (Figure 1, Structures VI and VII). The insecticidal properties of these chemicals were first described in 1945.[19] Chlordane, in particular, was a chemist's delight since it existed in two stereoisomeric, *endo* and *exo* forms and could have the chlorines on the five-membered ring on the same side or on the opposite side, giving use to *cis* and *trans* isomers, respectively. Technical chlordane, a complex mixture of compounds including heptachlor and hexachlor, was a potent insecticide. This led to the isolation and characterization of heptachlor, improvement upon the synthesis procedure to provide greater yields of this agent which was introduced for agricultural use in 1948 by the Velsicol Corporation. Heptachlor is converted to an epoxide, heptachlor epoxide (Figure 1, Structure VIII) in vivo and is stored in that form. The epoxide is much more potent than the parent compound.[20]

The most important cyclodiene insecticides are those containing four fused five-membered rings prepared by the Diels-Alder reaction between hexachlorocyclopentadiene with vinyl chloride with a subsequent reaction with cyclopentadiene to produce isodrin (Figure 1, Structure IX). When isodrin is peroxidated chemically or via metabolic pathways found in insects and various animals, endrin (Figure 1, Structure X), a commonly used and highly potent insecticide, is the product. The synthesis of these two chemicals and the ease with which the Diels-Alder reaction could be manipulated led to the synthesis of aldrin (Figure 1, Structure XI) (adduct of vinyl chloride with cyclopentadiene which is subsequently dehydrochlorinated and subjected to a second Diels-Alder reaction with hexachlorocyclopentadiene). The epoxide of aldrin, dieldrin (Figure 1, Structure XII), may be produced by chemical reaction in vitro or via metabolic pathways in vivo. Dieldrin and aldrin, named after Diels and Alder who discovered the process of diene synthesis, are the best known of the cyclodiene insecticides, being chemically stable, highly lipophilic, environmentally persistent agents which are excellent contact insecticides but possess very little, systemic action. They were introduced into the market in little over 5 to 7 years and, in spite of their wide application, little had been written about their chemistry or mode of action by 1955.[21]

Soloway stated that, to be biologically active, cyclodienes must contain two correctly separated electronegative centers which might associate with the biological site of action.[22] The molecular topogaphy of the agents was also important and, in spite of their appearance when drawn as flat structures, they do have similar shapes. The marked toxicity of the epoxides appears to be due to the orientation of the oxygen to the rings (dieldrin and heptachlor epoxide). Tragically, there are no better insecticides than the entire group of chlorinated hydrocarbons but their downfall was due primarily to the properties of (1) physical and chemical stability, (2) low rate of biodegradability, (3) lipophilicity, (4) persistence, and (5) bioaccumulation which gave them such distinct advantages as insecticides.[23]

B. Organophosphorus Esters

The story of organophosphorus esters began in the early 1800s when Lassaigne prepared these compounds by reacting alcohol and phosphoric acid.[24] Moschinin is said to have first synthesized tetraethylpyrophosphate (TEPP, Figure 2, Structure I) by heating the silver salt of pyrophosphoric acid with ethyl chloride while he was working in the laboratory of A. Wurtz. It was, however, Ph. De Clermont, in 1854, who synthesized, described, and even tasted TEPP, albeit without noting any of its toxic effects.[25] At the beginning of the 1900s, Michaelis in Rostock, Germany and Arbusov in Kazan, Russia were important figures in the field of organophosphorus chemistry, elucidating the fundamental reactions involved in the synthesis of many different derivatives. Nylen in Uppsala, Sweden reported the synthesis of TEPP by two different pathways, but was completely ignorant of the toxicity of this agent.[26] It has been noted that the yield of TEPP is quite low during synthesis unless great care is taken, and it may be this feature which (1) protected early chemists who inhaled or tasted the reaction product and (2) delayed the recognition of its toxicity for almost 80 years. In 1932, Lange in Berlin synthesized some compounds containing a phosphorus-fluorine bond (esters of monofluorophosphoric acid from silver salts and alkyl halides). During the synthesis of dimethyl- and diethylphosphorofluoridate, Lange and his graduate student, Gerda von Krueger, noted toxic effects of the vapors on themselves, the pertinent observations being included in a published chemical paper.[27] Lange was unable to convince the chemical industry, and I. G. Farbenindustrie in particular, that the alkyl esters synthesized might be useful insecticides.[26] In 1934, Gerhard Schrader was appointed by Otto Bayer to pursue the development of synthetic insecticides for I. G. Farbenindustrie, but it was not until 1936 that Schrader began working on phosphorus and sulfur acid fluorides in search of aphicidal and acaricidal compounds, initially discovering methane sulfonyl fluoride which was used as a fumigant. From 1938 to 1944, Schrader developed a series of fluorine-containing esters including DFP (di-isopropylfluorophosphate) and Sarin (1-methylethyl methylphosphonofluoridate), pyrophosphate esters including TEPP and OMPA (octamethylpyrophosphortetramide) and thio- and thionophosphorus esters including parathion (O,O-diethyl-O-[4-nitrophenyl] phosphorothioate) and its oxygen analog paraoxon (O,O-diethyl-O-(4-nitrophenyl) phosphate (Figure 2, Structures II to VI). He was aware of the toxic signs produced by these esters and, while the potency of some of these chemicals prevented their development and use as insecticides, they were of immediate interest to the German Ministry of Defense which recognized their value as chemical warfare agents. Production

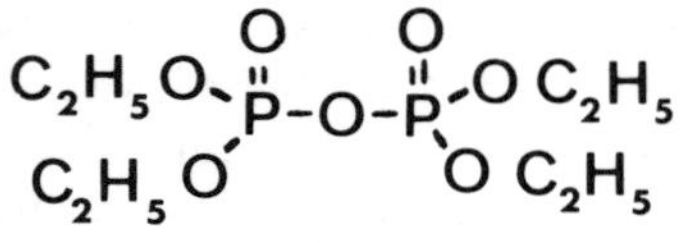

I Tetraethylpyrophosphate (TEPP)

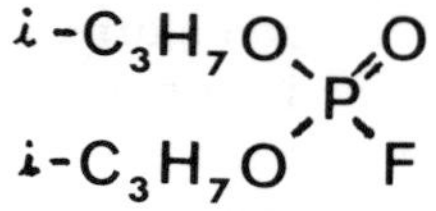

II Di-isopropyl-fluorophosphate

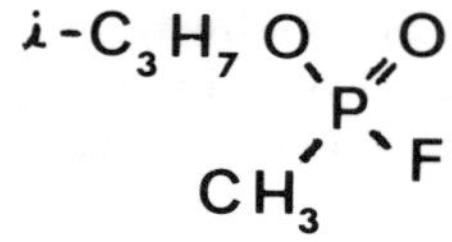

III Sarin

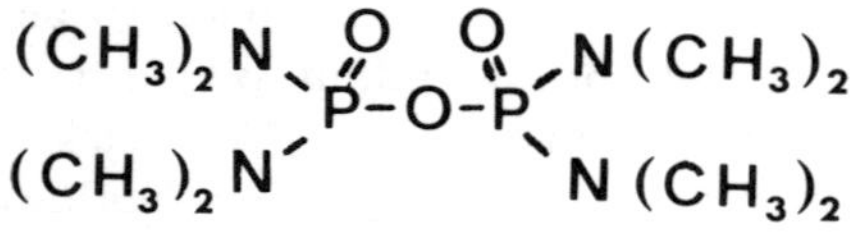

IV Octamethylpyrophosphoramide (OMPA)

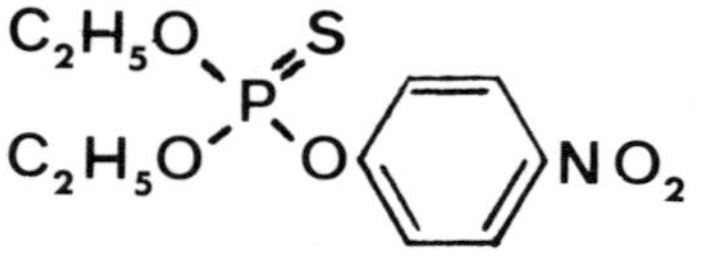

V Parathion

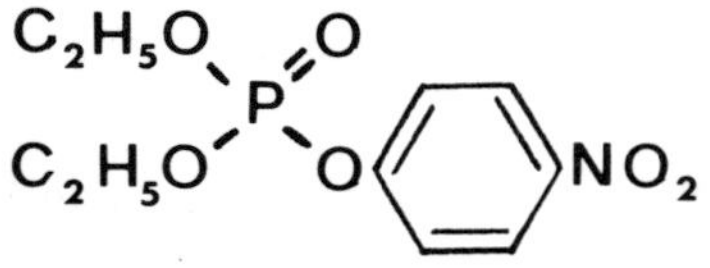

VI Paraoxon

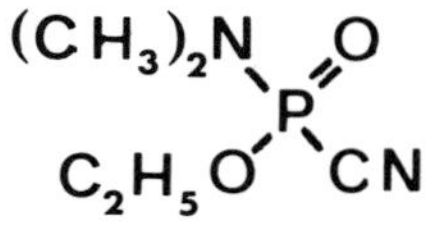

VII Tabun

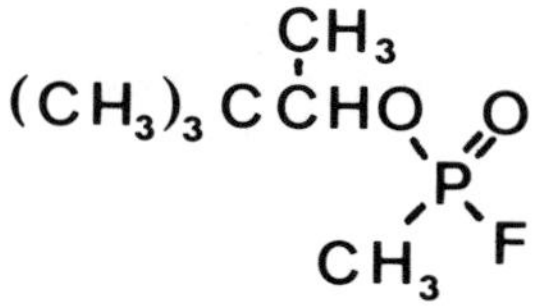

VIII Soman

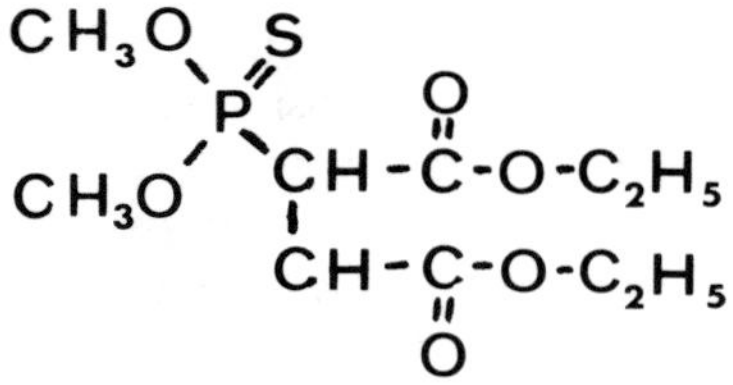

IX Malathion

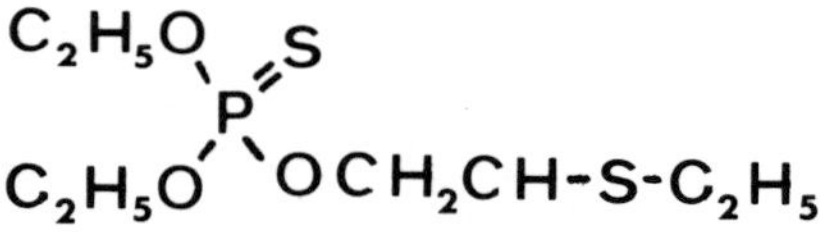

X Systox (mercaptophos)

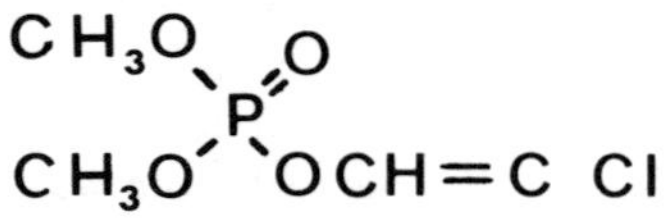

XI Dichlorvos

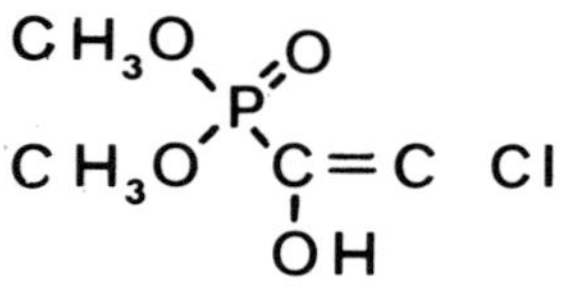

XII Trichlorfon

FIGURE 2. Structures of organophosphorus esters, showing the basic structure and the variations introduced to modify the physicochemical properties and the insecticidal potency.

of stocks of Tabun (ethyl-N, N-dimethyl phosphoramidocyanidate) (Figure 2, Structure VII) and Sarin was carried out in a factory outside of Dühernfurt, near Breslau. Soman (1,2,2-trimethylpropyl methylphosphonofluoridate) (Figure 2, Structure VIII), another "nerve gas", was also synthesized at this factory. The pharmacological and toxicological studies of these compounds were carried out in a number of industrial and military laboratories.[26]

British scientists had taken note of the comments of Lange and Krueger concerning the toxicity of acyl phosphorofluoridates, and during World War II, they were paying particular attention to fluorine-containing compounds. With this lead, it is interesting to note that studies conducted by these two protagonists were almost parallel, DFP and other alkyl phosphorofluoridates being the prime test chemicals.[28-31] A similar line of investigation was being followed at Edgewood Arsenal in the U.S., again DFP being a compound of choice in such studies.[32] Scientists on both sides of the Atlantic were well aware of the potent, irreversible, anticholinesterase properties of these esters,[33,34] When the structures and properties of the German nerve gases Tabun and Soman became known, it was realized that they were more potent than DFP by an order of two of magnitude.[33,35]

With the cessation of hostilities and the exchange of information in the post-war period, the chemistry of organophosphorus insecticides developed at a rapid rate. The decade from 1950 to 1960 can well be said to have been the era of the organophosphates. Malathion [diethyl(dimethoxyphosphinothioyl)thiobutanedioate] was introduced by the American Cyanamid Company in 1950, this ester (Figure 2, Structure IX) contains carboxy ester groups. In 1951, G. Schrader continued developing new insecticides including Systox® (demeton or mercaptophos, a mixture of the thiono- and thioloisomers of O,O-diethyl-2-ethylmercaptoethyl phosphorothioate) (Figure 2, Structure X), thereby introducing a new class of insecticides having a thioether group. In 1952, the Perkow reaction was first described in which alpha-halogen carbonyl compounds were reacted with triethyl phosphite, resulting in the synthesis of a number of new dialkylvinyl phosphate esters such as dichlorvos (2,2-dichlorovinyl dimethyl phosphate) (Figure 2, Structure XI) and trichlorfon (O,O-dimethyl[2,2,2-trichloro-1-hydroxyethyl]phosphate, Figure 2, Structure XII).[36] The thio- and thionophosphorus esters arising from parathion and containing substituted aryl and heterocyclic groups have also been synthesized. Today, a wide range of organophosphorus esters having a variety of biological properties are available for such equally diversified range of uses as insecticides, nematocides, acaricides, fungicides, etc.

Before the myriad of organophosphorus ester insecticides were introduced into agricultural practice, another class of phosphoric acid esters was known to be highly toxic. These chemicals, triaryl esters of phosphoric acid (the most common being tricresyl phosphates), have been used industrially as plasticizing agents, as additives to extreme-pressure lubricants in hydraulic systems, and as lead scavengers in gasoline. The earliest reference to the toxicity of these compounds was found to have been made in 1899, when severe polyneuritis was found among patients with pulmonary tuberculosis who were being treated with phospho-creosote.[37] Prohibition in the U.S. contributed greatly to our knowledge of these esters after an estimated 20,000 individuals who had imbibed a certain brand of alcoholic extract of Jamaica ginger suffered a debilitating peripheral neuropathy commonly known as "Ginger Jake Paralysis".[38,39] As a consequence of this epidemic, the extract was found to be adulterated with cresyl phosphate esters and the potent neurotoxin was identified as being one isomer: tri-ortho-cresyl phosphate or TOCP.[40,41] Despite our awareness of the biological effects of this chemical since 1931, epidemic-proportioned outbreaks of neuropathy continued to appear sporadically over the next 40 years, the source of TOCP usually being a contaminated cooking or salid oil. The most recent outbreak occurred in Viet Nam between 1970 and 1971, when a "black market" cooking oil was identified as a TOCP-containing military aviation lubricant supplied to South Vietnamese helicopter units.[42] TOCP will be discussed at length in a later chapter.

C. Carbamic Acid Esters

The story of "trial by ordeal", a crude form of justice practiced in West Africa during which the person suspected of being guilty was required to ingest a milky slurry of ground beans from the Calabar plant *(Physostigma venenosum)*, is well known.[43]

The unfortunate individual suffered certain death if he was unable to regurgitate the material. Studies of the seeds of this plant led to the eventual isolation and characterization of the active principle, (the alkaloid physostigmine or eserine) (Figure 3, Structure I), an understanding that the mechanism of action involved the inhibition of nervous tissue cholinesterase, and the use of this natural agent in medicine.[44,45] The active portion of the molecule was a portion of the ring structure and a monomethyl ester of carbamic acid which gave rise to a spatial and structural arrangement similar to that of acetylcholine. With the realization that the organophosphorus esters functioned because they possessed potent anticholinesterase activity, the search turned to the identification of agents with similar modes of action.

An initial but false start was made in the early 1930s by the E. I. duPont de Nemours Company with the synthesis of the dithiocarbamic acid esters, thiram, and nabam (Figure 3, Structures II and III).[46] These chemicals proved to be excellent fungicides but were not good insecticides when compared to the more potent chlorinated hydrocarbon and organophosphorus agents. The first potent carbamate insecticide synthesized was Dimetan® (Figure 3, Structure IV), a product of the Geigy Chemical Company of Switzerland. The insecticidal potency of this agent encouraged this firm to direct activity toward the development of carbamate insecticides: two dimethyl carbamic acid esters of heterocyclic enols (Isolan and dimetilan, Figure 3, Structures V and VI) being marketed in Europe in the 1950s.[43] In 1953, the Union Carbide Corporation in the U.S. introduced a different type of carbamate insecticide, carbaryl or sevin (Figure 3, Structure VII), in which an aryl (alpha-naphthol) group was substituted for the enol, and monomethyl carbamic acid replaced the dimethylcarbamate moiety.[47] This insecticide was and still is extemely popular because of its low mammalian toxicity, its environmental biodegradability, and its toxicity to a broad spectrum of insect pests. It has been used to a larger extent than any of the subsequently developed carbamates.[48]

A large number of carbamic acid esters were synthesized which inhibited nervous tissue acetylcholinesterase in vitro, but which exhibited minimal toxicity to insects when applied as a contact poison. It was discovered that all of these compounds were strong bases, highly ionized at a physiological pH, and that they could not penetrate the waxy cuticle of the insect or the ion-impermeable fatty sheath around the nervous tissue.[49] What was required was substituent groups which were less basic and more lipophilic. The group of Metcalf and Fukuto published what can be considered a classic paper dealing with the structure-activity relationships for this class of insecticides.[50] In the biological screening, several phenyl-substituted monomethylcarbamates were found to be highly toxic to insects. It was also established that the basic structure of potent carbamates should contain an aryl group. Once the optimal structure-activity relationships were established for carbamate toxicity toward insects, a large number of such esters were synthesized and tested but only 12 compounds have achieved any widespread use as contact or systemic insecticides. Primarily, these contain either an unsubstituted or substituted phenol or naphthol linked to monomethylcarbamic acid and include such agents as propoxur, Zectran®, carbofuran, pirimicarb and aminocarb (Figure 3, Structures VIII to XII). In attempting to synthesize N-methylcarbamates with a closer spatial similarity to acetylcholine, chemists of Union Carbide developed the oxime carbamate aldicarb (Temik®, Figure 3, Structure XIII) in 1965.[51,52] One additional oxime carbamate which is frequently used is methomyl (Lannate®; Figure 3, Structure XIV).

Carbamic acid esters have been playing a more significant role in the control of insect pests in the past decade and, although significant advances in carbamate chemistry have been made, two problems remain. Using the empirical approach to the design of such insecticides, useful potent agents have been discovered but often they lack

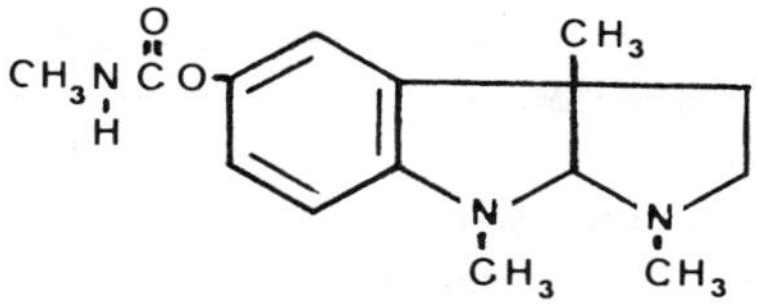

I Physostigmine

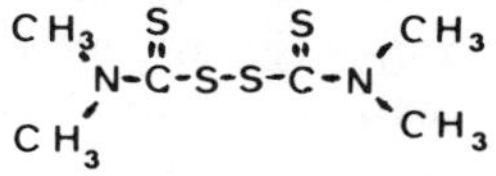

II Thiram

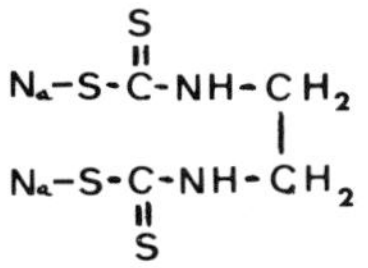

III Nabam

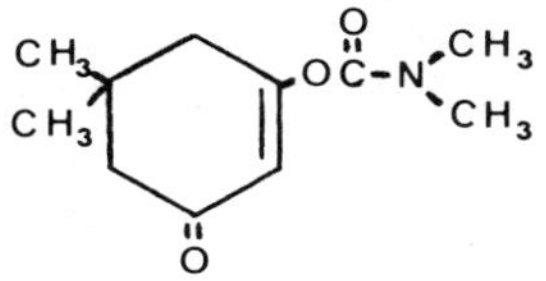

IV Dimetan

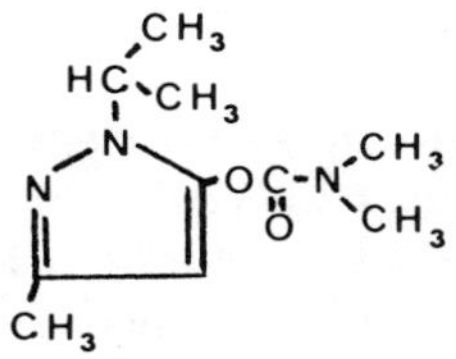

V Isolan

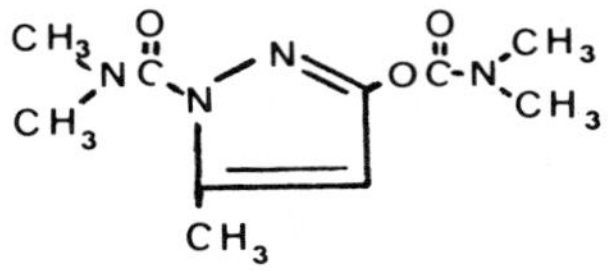

VI Dimetilan

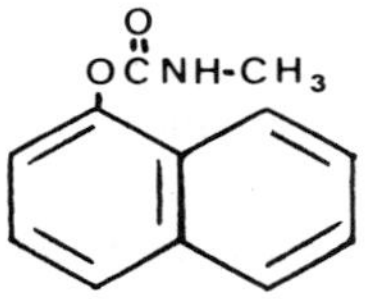

VII Carbaryl

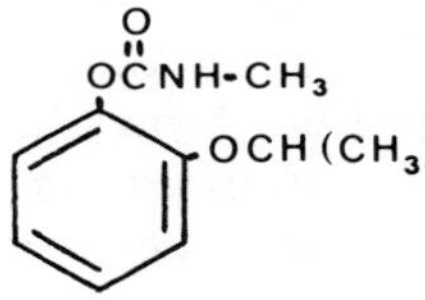

VIII Propoxur

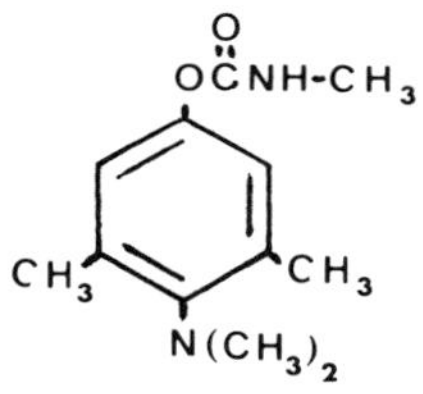

IX Zectran

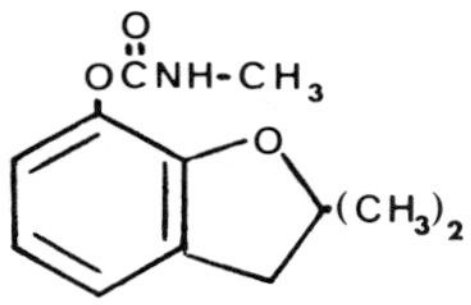

X Carbofuran

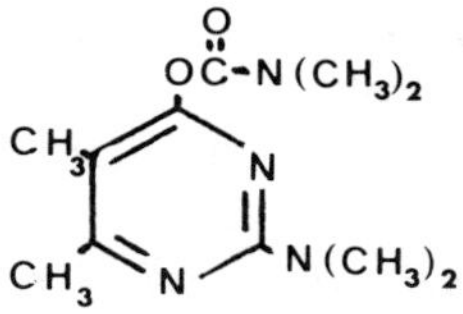

XI Pirimicarb

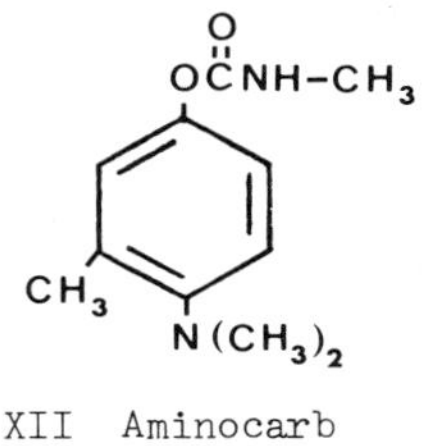

XII Aminocarb

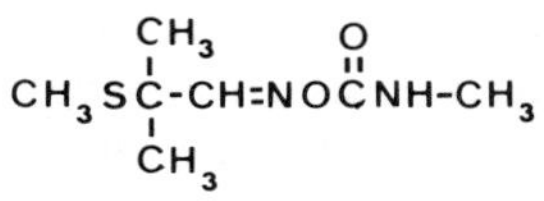

XIII Aldicarb

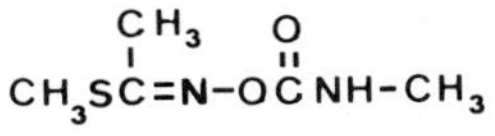

XIV Methomyl

FIGURE 3. Structures of carbamic acid esters, showing a "natural" drug (physostigmine) and the variations introduced into the basic structure to enhance the insecticidal and fungicidal activity of the molecule.

selectivity toward insect pests and, at the same time, are highly toxic to mammals. Aldicarb is a good example, having an acute LD_{50} in mice of 0.3 to 0.5 mg/kg body wt.[53] In a recent publication, Fukuto has demonstrated the necessity of abandoning the empirical trial-and-error techniques and of introducing "selectophore" groups into the molecule which will allow exploitation of the differences between the fundamental biochemistry of insects and mammals, thereby protecting the warm-blooded species.[54] This approach may contribute valuable alternatives to some of the indiscriminantly toxic agents now in use.

III. TOXICOLOGY

Toxicology is that branch of medical science that deals with the nature, properties, effects, and the detection of poisons.[55] It is the qualitative and especially the quantitative study of the injurious effects of chemicals and physical agents, as observed in alterations of structure and response in living systems, and it includes the application of such findings to the evaluation of safety and to the prevention of injury to man and all useful forms of life.[56]

Toxicology is a multidisciplinary field and may be differentiated into occupational or industrial, economic, medical, veterinary, forensic, or medico-legal and environmental toxicology. One may also partition toxicology into sub-disciplines according to the classification and/or use of the chemicals being studied, i.e., drugs, solvents or industrial chemicals, heavy metals, agricultural chemicals, pesticides, etc. The field of pesticide toxicology is a fascinating one in that it is composed of a wide variety of chemical classes having diverse physicochemical properties as well as very different and, at times, surprising mechanisms of action. The mechanisms by which some of these chemicals exert biological effects are known in great detail, but many others are still being studied in attempts to identify the biochemical or physiological event which initiates the toxic reaction. With proper use, these chemicals greatly aid man in eliminating unwanted pests and increasing crop yields, controlling disease, and improving man's own health as well as that of his domestic animals and wildlife. Used improperly, these same chemicals become a curse, causing overt, life-threatening toxicity or covert, subtle reversible or irreversible damage to organisms at every trophic level in the environment. We have witnessed the development of some of these chemicals as potent and deadly chemical warfare agents but out of these experiments came at least two classes of insecticides without which man would have been in dire straits. Other chemicals, once considered safe and a boon to mankind, we have watched pervade our environment, accumulating in food chains to levels resulting in disastrous effects to individual species occupying predatory niches near the top of these chains. The field of pesticide toxicology is an ever-changing one in which new chemicals are continually being introduced to supplant older, well-known agents as a consequence of: (1) observed resistance in pests as in the case of DDT and malathion resistance in mosquitos,[57] (2) identified covert toxicity or an unforeseen impact on the environment as in the instance of reproductive problems in birds exposed to chlorinated hydrocarbon insecticides,[58] (3) the development of a new concept in pest control, as has been seen with experiments using inhibitors of chitin synthesis or insect hormones or pheromones.[59-61]

While this volume is not to be a treatise on toxicology in general, it is important for the reader to be aware of a few definitions and concepts. Toxicology has been defined, but the discipline may be divided into acute and chronic toxicity. *Acute toxicity* denotes a situation where the effects observed are directly related to the ingestion of the poisonous substance. In contrast, *chronic toxicity* involves the continuous or intermittent ingestion of small quantities of the toxic substance over a prolonged period of time, with a slow accumulation of toxic concentrations resulting in symptoms of poisoning which may be similar to, or distinctly different from, those observed for acute toxicity. *Long-term toxicity*, which is actually a form of chronic toxicity, involves the manifestation of symptoms after a long latency period, as would be seen with carcinogens or mutagens. Chronic and long-term toxicities are of paramount concern in this day and age because of our chronic, continual exposure to low levels of a host of chemical substances in our environment (both at work and at home), in our air, food, and water. Another term, important in the field of pesticide toxicology, is *delayed lesion* which pertains to residual injury or the unexpected onset of related symptoms after a protracted period of months or years resulting from a single dose or brief exposure — the onset of symptoms requiring no further uptake of substance in the intervening interval.[62] In some instances, a single exposure (or dose) of the agent may be sufficient to initiate damage to the biochemical and/or physiological mechanisms which ultimately results in the appearance of toxicity persistent for months, years, or even a lifetime. The duration of exposure should be specified separately since the term "chronic poisoning" is somewhat confusing and, as has been suggested, should be restricted to chronic disease produced by a chemical or by chemical changes secondary to exposure.[56] Chronic poisoning may be caused by a single dose of agent and acute poisoning may follow repeated exposure to an agent.

Two independent factors, chemical and dosage, are important in acute toxicity situations. The latter parameter is most important since it has been found that every known chemical is toxic if enough of it is given. Three independent factors, chemical, dosage, and duration of dosage, plus a separate, measurable but dependent factor, storage, are intimately involved in chronic toxicity. In chronic toxicity, there is noessential relationship between the number of doses and the chronicity of the illness.
However, the mechanism of toxic action is important as is the storage of the toxic compound in the body. There is ample evidence in the literature of agents (heavy metals, carcinogens, tri-ortho-cresyl phosphate, paraquat) which indicates that these can produce chronic toxicity following the administration of a single dose. These have been called "hit and run poisons", signifying that the agent, once absorbed, has found a target tissue, has caused some finite form of cellular or subcellular damage, and then has been eliminated, leaving the cellular mechanisms crippled in some fashion and unable to cope with or repair the lesion.[63] There is also ample evidence in the literature to show that exposure to low doses of other chemicals can be just as dangerous as a consequence of poor or slow biotransformation or extensive storage in vivo; the body burden of this agent subsequently initiates the adaptation of biochemical and/or physiological mechanisms in the organism in attempts to deal with the foreign compound.

IV. PURPOSE AND SCOPE

It is not the purpose of this monograph to discuss the chemistry of pesticides. There are many excellent volumes on the subject of chlorinated hydrocarbon,[64] organophosphorus,[65-67] and carbamate[68] insecticides, fungicides, etc. and the reader is referred to these for complete discussions. Likewise, there are many fine volumes and monographs pertaining to the biochemistry, pharmacology, physiology, and toxicology of pesticides, including those of Holmstedt,[69] O'Brien,[65,70] Hayes,[71] Matsumura,[72] and Wilkinson.[73] It is not within the scope of this book to deal in depth with the pharmacological properties of pesticides other than as they pertain to the purpose of this treatise.

It is the purpose of this volume to discuss the toxicology of various chemical classes of pesticides in terms of the morphological, physiological and biochemical mechanisms related to one unique property of many of these agents, that is, the ability to cause subtle damage to the central and peripheral nervous system following either acute or chronic administration. Little attention has been paid to this aspect, perhaps since the symptoms often occur after the patient has been discharged from the physician's care or, if not severe enough, they were never recorded or were endured in silence by the patient. It is proposed to focus on pesticides and neurological sequelae, examining the consequences of acute and chronic exposure upon central and peripheral portions of the neuroaxis. It is hoped that we can convince investigators to convert acute and subacute experiments into chronic ones, to persist in observing "recovered" animals for subtle changes in behavior and performance, and to persuade clinicians to continue monitoring their patients well past the point of apparent recovery.

REFERENCES

1. **McEwen, F. L. and Stephenson, G. R.**, *The Use and Significance of Pesticides in the Environment,* John Wiley & Sons, New York, 1979, 8.
2. **Bakir, F., Damluji, S. F., Amin-Saki, L., Murtadha, M., Khalidi, A., Al-Rawi, N. Y., Tikriti, S., Dhahir, H. I., Clarkson, T. W., Amith, J. C., and Doherty, R. A.**, Methylmercury poisoning in Iraq, *Science,* 181, 230, 1973.
3. **Skerfving, S. B. and Copplestone, J. F.**, Poisoning caused by the consumption of organomercury-dressed seed in Iraq, *Bull. WHO,* 54, 101, 1976.
4. **Damluji, S.**, Mercurial poisoning with the fungicide Granosan M., *J. Fac. Med. Baghdad,* 4, 83, 1962.
5. **Schmid, R.**, Cutaneous porphyria in Turkey, *N. Engl. J. Med.,* 263, 397, 1960.
6. **Cam, C. and Nigogosyan, G.**, Acquired toxic porphyria cutanea tarda due to hexachlorobenzene, *J. A. M. A.,* 183, 88, 1963.
7. **Davies, J. E. and Edmundson, W. F.**, *Epidemiology of DDT,* Futura Publishing, New York, 1972.
8. **Davies, G. M. and Lewis, I.**, Outbreak of food-poisoning from bread made from chemically contaminated flour, *Br. Med. J.,* II, 393, 1956.
9. **Weeks, D. E.**, Endrin food poisoning. A report on four outbreaks caused by two separate shipments of endrin-contaminated flour, *Bull. WHO,* 37, 499, 1967.
10. **Karunakaran, C. O.**, The Kerala food poisoning, *J. Indian Med. Assoc.,* 31, 204, 1958.
11. **Baker, E. L., Zack, M., Miles, J. W., Alderman, L., Warren, McW., Dobbin, R. D., Miller, S., and Teeters, W. R.**, Epidemic malathion poisoning in Pakistan malaria workers, *Lancet,* I, 31, 1978.
12. **Milby, T. H., Ottoboni, F. and Mitchell, H. W.**, Parathion residue poisoning among orchard workers, *J.A.M.A.,* 189, 351, 1964.
13. **Cannon, S. B., Veazey, J. M., Jackson, R. S., Burse, V. W., Hayes, C., Straub, W. E., Landrigan, P. J., and Liddle, J. A.**, Epidemic kepone poisoning in chemical workers, *Am. J. Epidemiol.,* 107, 529, 1978.
14. **Pimentel, D. and Pimentel, M.**, The risks of pesticides, *Nat. Hist.,* 88, 24, 1979.
15. **Eliot, Michael**, The Future use of natural and synethetic byrethoids, in *The Future for Insecticides, Needs and Prospects,* Metcalf, R. L. and McKelvey, J. J., Jr., Eds., Wiley Interscience, New York, 1976, 1963.
16. **Brooks, G. T.**, *Chlorinated Insecticides, Technology and Application,* Vol. I, CRC Press, Boca Raton, Fla., 1974, 185.
17. **Brooks, G. T.**, *Chlorinated Insecticides, Technology and Application,* Vol I, CRC Press, Boca Raton, Fla., 1974, 7.
18. **Muller, P.**, Uber Zusammenhange zwischen konstitution und insektizider wirkung I, *Helv. Chim Acta,* 29, 1560, 1946.
19. **Kearns, C. W., Ingle, L., and Metcalf, R. L.**, A new chlorinated hydrocarbon insecticide, *J. Econ. Entomol.,* 38, 661, 1945.
20. **Davidow, B. and Radomski, J. L.**, The metabolite of heptochlor, its estimation, storage and toxicity, *J. Pharmacol. Exp. Ther.,* 107, 266, 1953.
21. **Metcalf, R. L.**, *Organic Insecticides,* John Wiley & Sons, New York, 1955.
22. **Soloway, S. B.**, Correlation between biological activity and molecular structure of the cyclodiene insecticides, in *Advances in Pest Control Research,* Vol. 6, Metcalf, R. L., Ed., Interscience Publishers, New York, 1965, 85.

23. Ecobichon, D. J., Chlorinated hydrocarbon insecticides: recent animal data of potential significance for man., *Can. Med. Assoc. J.*, 103, 711, 1970.
24. Fest, C. and Schmidt, K-J., *The Chemistry of Organophosphorus Pesticides. Reactivity, Synthesis, Mode of Action, Toxicology,* Springer-Verlag, New York, 1973, 12.
25. DeClermont, Ph., Note sur la préparation de quelques éthers., *Compt. Rond.*, 39, 338, 1854.
26. Holmstedt, B., Structure-activity relationships of the organophosphorus anticholinesterase agents, in *The Anticholinesterases, Handbuch der Experimentellan Pharmakologie,* Vol. 15, Koelle, G. B., Ed., Springer-Verlag, Berlin, 1963, chap. 9.
27. Lange, W. and von Krueger, G., Über Ester der Monofluorphosphorsaüre, *Ber. Dtsch. Chem. Ges.*, 65, 1598, 1932.
28. McCombie, H. and Saunders, B. C., Alkyl fluorophosphonates: preparation and physiological properties, *Nature (London)*, 157, 287, 1946.
29. McCombie, H. and Saunders, B. C., Alkyl fluorophosphonates, *Nature (London)*, 157, 776, 1946.
30. Saunders, B. C. and Stacey, G. J., Esters containing phosphorus. Part IV, diisopropyl fluorophosphonate, *J. Chem. Soc.*, 695, 1948.
31. Kilby, B. A., Kilby, M., The toxicity of alkyl fluorophosphonates in man and animals, *Br. J. Pharmacol.*, 2, 234, 1947.
32. Bodansky, O., Contributions of medical research in chemical warfare to medicine, *Science*, 102, 517, 1945.
33. Gilman, A., The effects of drugs on nerve activity, *Ann. N.Y. Acad. Sci.*, 47, 549, 1946.
34. Adrian, E. D., Feldberg, W. and Kilby, B. A., The cholinesterase inhibiting action of fluorophosphonates, *Br. J. Pharmacol.*, 2, 56, 1947.
35. Holmstedt, B., Synthesis and pharmacology of dimethylamidoethoxyphosphoryl cyanide (Tabun) together with a description of some allied anticholinesterase compounds containing the N-P bond, *Acta Physiol. Scand.*, 25 (Suppl. 90), 1, 1951.
36. Perkow, W., Umsetzungen mit Alkylphosphiten I, Mitteil: umlagerungen bei der Reaktion mit Chloral und Bromal, *Chem. Ber.*, 87, 755, 1954.
37. Lorot, C., Les combinaisons de la créosote dans de traitement de la tuberculose pulmonaire, *Thèse de Paris*, 1899.
38. Bennett, C. R., A group of patients suffering from paralysis due to drinking Jamaica ginger, *Southern Med. J.*, 23, 371, 1930.
39. Harris, S., Jr., Jamaica ginger paralysis (a peripheral polyneuritis), *Southern Med. J.*, 23, 375, 1930.
40. Smith, M. I., Elvove, E., Valaer, P. J., Frazier, W. H., and Mallory, G. E., Pharmacological and chemical studies of the cause of so-called ginger paralysis, *Public Health Rep.*, 45, 1703, 1930.
41. Smith, M. I., Elvove, E., and Frazier, W. H., The pharmacological action of certain phenol esters with special reference to the etiology of so-called ginger paralysis, *Public Health Rep.*, 45, 2509, 1930.
42. Dennis, D. T., Jake walk in Viet Nam, *Ann. Intern. Med.*, 86, 665, 1977.
43. Kuhr, R. J. and Dorough, H. W., *Carbamate Insecticides: Chemistry, Biochemistry and Toxicology,* CRC Press, Boca Raton, Fla., 1976, 2.
44. Stedman, E. and Barger, G., Physostigmine Part III, *J. Chem. Soc.*, 127, 247, 1925.
45. Englehart, E. and Loewi, O., Fermentative Azetylcholinspaltung im blut und ihre Hemmung durch Physostigmin, *Naunyn-Schmiedeberg's Arch Exp. Pathol. Pharmakol.*, 150, 1, 1930.
46. Tisdale, W. H. and Flenver, A. L., Derivatives of dithiocarbamic acid as pesticides, *Ind. Eng. Chem.*, 34, 501, 1942.
47. Weiden, M. H. J., Toxicity of carbamates to insects, *Bull. WHO*, 44, 203, 1971.
48. Matzumura, F., *Toxicology of Insecticides,* Plenum Press, New York, 1975, 6.
49. Kuhr, R. J. and Dorough, H. W., *Carbamate Insecticides: Chemistry, Biochemistry and Toxicology,* CRC Press, Boca Raton, Fla., 1976, chap. 4.
50. Kolbezen, M. J., Metcalf, R. L., and Fukuto, T. R., Insecticidal activity of carbamate cholinesterase inhibitors, *J. Agric. Food Chem.*, 2, 864, 1954.
51. Weiden, M. H. J., Moorefield, H. H., and Payne, L. K., O-(methylcarbamoyl) oximes: a new class of carbamate insecticide-acaricides, *J. Econ. Entomol.*, 58, 154, 1965.
52. Payne, L. K., Jr., Stansbury, H. A., Jr., and Weiden, M. H. J., The synthesis and insecticidal properties of some cholinergic trisubstituted acetaldehyde-O-(methylcarbamoyl) oximes, *J. Agric. Food Chem.*, 14, 356, 1966.
53. Fahmy, M. A. H., Fukuto, T. R., Myers, R. O., and March, R. B., The selective toxicity of new N-phosphorthioyl-carbamate esters, *J. Agric. Food Chem.*, 18, 793, 1970.
54. Fukuto, T. R., Carbamate insecticides, in *The Future for Insecticides. Needs and Prospects,* Metcalf, R. L. and McKelvey, J. J., Jr., Eds., Wiley-Interscience Publication, New York, 1976, 313.
55. DuBois, K. P. and Geiling, E. M. K., *Textbook of Toxicology,* Oxford University Press, Oxford, 1959, 302.

56. **Hayes, W. J., Jr.**, *Toxicology of Pesticides*, Williams & Wilkins, Baltimore, Md., 1975, 1.
57. **Matsumura, F. and Brown, A. W. A.**, Biochemistry of malathion resistance in culex tarsalis, *J. Econ. Entomol.*, 54, 1176, 1961.
58. **Stickel, L. F.**, Pesticide residues in birds and mammals, in *Environmental Pollution by Pesticides*, Edwards, C. A., Ed., Plenum Press, New York, 1973, 254.
59. **Metcalf, R. L., Lu, P-Y. and Bowlus, S.**, Degradation and environmental fate of 1-(2,6-difluorobenzoyl)-3-(4-chlorophenyl) urea, *J. Agric. Food Chem.*, 23, 359, 1975.
60. **Hajjar, N. P. and Casida, J. E.**, Insecticidal benzoylphenyl ureas: structure-activity relationships as chitin synthesis inhibitors, *Science*, 200, 1499, 1978.
61. **Wright, J. E.**, Environmental and toxicological aspects of insect growth regulators, *Environ. Health Perspec.*, 14, 127, 1976.
62. **SIPRI (Stockholm International Peace Research Institute)**, *Delayed Toxic Effects of Chemical Warfare Agents*, Almquist and Wiksell International Stockholm, New York, 1975, 1.
63. **Barnes, J. M.**, Poisons that hit and run, *New Sci.*, 38, 619, 1968.
64. **Brooks, G. T.**, *Chlorinated Insecticides*, Vols. 1 and 2, CRC Press, Boca Raton, Fla., 1974.
65. **O'Brien, R. D.**, *Toxic Phosphorus Esters. Chemistry, Metabolism and Biological effects*, Academic Press, New York, 1960.
66. **Heath, D. F.**, *Organophosphorus Poisons. Anticholinesterases and Related Compounds*, Pergamon Press, London, 1961.
67. **Eto, M.**, *Organophosphorus Pesticides: Organic and Biological Chemistry, Chemistry*, CRC Press, Boca Raton, Fla., 1974.
68. **Kuhr, R. J. and Dorough, H. W.**, *Carbamate Insecticides: Chemistry, Biochemistry and Toxicology*, CRC Press, Boca Raton, Fla., 1976.
69. **Holmstedt, B.**, Pharmacology of organophosphorus cholinesterase inhibitors, *Pharm. Rev.*, 11, 567, 1959.
70. **O'Brien, R. D.**, *Insecticides. Action and Metabolism*, Academic Press, New York, 1967.
71. **Hayes, W. J., Jr.**, *Toxicology of Pesticides*, Williams & Wilkins, Baltimore, Md., 1975.
72. **Matsumura, F.**, *Toxicology of Insecticides*, Plenum Press, New York, 1975.
73. **Wilkinson, D. F., Ed.**, *Insecticide Biochemistry and Physiology*, Plenum Press, New York, 1976.

Chapter 2

ENVIRONMENTAL DYNAMICS AND TOXICOKINETICS OF PESTICIDES

Donald J. Ecobichon

I. INTRODUCTION

As stated in the previous chapter, there is no question of stopping the annual application of pesticides. Pesticides are a necessity of life in order to prevent the extensive loss of food as well as the loss of "fiber" such as cotton, flax, forest products, etc. National regulations go to great lengths to protect people from secondary poisoning by foodstuffs contaminated with pesticides by establishing tolerance levels for residues above which legal action may be taken to restrict the marketing of contaminated produce, in addition to restricting the usage or effectively banning such chemicals. When one encounters statements that some 400 million lb of insecticides were used in 1964 in the U.S.[1] or that, in 1975, 1.3 billion lb of pesticides were applied to the environment of the U.S.,[2] it becomes apparent that we have not learned the lessons so eloquently presented by the late Rachel Carson in *Silent Spring,* published in 1962.[3] We have persisted in adding a variety of agents, not all of them pesticides, to our environment. Consider that the amount of insecticides applied per acre in southern Arizona in 1965 included 7.59 lb of toxaphene, 1.35 to 4.00 lb of DDT, 1.2 lb of endrin, and 0.58 to 1.02 lb of endosulfan covering 345,000 acres of cotton fields.[4] The amount of the organophosphorus insecticide fenitrothion applied to the New Brunswick forests by one company in 1971 constituted 43% of the total pesticides used in Canada that year.[5] In 1978, 295,000 kg of fenitrothion and 94,000 kg of aminocarb (carbamic acid ester) were applied to 1.54 million ha of New Brunswick forest in an ongoing attempt to check an epidemic of spruce budworm (*Choristoneura fumiferana,* Clemens).[6] While these are only isolated examples, they point out: (1) the continual contamination of the environment on a massive scale and (2) the chance of primary poisoning of the populace as a result of exposure to such chemicals. It has been estimated that, on a worldwide basis, accidental pesticide poisonings of humans (men, women, and children) number 500,000/year with a mortality rate in excess of 1.0%.[7] However, the need for more and more chemicals will persist as the human population continues to increase and the pressures for the production of more food, clothing, housing, etc. increase.

II. PESTICIDE ECONOMICS

The pesticide industry really dates from the end of World War II with the commercial introduction of synthetic chemical pesticides such as the chlorinated hydrocarbon and organophosphorus ester insecticides and the phenoxyacetic acid herbicides. It is of interest to examine the development and the pattern of usage of these chemicals since the late 1940s, in order to gain an appreciation of the global impact of pesticides on the environment.

Within 13 years of their introduction, some 275 million lb of chlorinated hydrocarbon insecticides were being produced annually in the U.S., as compared to 25 million lb of organophosphorus esters, 55 million lb of herbicides, and 75 million lb of fungicides (Figure 1).[8,9] From 1958 to 1967, the production of chlorinated hydrocarbons never exceeded 300 million lb annually but then began to decline, the production in

16 *Pesticides and Neurological Diseases*

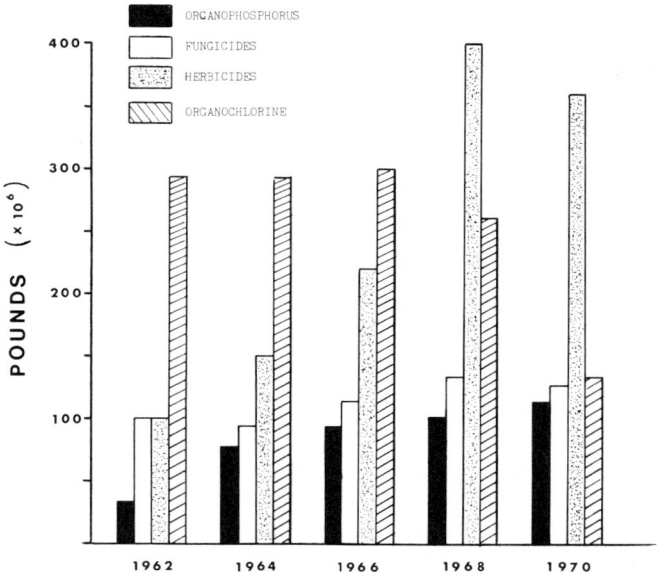

FIGURE 1. Pesticide production in the U.S. in terms of millions of pounds used per year. Data taken from Matsumura[8] and Kearney et al.[9]

1974 being slightly less than 150 million lb. The production of herbicides rose steadily from 60 million lb in 1961 to 220 million lb in 1966 and sharply increased to 400 million lb in 1968. Fungicide production remained relatively constant at 100 million lb from 1960 to 1965 but increased to a level of 130 million lb in 1971. The annual production of organophosphorus esters in the U.S. began to increase in 1962, reaching 50 million pounds in 1963 and 130 million lb in 1971. The production of herbicides remained relatively constant from 1968 to 1971 but has recently increased, production reaching 450 million lb in 1973 and 650 million lb in 1975. Today, herbicides are the most rapidly developing class of pesticides, sales being far in excess of those of insecticides. (Table 1).[10]

More recent data on world pesticide markets, including predictions for the future, can be gleaned from an examination of an economic survey reporting the market (user level) value of pesticides.[10] Data shown in Figure 2 predict that world pesticide sales will reach $8 billion by 1980 and $9.7 billion by 1984, a projected growth of 24% over the 5-year period. One can see from Table 1 that the trend in usage has continued from that shown in Figure 1, the quantity of herbicides produced and sold far outweighing that of insecticides both worldwide and in the U.S. Subdividing the insecticides into their respective chemical classes, the data shown in Figure 3 predict a marked worldwide increase in the usage of organophosphorus esters, a less dramatic expansion of the use of carbamate esters, a significant and substantial increase in the sale and use of synthetic pyrethroid insecticides.[10] In contrast, the use of chlorinated hydrocarbon insecticides will remain relatively constant on a worldwide basis. For comparison, the predicted sales of the major classes of insecticides in the U.S. market are shown in Figure 4, the trend being comparable to that predicted for world markets, except for the chlorinated hydrocarbon group of agents which will decrease sharply in 1980 in response to regulatory pressure and concern over environmental persistence.

It is of interest to look at the reasons for the distribution and usage of pesticides in the world. Herbicides will remain the major class of chemicals used in developed countries, whereas insecticides will be the most important commodities in the Third-World

Table 1
WORLD MARKETS (MILLIONS OF U.S. DOLLARS) OF VARIOUS CLASSES OF PESTICIDES

Pesticide	1974	1980	1984[a]
Herbicides	2190	3819	4668
Insecticides	1822	2575	3190
Fungicides	961	1418	1761
Soil fumigants	69	134	183
Defoliants and desiccants	19	49	68
Growth regulators	40	50	80
Phermones, attractants viruses	—	8	11
Totals	5138	8053	9961

[a] Estimated sale values from confidential survey of leading international marketers carried out and published by Farm Chemicals[10]

developing countries. Important factors contributing to the extensive use of herbicides must be reflected in the high degree of mechanization in agriculture and in high labor costs in developed countries. Pesticide sales in the U.K. can be subdivided into 66% herbicides, 20% fungicides, 10% insecticides, and 4% miscellaneous agro-chemicals.[11] Tropical and subtropical countries spend proportionally more on insecticides. It is in these regions that there is the greatest need for plant protection chemicals since they contain 49% of the world population, 46% of the cultivatable land of the world, and suffer the most extensive crop losses (40%) from pest infestation.[11] The public health problems of tropical areas also necessitate the use of large quantities of insecticides, primarily of the cheapest sort which signifies the continued use of the persistent chlorinated hydrocarbon class of chemicals rather than the more costly and short-lived organophosphorus and carbamate ester insecticides. This usage is reflected in the fact that while little DDT has been used in the U.S. since 1973, it and similar agents are still manufactured on a large scale for export to developing countries (Figures 3 and 4). The inexpensive chlorinated hydrocarbon insecticides can be used to kill adult insects by spraying dwellings and crops, whereas the newer chemicals must be used as larvicides with all of the accompanying inherent problems of environmental contamination, uneven distribution, and a lower order of effectiveness.[12] In balancing risks against benefits, the benefits of continued use of chlorinated hydrocarbon insecticides in increased food production and reduction of insect-borne disease far outweigh the risks to human health and to the environment.

III. ENVIRONMENTAL DISTRIBUTION

Despite the large amounts of pesticides manufactured and used, widespread contamination of the environment has occurred with only a few agents. With DDT, we have achieved significant environmental levels capable of causing chronic toxicity. Westlake and Gunther estimated that if the cumulative total amount of DDT produced in the world (approximately 3.5 billion lb) were spread over the total amount of arable land in the world, there would be approximately 1.5 lb of DDT/acre.[13] Considerable interest has been generated in global contamination with insecticides and this subject has been extensively reviewed.[14-16] Only the extensive use of the more persistent chlorinated hydrocarbon class of insecticides during the 1950s and early 1960s resulted in detectable atmospheric levels which assisted scientists in their development of the technology necessary for such studies.

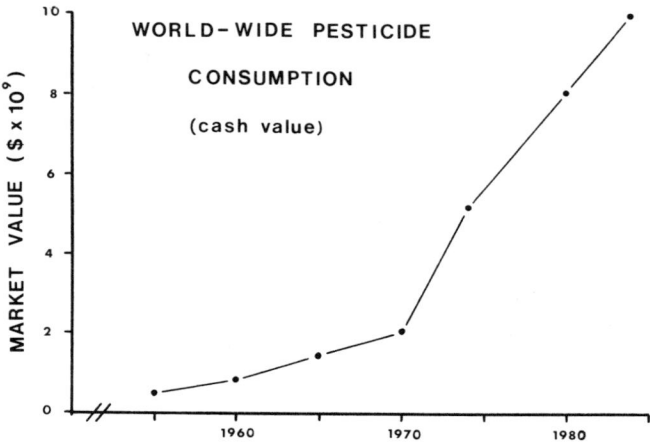

FIGURE 2. World sales of pesticidal chemicals including estimated sales in 1980 and 1981. Data originally from a market report.[10]

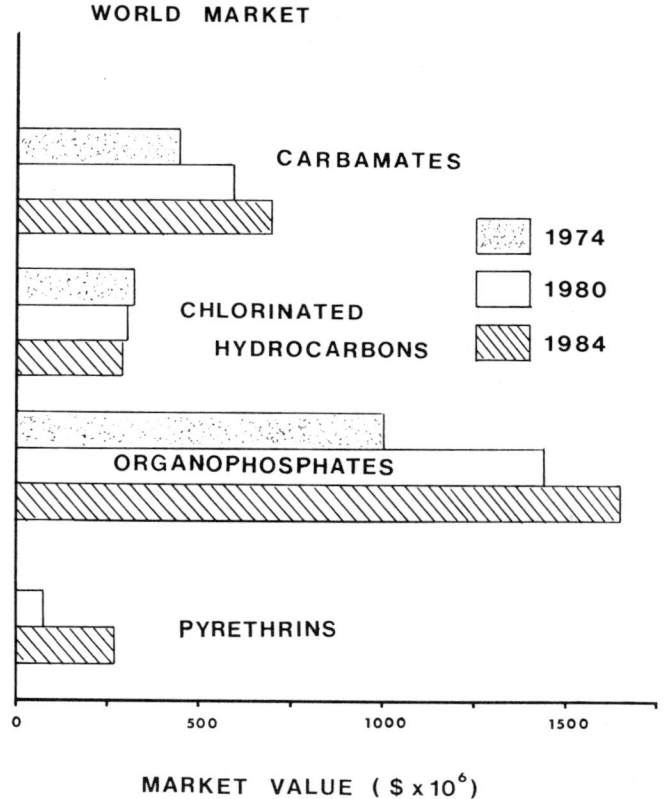

FIGURE 3. World consumption of carbamate, chlorinated hydrocarbon, organophosphorus, and pyrethrin insecticides in 1974, and estimated for 1980 and 1984. Consumption is estimated in terms of the market value in U.S. dollars.[10]

Pesticides are introduced into the environment as a consequence of their application to crops and forests from ground-based spraying equipment or from low-flying air-

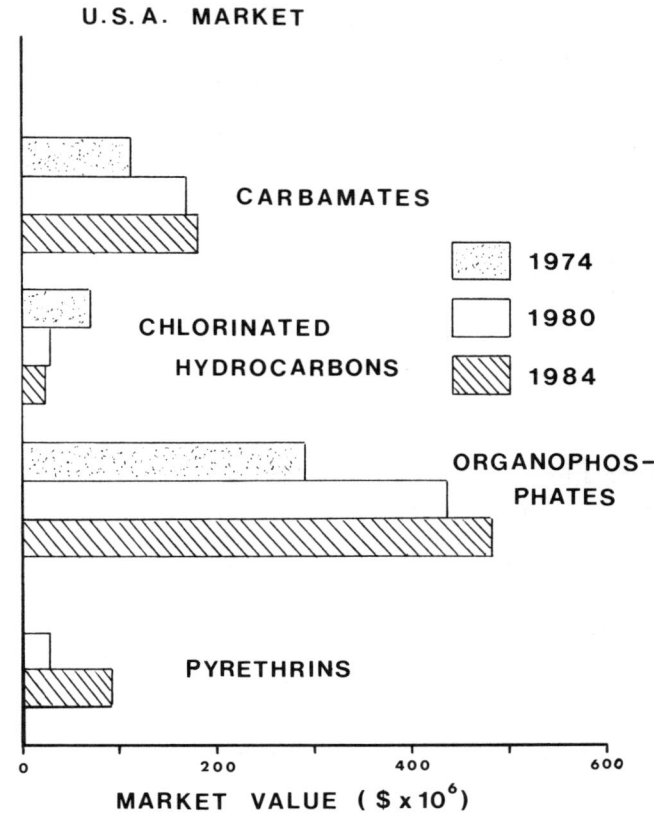

FIGURE 4. Consumption of carbamate, chlorinated hydrocarbon, organophosphorus and pyrethrin insecticides in the U.S. market in 1974 with estimated consumption for 1980 and 1984. Use is estimated in terms of the market value in U.S. dollars.[10]

craft. The off-target losses for aircraft spraying have been estimated as being between 50 to 75%.[17-19] Appreciably more drifting of pesticide occurs from mist-blowers than from aircraft.[20] The dispersal of pesticide into the air during application is dependent upon the size of the droplet produced by the equipment.[21] While the larger droplets will impinge on the nearby foliage or ground, smaller droplets will drift. Under conditions of relative low humidity, a spray droplet 125 μm in diameter loses 75% of its volume within a short distance (36 ft) of the mist generator.[22] With the dramatic decrease in volume observed with time, small droplets of 5 to 10 μm in diameter can be formed within 1 to 2 min of release from the sprayer and these will remain airborne and travel considerable distances (Table 2).[23] The smaller droplets resulting from the evaporation of the water will consist of almost pure insecticide plus emulsifiers. These small particles will be picked up by air currents and carried upward where they become diluted in the troposphere.[24]

Pesticides may reach the atmosphere by volatilization during or within hours of application, the amount lost being directly related to the vapor pressure (a measure of the volatility) of the particular compound and the ambient air temperature.[25,26] This material will be in a vapor state. Studies carried out during aerial forest spraying with DDT revealed that the highest concentration (19 ng/m^3) of DDT in the air occurred 150 m above the forest canopy some 2 hr after application.[27] Of this DDT, 20% was in the vapor state, the remainder being small particles.[27] Recent field studies with the organophosphorus ester fenitrothion have revealed that a sustained vapor plume can

Table 2
AERIAL DRIFT OF SPRAY DROPLETS OF DIFFERENT SIZES[a]

Droplet size (microns = μ)	Drift (ft)
450	8.5
150	22.0
100	48.0
50	178.0
20	11,000
10	44,000
2	21 miles

[a] Drift conditions were standardized for droplets falling a vertical distance of 10 ft in a 3.0 mph air current.[23]

be generated by volatilization from the forest for up to 10 hr.[28] It would appear that some material may be volatilized from treated surfaces, including water, the amount released being governed by ambient temperature, wind velocity and, of course, the inherent volatility of the agent. The question of codistillation of pesticides from soils along with evaporating water has been equivocal, some investigators concluding that it does not apply to pesticides below the boiling point (bp) of water, while others insist that, for agents of low solubility, the liquid phase controls the rate of vaporization and that this is a major factor in pesticide loss.[16,29]

Wind erosion of small soil particles carrying adsorbed pesticide may be an important source of atmospheric residues. Few studies have documented the significance of this source of atmospheric contamination, though one has demonstrated that insecticides (DDT, DDE, chlordane, dieldrin, and heptochlor epoxide) sprayed in western Texas on cotton crops were picked up in a dust storm and deposited the following day along with rain in a swath 1500 miles long and 200 miles wide stretching from the Gulf Coast through Cincinnati, Ohio to Lake Erie.[30] Several investigators have used oil-coated nylon mesh screens to trap air particles of 1.0 μm in diameter, presumably being blown offshore from the U.S.[31-33] Significant quantities of DDT (0.078 to 0.234 $\times$ 10^{-12}g), were measured in such dust particles. More sophisticated techniques which allowed the trapping and measurement of vapor phase as well as particulate phase insecticide (DDT) yielded much higher levels, 1.0 to 100 $\times$ 10^{-12} g/m^3 being detected.[33,34] Studies carried out in agricultural regions have measured the persistent presence of chlorinated hydrocarbon insecticides in the air in all samples collected over a 3-year period, whereas the organophosphorus esters were detected during the growing season.[35] As DDT was phased out of use, the levels declined from 99.5 $\times$ 10^{-9}g/m^3 in 1972 to 11.9 $\times$ 10^{-9}g/m^3 in 1974. While the inhalation of pesticides can be toxic, the levels of such agents as DDT found in air would lead to an estimated intake of 0.2 to 32 μg of DDT, a quantity well below that estimated to be toxic by this route.[36] Similar calculations can be made for other insecticides. It should be recognized that field workers would be exposed to far higher levels resulting from volatilized pesticide from crops. These aspects have been reviewed recently by McEwen and Stephenson.[37]

By far, the pesticide to receive the most attention concerning environmental dispersion has been DDT, partly because it has been environmentally persistent, easily detected, and for some 20 years was the most widely used agent. Based on the assumptions that (1) the amount of DDT produced up to 1974 was some 2.8 $\times$ 10^{12}g and (2) approximately 2.7 $\times$ 10^{10}g were used annually (0.28 lb/acre), Woodwell et al.[14] have developed an elaborate atmospheric model for the mass balance of this insecticide (Figure 5). How DDT reaches the soil or becomes localized in the atmosphere has been mental concentrations to initiate selective toxic effects.[14] Was the plateau ever reached with DDT or would the residues have continued to climb with continued use? We will never know since, when the apparent environmental problems of this chemical were finally recognized, its use was drastically curtailed in the mid-1960s. With the possible exception of dieldrin, the environmental persistence and chemical stability of other chlorinated hydrocarbons as well as organophosphorus and carbamate insecticides has been considerably less than that of DDT and, except in regions of high intensity use, these chemicals have created fewer environmental problems. This does not mean that incidents of environmental contamination and subsequent serious poisoning have not occurred with other agents, as has been observed recently involving parathion applied to crops in Texas and poisoning in a breeding colony of laughing gulls.[44] The absence of global scale problems with these other agents is also reflected by the fact that, during the mid-1960s at the peak of the chlorinated hydrocarbon usage, the concentrations of DDT being used far exceeded those of the organophosphorus and carbamate insecticides combined (Figure 1).[1,9]

The "bioaccumulation" of an agent depends upon the dose, the dose interval, and the half-life of the agent. If the half-life is small in relation to the dose interval, the chemical may be completely eliminated before the next exposure occurs. When the half-life is of the same order of magnitude as the dose interval, a quantity of chemical will still be present at the end of each dose interval. A subsequent dose of agent then results in an appreciably higher blood plasma and eventual tissue concentration than did the previous dose. Without terminating the exposure, there is no way in which the organism can be rid of the agent in the body tissues and the organism is said to have a "body burden" of the chemical. This concept was shown diagrammatically in Figure 6. Another striking example of bioaccumulation which caused marked toxicity in Lake Michigan gulls is shown in Figure 8.

A dramatic increase in the number of deaths of herring gulls at nesting colonies in Lake Michigan led Hickey et al.[45] to measure the pesticide residues at different trophic levels in the ecosystem in the Green Bay area since the neighboring region of Wisconsin contains the bulk of the cherry orchard industry of that state. The annual insecticide usage in this region was high, some 30 tons of DDT, 15 tons of methoxychlor, 15 tons of DDD, 7.5 tons of parathion, and 22.5 tons of malathion being applied annually (1 ton = 1016 kg). The results demonstrated that the insecticides did not stay on the land but gained access to the ecosystem, most likely by runoff, and concentrated in the various layers of the Lake Michigan animal pyramid. Mortality was observed in two species of predaceous gulls at the end of the food chain. In addition, a more subtle form of toxicity, equally as important as death, was the effect on reproductive success with large numbers of unhatched eggs and dead chicks being found in nests. It should be realized that it is not necessary to kill animals to make a species extinct, interference in the delicate balance of reproduction serving just as well. The body burden of pesticides in the gulls, a fraction of which would be mobilized from adipose tissue at breeding time, would consequently be transferred to the developing embryo and result in embryotoxicity or decreased viability of the newly hatched chick. It is important to note that the chlorinated hydrocarbon insecticides appeared to be the causative agents in this environmental problem. The organophosphorus esters, parathion and malathion, while used in high concentrations, did not persist in the environment and were not detected in body tissues. However, we should not ignore the potential of some organophosphorus esters to persist in the environment, to bioaccumulate, and to cause serious toxicity, given the proper environmental setting and conditions, as has been demonstrated for parathion in at least two separate episodes.[44,46] Recent studies have demonstrated that, compared to some other organophosphorus esters, parathion is relatively stable with a biological half-life (the hydrolysis rate measured at pH 7.4 and 20°C) of 130 days.[47]

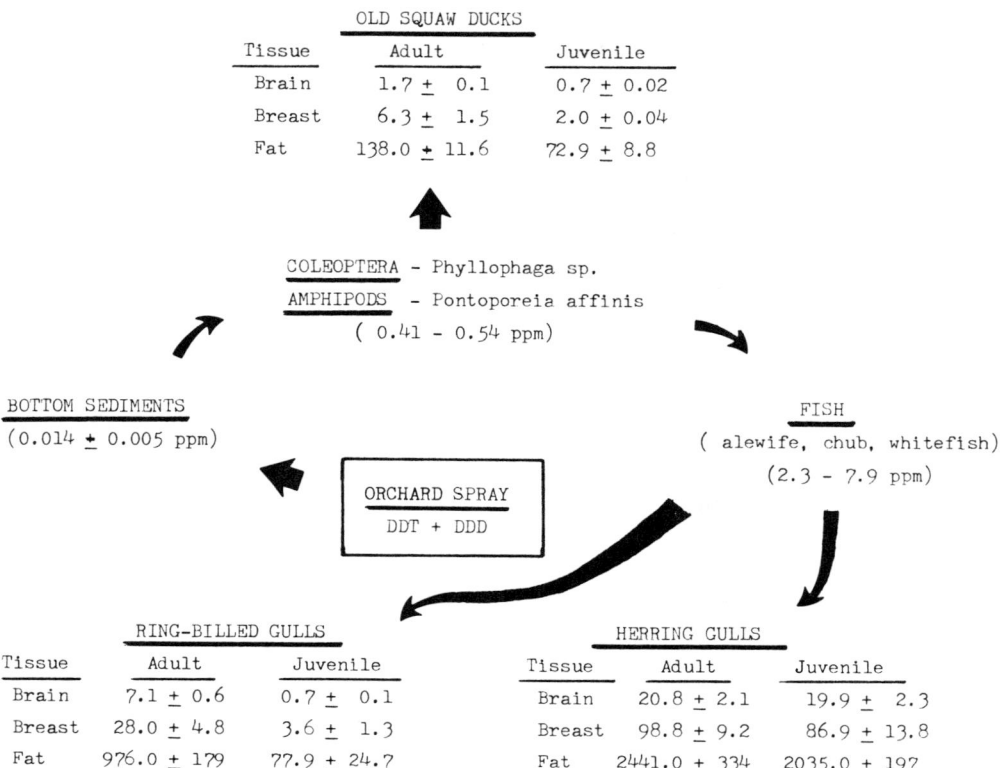

FIGURE 8. The bioaccumulation of DDT in various trophic levels of wildlife in Lake Michigan (Green Bay) as a consequence of repeated annual spraying of Wisconsin fruit orchards. The results taken from Hickey et al.[45] show not only the bioaccumulation as one ascends the food chain but also the tissue distribution of the insecticide residues in the upper level predaceous species.

The terms "bioaccumulation", "biological magnification", or "bioconcentration" have been used to describe the phenomena shown in Figures 6 and 8 and may be defined as "the amount of agent accumulated by an organism by absorption and adsorption via whatever route of entry which results in an increased concentration of that agent by the organism or specific tissues."[48] The initial concentration of the agent in water or soil is generally quite low but, as one progresses upward through a food chain or web, the agent accumulates, the chemical being stored in particular tissues (soft tissues such as liver, kidney, adipose tissue, nervous tissue) with the concentration usually bearing a direct relationship to the lipid content of the tissue or organism. By this mechanism, the persistent chemical can be selectively stored, can escape degradative processes in vivo and lead to a body burden of the agent. Thus, individuals at each level in a particular food chain generally harbor a higher body burden of the agent than those upon which they are dependent for food.

If the use of an agent is stopped abruptly, there will be a corresponding change in the dynamics of the chemical in the food chain. The disappearance of pesticides from the water is quite rapid, being of the order of 1 to 3 days and governed by the rates of hydrolysis, photolytic, and chemical decomposition, natural filtration or extraction by various species of aquatic biota. Disappearance from various animal species or plants will be much slower, of the order of 3 to 18 months, depending upon the nature of the chemical, the extent of the storage, and the ability of the organism to mobilize the agent from storage sites and biotransform it into readily excreted products. Soil can act as a significant reservoir for such chemicals, the persistence of the highly lipid-soluble, chlorinated hydrocarbon insecticides (dieldrin/aldrin, chlordane, lindane, endrin, DDT) being measured in terms of 3 to 10 years for the disappearance of 95% of the quantity applied annually.[43,49-51] As a class, the organophosphorus and carbamic acid esters are much less persistent in soil and one is talking in terms of weeks or months rather than years for their disappearance, the agents having been destroyed before the next planting season.[52-54] The herbicides vary greatly in their persistence in soil, 75% of the residues of the urea, triazine, and picloram type agents disappearing in 3 to 18 months; in 1 to 6 months for the phenoxy, toluidine, and nitrile group; 2 to 12 months for the benzoate and amide classes of compounds; and 2 to 12 weeks for carbamate and aliphatic acid herbicides.[53] It is beyond the scope of this text to discuss the environmental problems of pesticidal agents fully and there have been any number of pertinent papers and excellent chapters devoted to this subject to which the reader is referred to obtain an overview of this topic.[1,2,55]

IV. ENVIRONMENTAL KINETICS

If one reduces the scale of time shown in Figure 7 from years to days or weeks, one can apply this dose-response relationship to the pharmacokinetic or toxicokinetic parameters of an agent in a food pyramid or to an individual species or organism occupying a niche in that ecosystem. In this section, we shall initially examine the ecosystem and its components, proceed to a consideration of kinetic models, and then discuss these models in terms of the fit of experimental data.

In the simplest of terms, the "toxicokinetics" of an environmental system can be considered to consist of a series of two-compartment models as is shown in Figure 9, consisting of central and peripheral compartments. The central compartment is essential for biotransformation and elimination while the peripheral compartment is associated with storage or sequesration. Theoretically, the organisms at each level can be considered to be composed of an infinite number of compartments, each of which functions as an independent unit in relation to the tissue concentrations of agent. Beginning at the base of the food chain, one finds small organisms which can easily equilibrate with their environmental surroundings (i.e., the water), and have a low lipid content and store very little agent. As one ascends the food chain, the biomass of the predators increases, with the peripheral compartment becoming larger and more important in that more extensive storage of the agent occurs in tissues. The lipid fraction of the biomass contributes greatly to the size of the peripheral compartment. Toward the top of the food web, one finds little of the agent escaping or being eliminated from the particular organisms and the peripheral compartment, consisting of skin, subcutaneous fat, abdominal adipose tissue, and muscle, may constitute a significant portion of the body mass available for storage. Toxic levels of lipid-soluble chemical may be found, at least toxic to the predaceous species at the apex of the ecosystem.

The rate(s) of appearance or disappearance of the chemical from a particular trophic level, as is shown in Figure 10, is dependent upon a number of factors including the number of available mechanisms of biotransformation (degradation), the rate of degradation, the rate of storage in and depletion from specific compartments, and, in the case of animals of all sorts, the capacity of the organism to form degradation products which can be eliminated. In many instances, whether accumulating or disappearing from an ecosystem, the chemical will appear to follow the law of "first order kinetics", where the rate of change of the chemical is proportional to the concentration. This can be described mathematically by the following linear differential equation

$$dc/dt = kc \qquad (1)$$

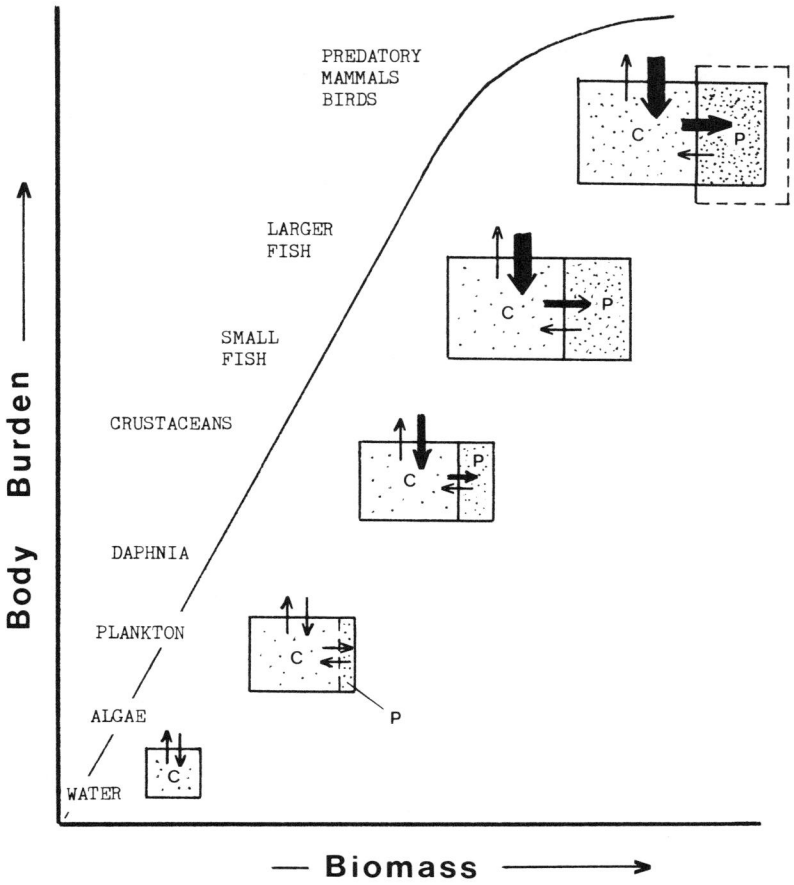

FIGURE 9. A schematic diagram illustrating bioaccumulation of a persistent chemical in a food chain using the concept of open two-compartment models at each level to demonstrate uptake and elimination from a central compartment "C" and storage in a peripheral compartment "P", presumably adipose tissue and muscle, which becomes larger and plays a more dominant role at the higher levels.

where k is a constant of proportionality, c is the concentration of the agent and dc/dt is the rate of change of c at any time t. If the agent is accumulating (Figure 10), dc/dt will have a positive sign. The constant k may be multifactorial and involve parameters of biotransformation and degradation, deposition and mobilization from body compartments or trophic levels, etc. One can think of "environmental elimination" in terms of the pharmacokinetic principles studied extensively in mammals, but one must recognize that the compartmentalization becomes more complex and includes chemical photolysis and degradation, evaporation or volatilization, and transport from one compartment to another in addition to biodegradation by animals, microorganisms, plants, etc. and the eventual disappearance or dilution into vast environmental compartments such as the air, oceans, etc.[14]

The above equation becomes extremely useful in assessing how long a period of time is necessary to eliminate a chemical from a particular "compartment". If c_o is the concentration of c at some instant in time t_o, the above equation can be integrated to form the exponential equation

$$c(t) = c_o e^{-kt} \qquad (2)$$

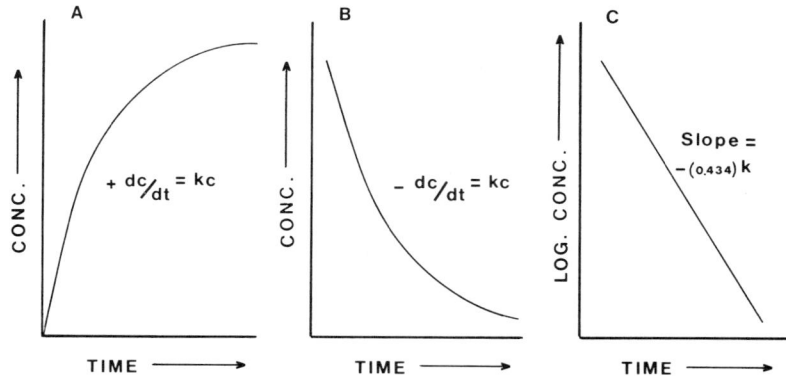

FIGURE 10. Schematic diagrams showing (A) the rate of appearance of an agent in a body "tissue" accompanied by the first-order kinetic equation; (B) the rate of disappearance or elimination of an agent from a body "tissue" and the appropriate first-order kinetic equation; and (C) the semilogarithmic transposition of the data thereby establishing the linear relationship characteristic of a first-order reaction. The values "c", "t" and "k" are the concentration of agent, time and the constant of proportionality (rate constant) respectively.

expressing c explicitly as a function of time. The concentration of c can be calculated at any point in time if one knows the value of c_o and of k. By using the logarithms of the values on both sides of the equation, one can transform the exponential equation into a linear one

$$\ln c(t) = \ln c_0 - kt \quad \text{(natural logarithms)} \qquad (3)$$

$$\log c(t) = \log c_0 - (0.434)kt \quad \text{(base 10 logarithms)} \qquad (4)$$

Taking the data shown in Figure 10B and plotting the logarithm of the concentration of c as a function of time, a linear relationship would be found with an intercept log y_o and a slope (0.434)k (Figure 10C). The rate of disappearance is traditionally characterized by the half-life ($t_{1/2}$ which can be determined easily from the linear relationship just established (Figure 11). If t_1 and t_2 are two points in time such that the concentration of c_2 is ½ that of c_1, then Equation 2 becomes

$$\frac{c(t_2)}{c(t_1)} = \frac{1}{2} = e^{-kt(t_2 - t_1)} \qquad (5)$$

which when converted to logarithmic units becomes

$$t_2 - t_1 = \frac{\ln 2}{k} = \frac{\log 2}{0.434k} = \frac{0.693}{k} = t_{1/2} \qquad (6)$$

In a practical way, point t_1 can be selected arbitrarily, making note of the concentration of c_1 at that particular time. Taking the value c_2 as being 50% of c_1, time t_2 can be estimated from the line, the interval between these two time points being the half-life. As can be seen from the above equation and Figure 11, $t_{1/2}$ is a constant whose value is independent of the concentration of chemical but is dependent on the rate constant k. A complete and easily understood description of these pharmacokinetic principles has recently been published by Gladtke and von Hattingberg[56] or in considerably more detail by Filov et al.[57]

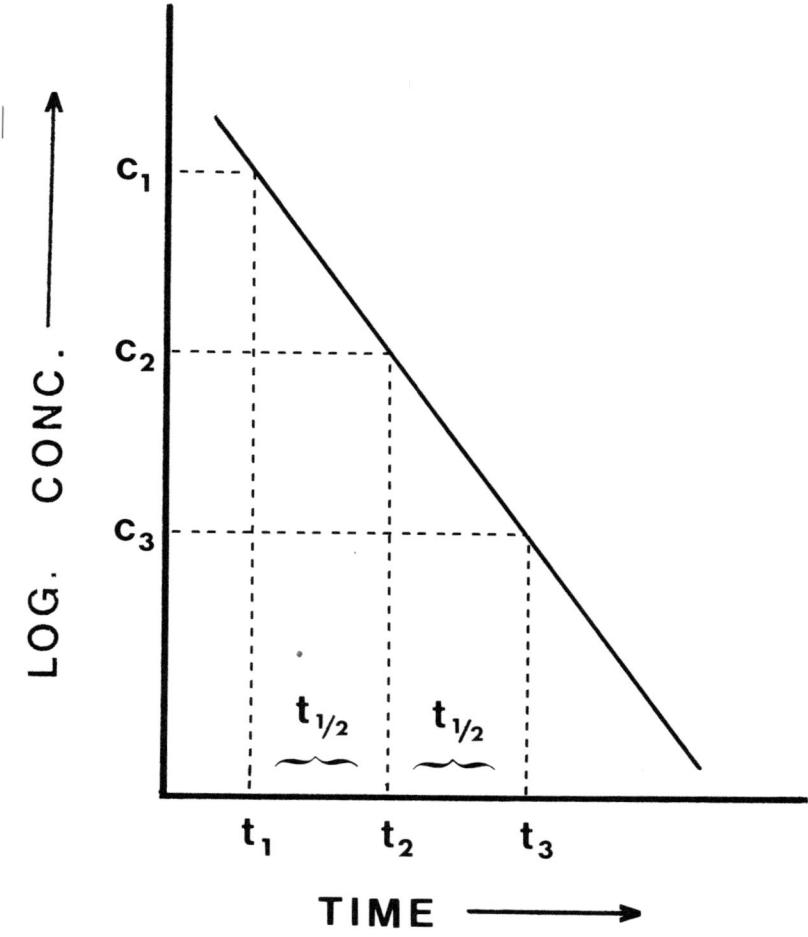

FIGURE 11. A schematic diagram illustrating the method of determining the biological half-life of an agent in a body tissue from a semilogarithmic plot of concentration of agent "c" against time "t". The linear relationship insures that the disappearance is a first-order reaction.

Taking a closer look at the exponential disappearance curve, one can readily see that it is a biphasic curve (Figure 12A). The initial, apparently rapid phase represents the redistribution or dissipation of the readily available, loosely associated residues in the compartments. The slower phase of elimination represents those residues which have been sequestered in storage compartments (adsorbed to soil particles, absorbed in plant material, deposited in adipose tissue, etc.) and which have escaped from the factors which contribute to the destruction and/or dissipation of the chemical.[48] Invariably, unless all of the mechanisms associated with elimination (disappearance), are saturated, the elimination will follow a curvilinear relationship as is shown. Replotting the concentrations of chemical as logarithmic functions, a linear relationship will be observed, signifying a first-order kinetic relationship (Figure 12, insert A'). If the environmental concentrations of chemical are sufficiently high enough to saturate the sum total of the elimination mechanisms, a curvilinear relationship will be observed (Figure 12B) which will remain so when plotted on a semilogarithmic scale (Figure 12, insert B') until the concentration of agent decreases to a value at which the rate of elimination becomes somewhat more simplified in terms of the constant k.

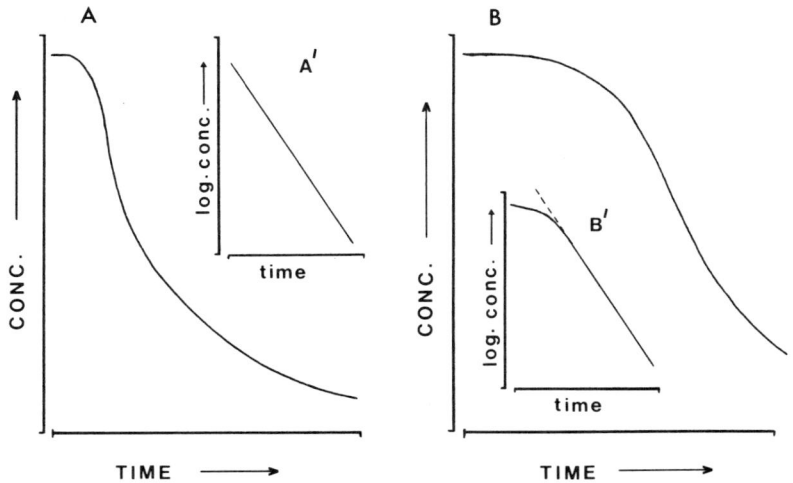

FIGURE 12. Schematic diagrams illustrating the elimination of agents by (A) a first-order mechanism and (B) by a mechanism which is initially saturated (zero-order) and subsequently becomes first-order in nature.

Prior to examining some practical samples of what has become known as "toxicokinetics", it would be valuable to look at some simplified compartmental models frequently employed in such studies which could be adapted for use in the environment. Toxicokinetic models depend on the ability to fit theoretical curves to experimental data obtained by the analysis of chemical residues in serial samples (blood, tissues, air, soil, plants, etc.) collected from "compartments" of a body, ecosystem, etc. The pharmacokinetic literature dealing with compartmental models is extensive and the reader is referred to Wagner,[58,59] Levy and Gibaldi,[60] Gehring et al.,[61] and Gladtke and von Hattingberg[56] for detailed accounts. A paper by Tuey in a recent book edited by Hodgson and Guthrie[62] has provided an interesting and simplified account of various toxicokinetic models.

If we make the assumption that the rate of absorption of a chemical is first-order (being dependent upon a rate constant and the concentration of chemical), the simplest model would be one-compartment model with the chemical uniformly distributed within this compartment (Figure 13A). The constants k_a and k_e are the rate constants for adsorption and elimination, respectively. The kinetic picture is shown for a single dose, $c_a(t)$ being the concentration of chemical at the adsorption site at time t, $c_a(t)$ being that concentration in the compartment at t and $c_e(t)$ being the cumulative concentration eliminated with time t. If a chemical is ingested chronically and is slowly eliminated from the body, the dose can be treated as though it was a constant input, giving rise to the model and kinetic picture shown in Figure 13B. Following a time interval equivalent to 5 or 6 half-lives, a plateau concentration of agent in the compartment will be reached whereby input will be equivalent to output at that particular dose.

Many chemical toxicants do not distribute evenly throughout the body as a consequence of their physicochemical properties or physiological properties of the organisms (i.e., blood flow or perfusion, lipid content, capillary bed, etc.) which cause the agent to be sequestered in certain organs. Such observations have led to the development of the *two-compartment open model* (Figure 13C). In animals, the central compartment consists of the blood and tissues which are highly perfused and which rapidly equilibrate with the blood (viscera, heart, liver, kidney). The second compartment, often called the peripheral compartment, corresponds to poorly perfused tissues such as the

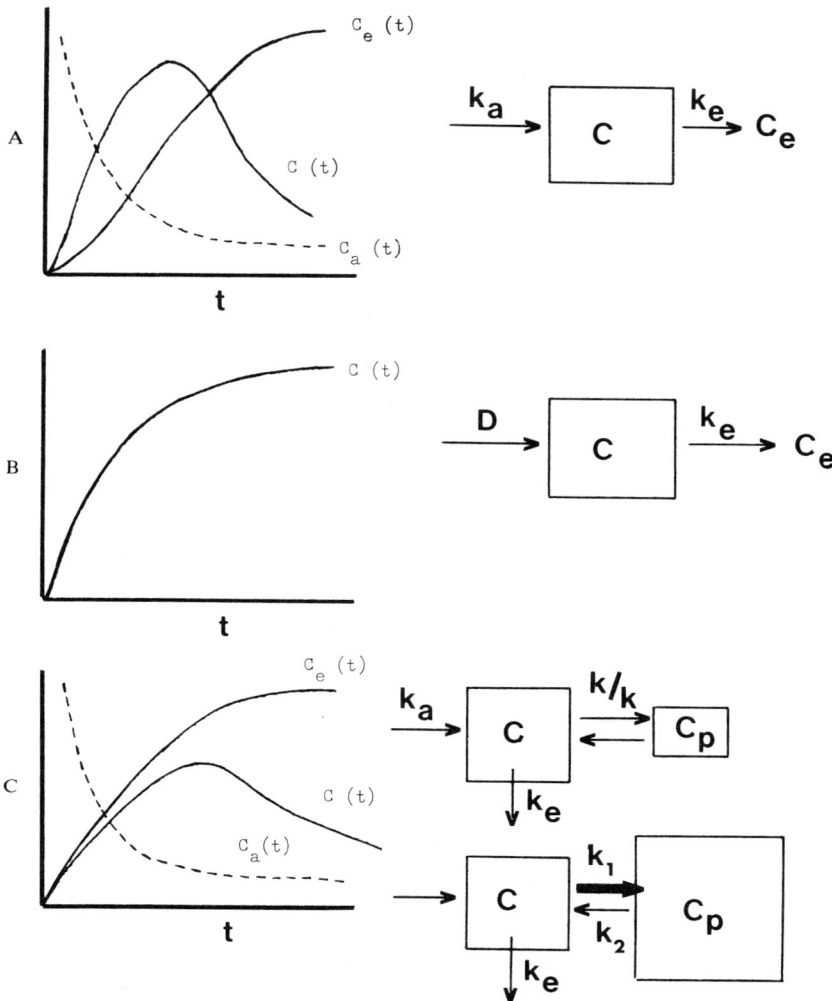

FIGURE 13. Schematic diagrams showing one-compartment models and plots of the dynamic relationships for (A) first-order absorption; (B) constant input. Diagram (C) shows 2 two-compartment models, the second having a large peripheral compartment. C is the central compartment concentration, C_p the peripheral compartment concentration, C_e the concentration eliminated. D is the dose, k_a and k_e are rate constants for absorption and elimination, respectively, while k_1 and k_2 are the rate constants for entry into and exit from the peripheral compartment, respectively.

muscle, lean tissues, and adipose tissue. As is shown here, the material found in the peripheral compartment must be returned to the central compartment for elimination from the body. In addition, the size of the peripheral compartment may often depend upon the physicochemical properties of the agent. The kinetic picture of a highly lipophilic chemical which might be extensively stored in the body fat would give the impression that the volume of the peripheral compartment (dotted line) was very large in comparison to that of the central compartment. The equilibrium between the central and peripheral compartments would be unbalanced, favoring "sequestration" in the adipose tissue with only a very small amount of the agent finding its way slowly back into the central compartment for eventual elimination. In effect, the dotted line in the kinetic profile indicates bioaccumulation of the material in the peripheral compart-

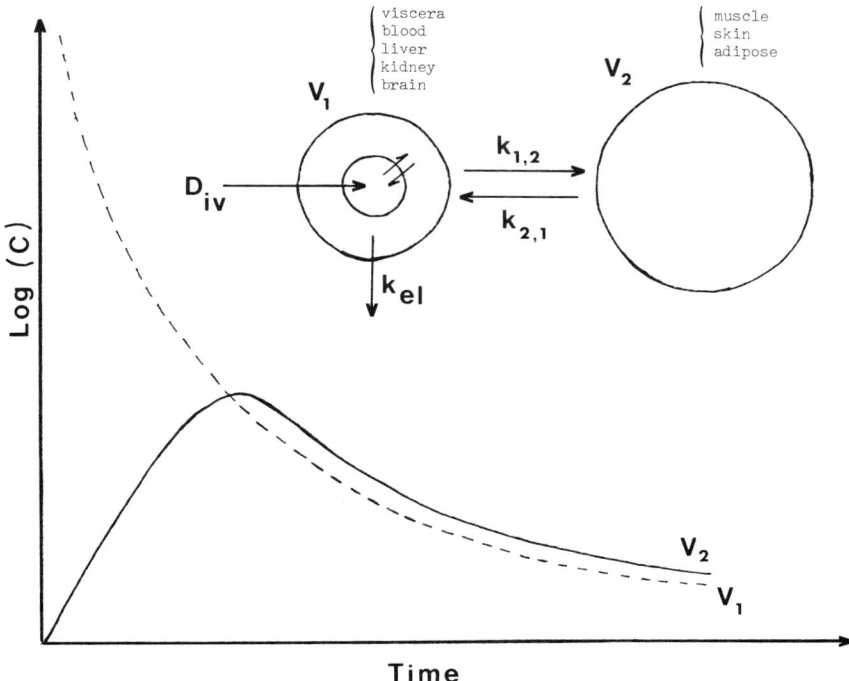

FIGURE 14. A schematic representation of a two-compartment open model with the concentration/time profiles of an agent in the two compartments.

ment, a situation often found in wildlife species continuously exposed to lipophilic agents such as chlorinated hydrocarbon insecticides, methylmercury, etc. Again, a relatively constant intake by any organisms at a particular trophic level in an ecosystem will result in a plateau effect similar to that shown in Figure 13B.

Considering the higher forms of life with their complex vasculature and infinite number of compartments, the toxicokinetics of an exogenous contaminant can be simplifed somewhat. Once a chemical has entered the bloodstream, it is distributed to the body tissues. Figure 14 depicts a schematic representation of a commonly used, two-compartment open model with the theoretical concentration-time profiles of an agent in these two compartments following intravenous administration. (The only difference between these profiles and those for orally or dermally absorbed agents is the absence of the initial absorptive phase. Once peak absorption is achieved, the subsequent concentration-time relationship will be identical to that shown.) The central compartment (V_1) is highly perfused with blood and accounts for the early decrease in blood levels of agent as the chemical is distributed to the tissues within this compartment. Within a certain period of time, the peripheral compartment (V_2), more poorly perfused but representing a larger mass of tissue, begins to sequester or store the agent. Elimination occurs in the reverse order, the agent being slowly removed from V_2 and entering V_1 to be eventually degraded by the enzymatic processes of the liver into readily excreted products via the kidneys or the gastrointestinal tract.

The tissue distribution of lipophilic chemicals is dependent upon the blood flow to the organs, i.e., the perfusion rate of the body tissues. Table 4 shows the blood flow (ℓ/min) to tissues compared with the tissue mass (as a percent of the body weight).[63] In addition, other factors which contribute to the initial localization of a lipid-soluble agent include (1) the lipid content of the tissue and (2) the nature of the circulation in the tissue, i.e., a slow perfusion through an extensive capillary network or a rapid

Table 4
BLOOD FLOW TO VARIOUS ORGANS AND TISSUES[a]

Tissue	Blood flow (l/min)	Tissue mass (% body wt)
Brain	0.75	2.0
Liver	1.55	3.5
Kidney	1.20	0.5
Cardiac muscle	0.25	0.5
Skeleton muscle	0.80	48.0
Skin	0.40	6.5
Fat	0.25	14.0
Skeleton	0.20	17.0

Note: Values should be considered as approximations.

[a] Data from R. Levine[63]

Table 5
TISSUE INSECTICIDE LEVELS IN RELATION TO AVERAGE LIPID CONTENT AND BLOOD FLOW IN RABBITS

Tissue	Lipid (%)	Blood flow (ml/min/kg)	Mean DDT (μg/g)
Muscle	1.7	250—300	26.3
Brain	8.3	615	40.6
Kidney	3.2	3330	87.1
Heart	2.7	833	104.7
Liver	3.8	680	207.2
Fat	90.9	25	39.3

From Black, W. D. and Ecobichon, D. J., *Can. J. Physiol. Pharmacol.*, 49, 45, 1971. With permission.

transport. Table 5 shows the tissue distribution of technical DDT in rabbits killed 90 min after the intravenous administration of an emulsified preparation at a dose of 100 mg/kg body weight.[64] The residue level, blood flow, and lipid content of the tissues are also shown. While muscle had a reasonable perfusion rate, it contained the lowest concentration of DDT because of the low lipid content though it would represent a large fraction of the body mass. By comparison, the perirenal adipose tissue contained a similar amount of DDT but the low perfusion rate was offset by a high lipid content. The residue levels in the liver, kidney, and heart were comparable, these tissues belonging to the visceral or central compartment which is highly perfused. They also contained comparable amounts of lipid. At 24 hr after administration, most of the body burden of DDT was sequestered in the adipose tissue as a consequence of residue redistribution, the large total amount of this tissue, and the percent fat content (See Table 4). Relatively high concentrations were also found in the "soft" tissues (liver, kidney) of the central compartment.

As was mentioned earlier, it is known that most lipophilic environmental contaminants such as chlorinated hydrocarbon and organophosphorus insecticides, polychlorinated and polybrominated biphenyls, hexachlorobenzene, halogenated phenols, dioxins, etc. do not keep accumulating when continuous exposure to a relatively constant amount of the agent(s) occurs. The tissue levels rise to a plateau as was described earlier (Figure 7) and then generally decline somewhat despite continual and prolonged exposure.[61,65] Three experiments, among many conducted, may serve to demonstrate this phenomenon. Blood levels of dieldrin in dogs daily receiving this insecticide at dietary levels of 0.005, 0.05, and 0.20 mg/kg body weight for 2 to 5 years were found to reach a plateau after approximately 100 weeks of exposure at the two lower doses but, at the highest dose, to reach a plateau after 20 weeks of feeding and then to subsequently decline slowly during the remainder of the study.[65] In another study, shown in Figure 15, dogs receiving 0.6 mg of technical aldrin/kg body wt/day in their diet for 10 consecutive months never reached plateau tissue levels during the study period with the exception of those in the blood.[66] In dogs fed DDT at 24 mg/kg body wt/day for the same period, plateau levels of the parent compound were attained in the liver within 4 months of initiating treatment but began to decline before termination of treatment at 10 months. In the adipose tissue of these animals, DDT continued to accumulate during feeding but levels of the metabolite (degradation product or contaminant), DDE had reached plateau levels early in treatment and were declining before termination of the feeding experiment.[66] Upon terminating the daily feeding of these two chemicals, there was a slow decline of the metabolites dieldrin and DDE but a rapid loss of the parent compounds aldrin and DDT. Another example, in which steers were fed a fattening ration containing 0.19 ppm heptachlor, revealed that almost 60 weeks of continuous feeding was required before a plateau level in body fat was achieved, though this concentration declined slightly with continued administration.[67] This phenomenon of bioaccumulation to a certain plateau concentration appears to hold in other species as well as has been recently demonstrated for hexachlorobenzene residues in coho salmon and rainbow trout in Lake Ontario.[68]

The important point to be gleaned from such studies is that accumulation processes follow concentration-dependent, first-order kinetics, the relationship being identical to that shown in Figure 10A and expressed as $+ dc/dt = kc$ where c is the concentration of the insecticide and k is the first-order reaction constant. This relationship, established for each body tissue, makes the assumption that the concentration of agent within the "compartment" is constant and that changes occur simultaneously in all parts of the compartment. The whole animal may be considered to be one compartment if the system behaves as one unit, ideal in the situation of small organisms with poorly defined peripheral compartments of minimal size. Accumulation in a tissue such as adipose tissue may appear to be relatively linear with time when the residue concentration is plotted as a logarithmic function whereas other tissues (blood, liver, brain, etc.) may show a more complex, curvilinear relationship when plotted in a similar fashion.[65] The chronic administration of different dietary levels of DDT to White Leghorn cockerels revealed, as is shown in Table 6 that there was a wide range of tissue residue levels found with no apparent obvious dose-relationship in these healthy birds.[69] It should be pointed out that, in this particular study, 250 ppm of technical DDT in the diet of the birds was without obvious effect, body weight gain being no different from that of control birds. In tissues such as brain where there were obvious regional differences in lipid content, there were distinctive regional differences in residue concentration following the acute intravenous administration of DDT (100 mg/kg body wt) to rabbits.[64] Considerably higher levels of DDT were found in regions of predominantly gray matter than in regions of extensive myelination, suggesting that in such an acute experiment the myelin actually prevented penetration of the lipophilic insecticide, the agent easily crossing the blood capillary barrier into the gray matter (Figure 16).

The efficiency of absorption and storage of a highly lipophilic agent such as DDT

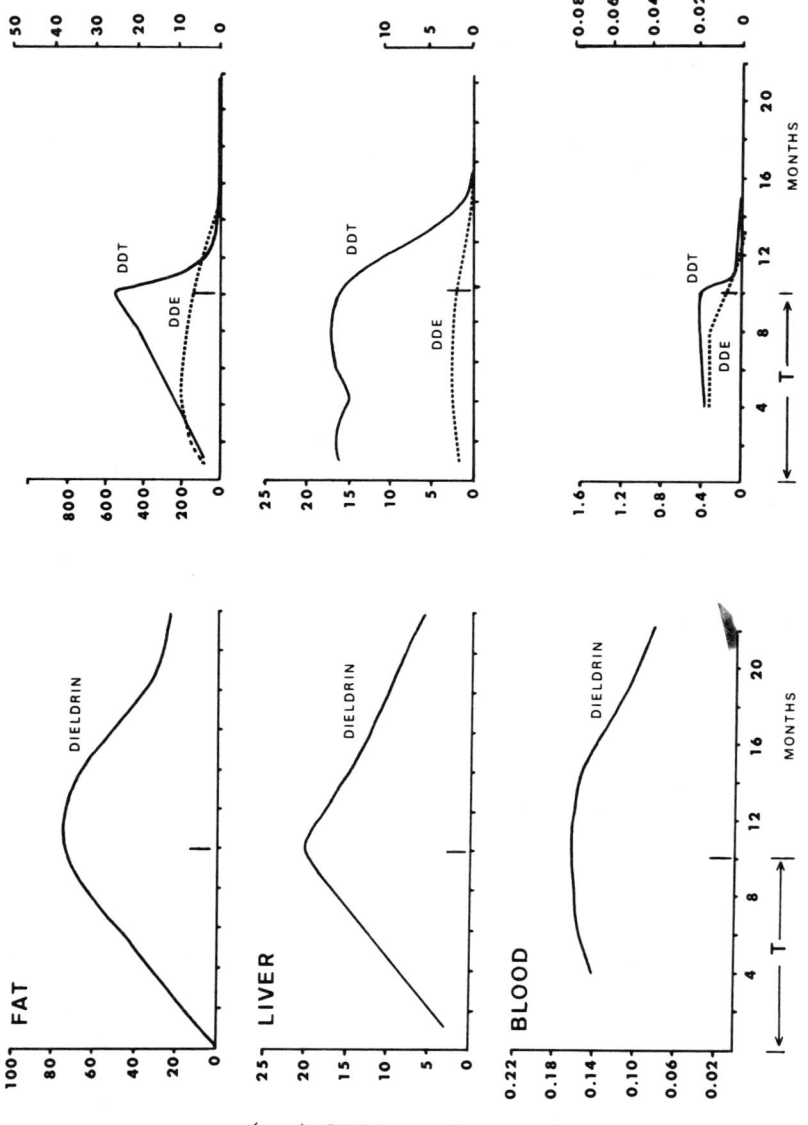

FIGURE 15. Levels of dieldrin, DDT, and its metabolite DDE in tissues of dogs fed 0.6 mg of technical aldrin or 24 mg of technical DDT/kg body weight/day for 10 months. A plateau in tissue concentration was attained only with DDT in the liver and with DDE in liver and fat and levels had begun to decline before the termination of feeding. (Redrawn from Deichmann, W. B., Keplinger, M., Dressler, I., and Sala, F., *Toxicol. Appl. Pharmacol.*, 14, 205, 1969. With permission.)

Table 6
TISSUE RESIDUES OF DDT AND DDD IN HEALTHY LEGHORN COCKERELS FED TECHNICAL DDT AT DIFFERENT DIETARY LEVELS FOR COMPARABLE PERIODS OF TIME

Tissue	Mean Toxic Residues (p.p'-DDD + p,p'-DDT) ± SEM[a]		
	250 ppm	500 ppm	1000 ppm
Brain	22.6 ± 5.5	20.1 ± 4.3	38.3 ± 9.9
Plasma	6.1 ± 1.5	7.8 ± 2.9	13.0 ± 4.6
Liver	8.8 ± 3.7	110.7 ± 35.8	263.8 ± 174.8
Kidney	102.3 ± 29.9	111.7 ± 35.2	97.3 ± 14.6
Heart	139.2 ± 26.4	87.8 ± 10.9	321.5 ± 83.5
Breast muscle	11.9 ± 4.1	6.6 ± 1.2	11.2 ± 4.0
Skin	782.5 ± 168.8	689.7 ± 210.3	2793.4 ± 642.7
Fat	3096.8 ± 563.9	3652.7 ± 449.9	—

[a] Results are presented as ppm (μg/g of tissue, wet weight). Data are from Ecobichon and Saschenbrecker.[69]

has been estimated for the human.[70] Minimum absorption was evaluated by comparing increments in fat storage with amounts ingested in the course of dosing. Two estimates of storage can be obtained: (1) the slope of DDT concentration in fat regressed on time during dosage can be multiplied by the amount of body fat; (2) the maximum concentration of DDT attained in body fat during or after dosing can be compared with the total amount of DDT ingested. The practical results revealed that very substantial fractions of the ingested p,p'-DDT (50 to 63%) could be accounted for in body fat during and after dosing. The other isomer, o,p'-DDT, found in technical grade insecticide was poorly stored (29 to 33%) in the body fat, presumably because of its rather different physicochemical properties. A very slow rate of elimination of p,p'-DDT and the metabolite p,p'-DDE from storage was encountered in the human studies, a very prolonged half-life being measured. The half-life for the o,p'-isomer was considerably shorter.

The elimination of body burdens of highly lipophilic agents from storage depots in animals has been extensively investigated using a number of species, including man. The general rule of thumb is that any agent which can be extensively stored in vivo usually is also very persistent and is excreted only slowly. The kinetic models described earlier (Figures 10 and 12), can be utilized, the elimination of most lipid-soluble chemicals following first-order processes. We shall examine some examples for pesticides and also discuss some newer models which have been applied to lipid-soluble environmental contaminants such as halogenated biphenyls and mirex (dodecachlorooctahydro-1,3,4-methano-2H-cyclobuta[cd] pentalene).

Since they have been in use longer than other classes of insecticides, it is not surprising that there is considerable data on the elimination of chlorinated hydrocarbon compounds in the literature. In commercially important domestic animals, both Bovard et al.[71] and McCully et al.[72] demonstrated that the elimination of residues of DDT and/or its degradation products was a slow process indeed. At a body burden (adipose tissue) of approximately 80 ppm, a period in excess of 600 days was required to reduce the adipose tissue residues to 7.0 ppm, at that time the tolerance level for DDT.[71] The biological half-life ($t_{1/2}$) was estimated at approximately 100 days. For steers fed forage for 83 consecutive days which had been sprayed with technical DDT at the rate of 1.5 lb/acre, McCully et al.[72] reported body fat burdens of approximately 700 ppm which decreased so slowly that, even at 331 days post-feeding, adipose tissue levels were still

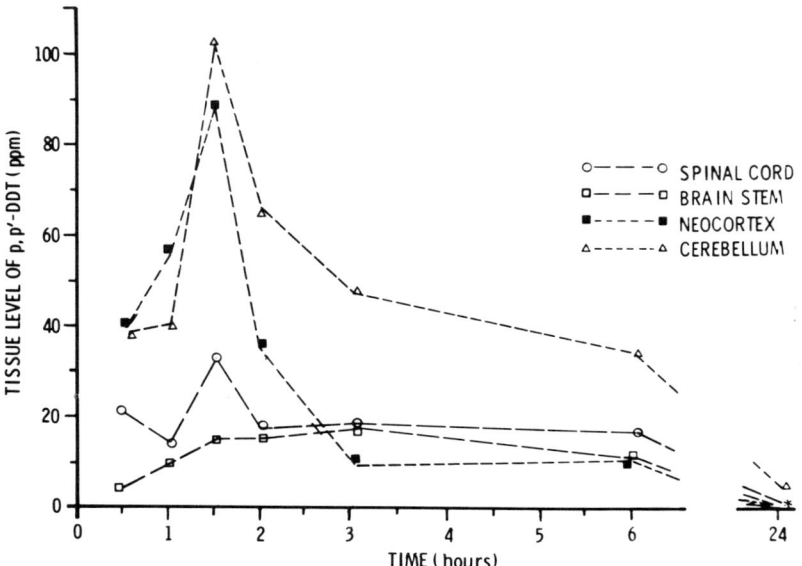

FIGURE 16. The levels of p,p'-DDT detected in regions of gray matter (cerebellum, neocortex) and white matter or mixed tissue (brainstem and spinal cord) of rabbits receiving 50 mg p,p'-DDT/kg bw intravenously. The data are plotted against the duration of the experiment. Each point is the mean value of four individual rabbits. (From Black, W. D., and Ecobichon, D. J., Can. J. Physiol. Pharmacol., 49, 45, 1971. With permission.)

in excess of 100 ppm. Unlike the first study mentioned, excretion in these forage-fed cattle appeared to involve some other complex factors since a linear relationship was established only when the logarithmic functions of tissue concentration and time were plotted against each other, suggesting something other than a first-order kinetic process.

Using an open two-compartment model similar to those shown in Figures 13 and 14, Robinson et al.[73] studied the elimination of dieldrin from adipose tissue, brain, liver, and blood of rats fed a diet containing 10 ppm of this chemical for 8 weeks. The mean tissue concentrations of insecticide were not related in a semilogarithmic manner to the time after feeding was terminated, even the log-log relationships containing one exponential term in the case of adipose tissue and brain and two exponential terms in the case of liver and blood were fitted to the results. Initial rapid periods of elimination followed by slower rates of elimination were established for all tissues, the estimated half-lives of dieldrin in blood and liver being 1.3 and 10.2 days for the rapid and slow periods, respectively. In contrast, the half-lives for dieldrin in adipose tissue and brain were 10.3 and 3.0 days, respectively.

Lindstrom et al.[74] proposed an eight-compartment nonlinear model for dieldrin distribution in mammals in which compartment volumes and transfer parameters were based upon the size of tissues and the flow rates of the blood perfusing those tissues, respectively (Figure 17). Based on the evidence that the dieldrin mass in a tissue region is roughly proportional to its lipid content, the relationships of the compartment lipid fractions of an animal were used to form a conceptual scheme for dieldrin distribution. Certain other assumptions were also made, these including:

1. Dieldrin was transported almost exclusively via lipids and does not interact strongly with other molecules.

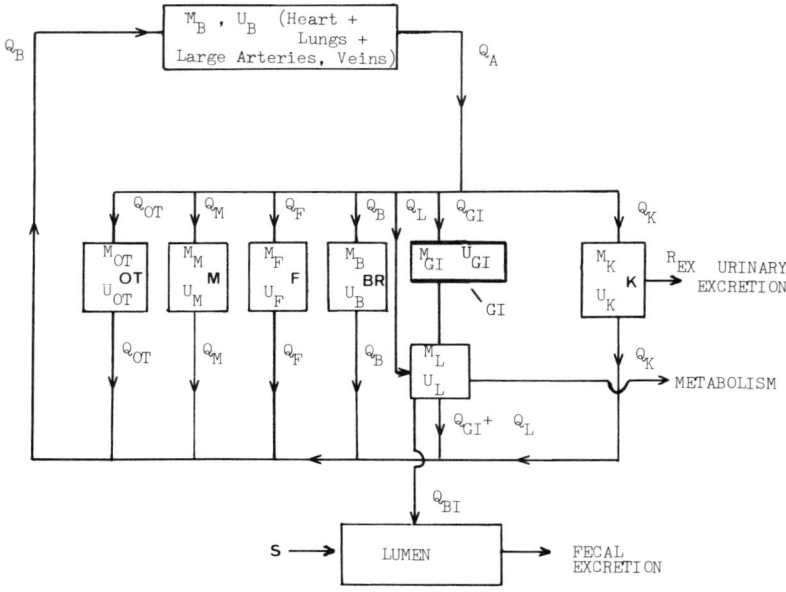

FIGURE 17. A schematic diagram representing an eight-compartment nonlinear model for dieldrin distribution in mammals in terms of blood lipid flows (Q, kg of lipid/hr) for the tissue compartment lipid mass (M, kg lipid/tissue) containing the mass of dieldrin (U, g of dieldrin/tissue). The subscripts represent B (general blood pool), L (liver), K (kidneys), BR (brain), GI (intestinal tract), F (adipose tissue), M (muscle, including associated fat) and OT (other, including skin, bone, etc.). (From Lindstrom, F. T., Gillette, J. W., and Rodecap, S. E., Arch. Environ. Contam. Toxicol., 2, 9, 1974. With permission.

2. The transfer of dieldrin from blood to lipid-containing tissues is very rapid.
3. Metabolism of dieldrin is carried out exclusively in the liver at rates much slower than the rates of transfer of dieldrin from one lipid compartment to another.
4. Dieldrin is almost totally absorbed from the intestine via lipid absorption and is not readily excreted in the feces or urine.
5. Tissues can be assigned to large compartments of similar constitution.
6. Each compartment behaves as a well-stirred chemical reactor.
7. The complete open system is representable by a system of mass balance equations.
8. The blood lipid flow and the compartment lipid mass functions in the equations developed are continuous functions of time ($0 \leq t < \infty$).

To the extent that other agents share properties similar to those of dieldrin, such a concept may be used to explain and predict their pharmacodynamics.

In human volunteers who had received dietary DDT, the stored p,p'-DDT in adipose tissue declined in a manner which suggested first-order kinetics, i.e., a rate of decay proportional to the remaining store.[70] Disappearance was found to be rapid if the level of stored material was high but became slower as the tissue residues decreased. This characteristic pattern of loss of chlorinated hydrocarbon insecticides from fat has been reported by Hayes.[75] In effect, the lower the level of stored material, the longer the estimate of biological half-life, all of which explains why most people over the age of 30 harbor trace amounts of p,p'-DDT or the metabolite p,p-DDE in their body fat. This slow loss implies a decay process in which the constant relating storage loss to remaining tissue concentration is a function of the prevaling store.

The importance of the peripheral compartment in retaining a lipid-soluble agent can be observed even with chemical such as methyl parathion which do not have a prolonged half-life in tbe body. Figure 18 shows the two- and three-compartment models used by Braeckman and colleagues to describe the observed toxicokinetics of methyl parathion administered intravenously at different doses (1.0, 3.0, 10.0, and 30.0 mg/kg body wt) to dogs.[76] With the lower doses, disappearance of the agent from the serum was more rapid than at the higher doses where the rate of decrease tended to become slower with time. The presence of "deeper" peripheral compartments where the toxicant can be stored is revealed with these higher doses, allowing use of the three-compartment model shown. Simulation studies using kinetic parameters from the 10.0 mg/kg dose of methyl parathion revealed that, while elimination was rapid, after 5 hr following injection, 30% of the dose remained in the body and was located in peripheral compartments. Given a highly lipid-soluble chemical, this three-compartment model would be significantly more important than the two-compartment model shown.

Serious toxicological problems have been encountered with a group of compounds known as polyhalogenated biphenyls. The accidental leakage of a heat transfer medium composed of polychlorinated biphenyls into a cooking oil rendered from rice bran resulted in extensive poisoning in Japan, the condition being called "Yusho oil disease".[77] The inadvertent mixing of a polybrominated biphenyl fire retardant (Firemaster BP-6 or FF-1) into animal feed in the state of Michigan resulted in widespread chronic toxicity in cattle and other domestic species.[78] The polyhalogenated biphenyls are produced by the controlled halogenation of biphenyl at one or more of five possible positions on each phenyl ring, resulting in a complex mixture of congeners, the commercially available chloro-compounds being much more complex than the brominated biphenyls.[78,79] Toxicological assessment of these mixtures has been complicated by the heterogeneity of the congeners, by marked differences in physical and chemical properties influencing rates of absorption, distribution, biotransformation, and elimination. As a consequence of this challenging problem, some unique studies concerning the pharmacokinetics of these polyhalogenated congeners have been carried out, with the development of some interesting models using isomerically pure congeners.

In 1974, it was suggested that the position of chlorination might exert a directing influence on the rate of biotransformation, on the persistence of the individual chlorobiphenyl in the body, and hence on the biological effects produced.[80,81] Studies with isomerically pure PCB and PBB congeners demonstrated that the position of chlorination on the ring was as important as the degree of chlorination when the capability of hepatic enzyme induction was compared with the rate of elimination from this organ.[82,83] An example of the rate of disappearance of selected congeners from rat liver following single oral doses (0.2 mmol/kg) is seen in Figure 19. It was found that 4,4'-di-, 2,4,2',4'-tetra-, and 2,4,5,2',4',5'-hexachlorobiphenyls were excellent inducers of hepatic microsomal monooxygenases, whereas the 2.5,2',5', -tetrachlorobiphenyl had no effect.[83] Detailed pharmacokinetic studies by Matthews and colleagues resulted in the development of flow diagrams for models to describe the elimination of selected congeners from mammalian systems.[84-86] One such model is shown in Figure 20 and consists of two parts: (1) a simple flow-limited model used for PCB distribution which operates under the assumption that the tissue concentration of parent compound and metabolite(s) maintains equilibrium with the venous blood (implying that uptake by tissue is limited primarily by tissue perfusion rates), and (2) the full model and a set of differential equations which represent individual mass balances for both the parent compound and the metabolite(s) in each compartment.[85] According to this model, the rate of accumulation (or depletion) of each species in each compartment is equal to the rate of influx with arterial blood minus the rate of efflux with venous blood minus the rate of metabolism or clearance by any physiologic process.[85] The resulting solved

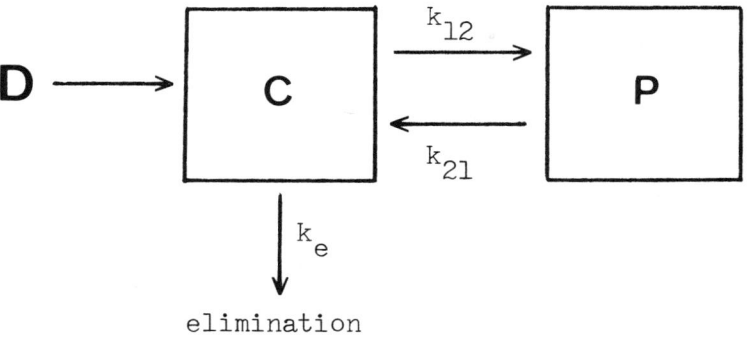

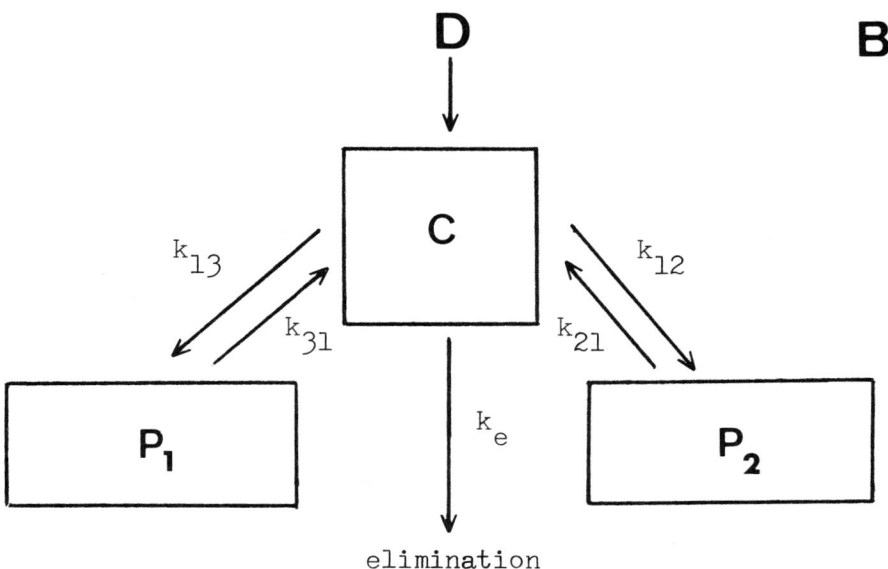

FIGURE 18. Schematic diagrams of (A) a linear two-compartment open model with its two exponential model equation and (B) a linear three-compartment open model with its three exponential equation of the concentration of agent in the central compartment "C". (From Braeckman, R. A., Godefroot, M. G., Blondeel, G. M., Belpaire, F. M., and Willems, J. L., *Arch Toxicol.* 43, 263, 1980. With permission.)

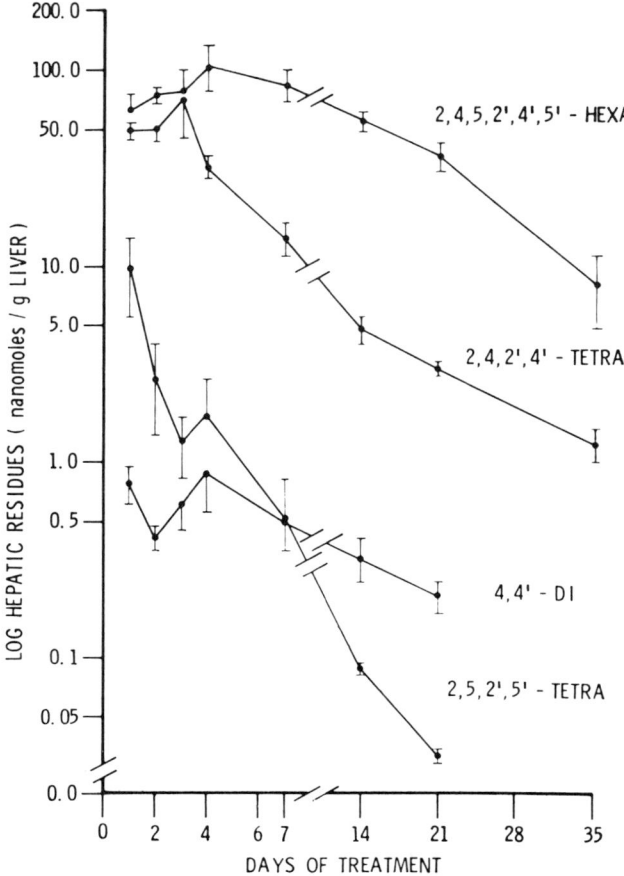

FIGURE 19. The disappearance of residues of 4,4′-di-, 2,4,2,′4′-tetra-, 2,5,2,′5′-tetra- and 2,4,5,2,′4,′5′-hexa-chlorobiphenyl from rat liver as measured by gas-liquid chromatography following a single ip injection (0.2 mmoles of agent/kg body wt). Each point represents the mean hepatic residue concentration ± SD (vertical bars) from 4 to 6 treated rats. (From Hansell, M. M., Ecobichon, D. J., Comeau, A. M., Cameron, P. H., *Exp. Mol. Pathol.*, 26, 1977. With permission.)

equations give the predicted parent compound and metabolite concentrations in the tissues as well as the total metabolite eliminated via the urine and the feces for a given dose of agent. Considering the longterm storage and excretion of an agent such as 2,4,5,2′,5′-pentachlorobiphenyl, analysis of the data revealed, as is shown in Figure 21, that 24 hr after the intravenous administration of the agent less than 5.0% of the total dose remained in any tissue other than adipose tissue and skin.[87] The PCB content of these two tissues was approximately 30% of the total dose and declined slowly over the next 42 days after treatment. Characteristically, at this low dosage (0.6 mg/kg body wt), a biphasic curve was observed, with a rapid linear decline occurring within the first 7 days and a second slow and linear phase decline starting on day 14 and persisting past day 42, the termination of the experiment. These results signify that PCBs, like chlorinated hydrocarbon insecticides, are extensively stored in tissues having a high lipid content and are only slowly released.

Morales et al.[88] studied the distribution of multiple doses of selected PCB congeners of known degree and position of chlorination in mice and found that bioaccumulation

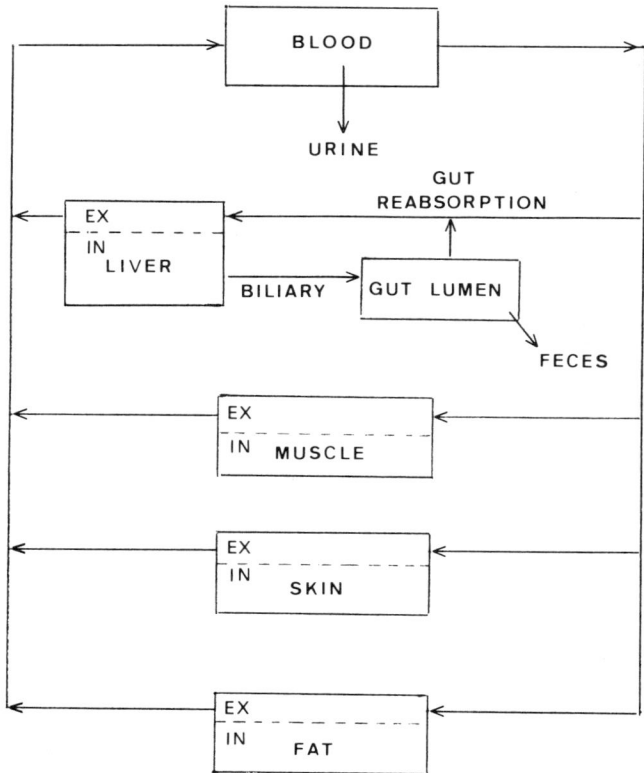

FIGURE 20. Schematic diagram for a flow-limited pharmacokinetic model for chlorinated biphenyls in the rat, useful for highly lipophilic agents. The scheme is redrawn from Lutz et al.[85] The major compartments are subdivided into an extracellular space (EX) and an intracellular space (IN) combined into what the authors call a "lumped" compartment. The differential equations which represent the individual mass balances for parent compound and metabolite for each compartment will be found in the above reference. The equations state that the rate of accumulation (or depletion) of each species in each compartment is equal to the rate of influx with arterial blood minus the rate of efflux with venous blood minus the rate of biotransformation or clearance by any physiological process. The resultant solutions give the predicted parent compound and metabolite concentrations in the tissues and the total metabolite excreted in the urine and feces.

occurred mainly in adipose tissue, skin, and muscle plus the fact that the tissue distribution for any particular PCB was similar regardless of the route of administration and the number of doses. They reported that the fecal excretion rates following multiple doses could be predicted from the excretion rates obtained after a single intravenous dose. Estimates of minimum body burden were based on a two-compartment open model, whereas estimates of the maximum body burden were based on a model that included a third permanent retention, or a "deep" storage compartment. The actual experimental body burdens fell between the predicted upper and lower limit values. More recently, a blood flow-limited physiological compartment model has been constructed for one particular chlorobiphenyl congener (2,4,5,2′,4′,5′-hexachlorobiphenyl) and used to stimulate the time course of this chemical in rat tissue and excreta following single intravenous or multiple oral doses.[89] This model was basically a mod-

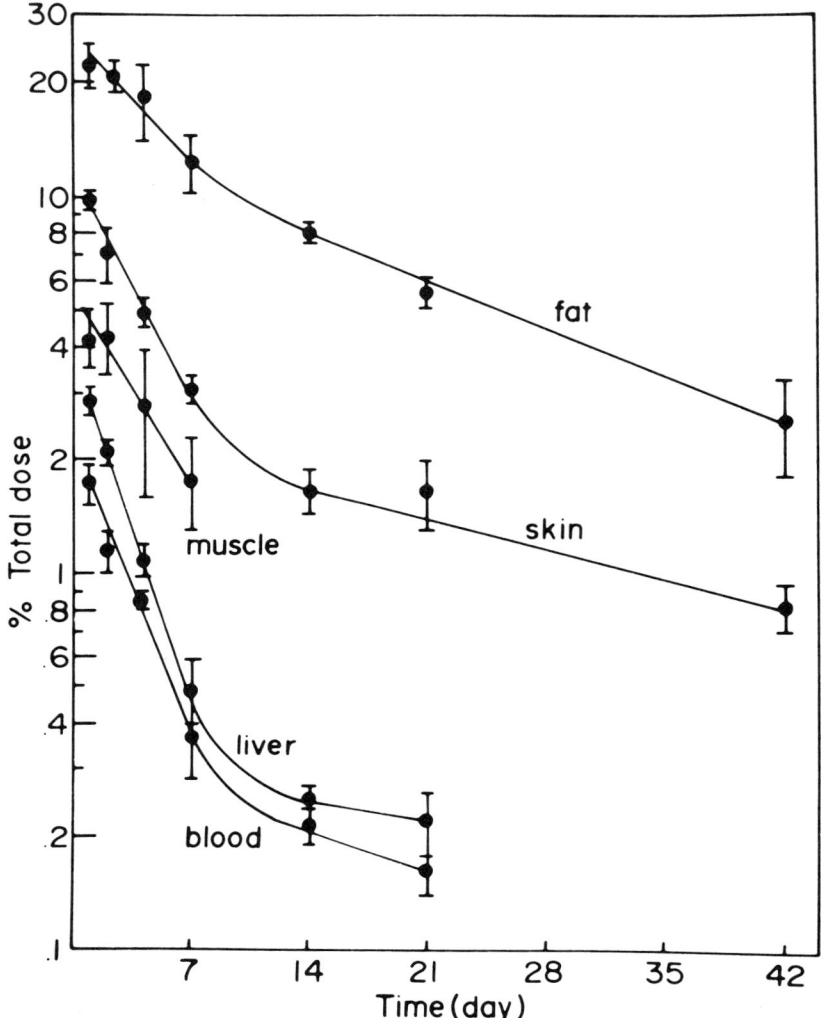

FIGURE 21. Disappearance of radiolabeled pentachlorobiphenyl from the blood, liver, skin and fat of rats held for up to 42 days following a single iv administration of 0.6 mg/kg body wt. Each point represents the average value (±SD) obtained for 3 animals. (From Matthews, H. B. and Anderson, M. W., *Drug Metab. Dispos.*, 3, 211, 1975. With permission.)

ification of that shown in Figure 20, but it was realized that a "growing" adipose tissue compartment was needed to properly describe the pharmacokinetics of 2,4,5,2′,4′,5′-hexabromobiphenyl in growing rats. The model was adapted to man by adjustment of tissue volumes and blood flow for a 250 g rat and a 70 kg man, the actual data being shown in Table 7.

Table 8 shows the clearance and rate constant estimates for biliary clearance, permeability, stomach and fecal transport of this agent and one can readily see the close agreement between the two species for all parameters with the exception of the biliary clearance. This useful model may be adapted for use with other lipophilic, persistent chemicals including pesticides.

Hammer and Bozler[90] proposed that the administration of a single radioactive tracer dose of an agent to control or treated animals at the end of subacute or chronic toxicity

Table 7
TISSUE VOLUME AND BLOOD FLOW
PARAMETERS FOR A PHARMACOKINETIC
MODEL FOR THE RAT AND FOR MAN[a]

Compartment	Volume (ml)		Blood Flow (ml/hr)	
	Rat	Man	Rat	Man
Blood	22.5	5,400	—	—
Liver	10.0	1.700	960	87,000
Muscle	125.0	30,370	450	36,000
Skin	40.0	7,880	10	7,800
Adipose	17.5	10,123	24	12,000
Intestinal tissue	7.0	1,000	336	66,000
Intestinal lumen	7.7	1,700	—	—

[a] Values are based on a standard 250-g rat and a 70-kg man.

From Tuey, D. B. and Matthews, H. B., *Toxicol. Appl. Pharmacol.*, 53, 420, 1980. With permission.

Table 8
CLEARANCE AND RATE CONSTANT
ESTIMATES FOR
HEXABROMOBIPHENYL DISPOSITION
KINETICS IN THE RAT AND MAN[a]

Parameter	Rat	Man
Biliary Clearance k_b ml/hr	0.074	5.06
Permeability constant k_g/hr^{-1}	0.70	0.70
Stomach transport k_s (hr^{-1})	0.87	0.82
Fecal transport k_f (hr^{-1})	0.095	0.05

[a] These estimates can be applied to the equations shown in Figure 19.

From Tuey, D. B. and Matthews, H. B., *Toxicol. Appl. Pharmacol.*, 53, 420, 1980. With permission.

studies could be used to determine possible self-induced changes in the pharmacokinetics of compounds due to their own toxicity or influence on cellular mechanisms. This technique assumes that the pharmacokinetic profile of the last dose of a repeated treatment period would be identical to the profile of a dose given to untreated animals, i.e., the last-in, first-out phenomenon. In testing this concept experimentally, Colburn and Matthews[91] used diphenylhydantoin as one agent and, 2,4,4,2′,4′,5′-hexachlorobiphenyl (HCB) as the other chemical. The mathematical models were based on the assumptions that compartments were well stirred and that the test compounds are instantaneously distributed throughout the compartments. The experiments with diphenylhydantoin supported the proposed theory, but the pharmacokinetics of HCB in vivo did not meet the criteria. From the experiments, it appeared that HCB is accumulated and released from body fat relative to the time of administration. The theory was applicable to compounds that were distributed rapidly throughout the organism and could be handled by a simple two-compartment open pharmacokinetic model. Such a theory should be applied with caution and experimentally verified if chemicals which are slowly distributed, redistributed, and eliminated from the body are to be studied.

Mirex (dodecachlorooctahydro-1,3,4-methano-2H-cyclobuta (c,d) pentalene) is a complex "cage-structured" insecticide used in the form of a ground bait for the control of fire ants in the southern U.S.[92] Long-term feeding studies in three species (fish, quail, rats) showed a continual uptake in the body fat with no plateau being attained even after 16 months of continuous exposure.[93] Little evidence of extensive biotransformation of this compound has been obtained. Mirex accumulates and persists in animal tissues, particularly in fat.[93,94] Accumulation and tissue storage were found to be directly related to the dose administered over a 100-fold range of dose.[93,95] Elimination is an extremely slow process, the level in the fat of female rats decreasing by only 40% some 10 months after a return to normal diet.[93] In female rhesus monkeys, the cumulative fecal excretions of radiolabelled mirex was only of the order of 7% of the dose administered after 350 days.[96] Because of these properties, mirex may be an excellent model compound with which to study long-term kinetic processes of chlorinated hydrocarbons in individual animals and to develop models useful for predicting the behaviour of such chemicals during many years of chronic ingestion.[97]

The studies of Pittman and colleagues[96,97] demonstrated that the analysis (resolved by the method of residuals) of the semilogarithmic plots of plasma concentration-time data for monkeys receiving a single intravenous dose (0.86 to 1.05 mg/kg body wt) resulted in more than four linear components. The earlier components had half-lives of fractions of an hour and were thought to represent distributional phases which occurred with great rapidity. It was thought that these rapid phases would provide little information on the long-term fate of mirex. The last four components had half-lives ranging from a few hours to many days and represented distribution and excretion phases which were slow. Since a slow fecal elimination was shown to be the major excretory route and that mirex levels in liver, bile, and intestine were higher than in the plasma, models allowing for excretion from a peripheral compartment as well as from a central compartment were constructed. Two such models, a three- and a four-compartment model, which were explored are shown in Figure 22. Compartment 1 in each case is the central compartment containing plasma and any tissues for which the distribution phases are very rapid. Compartments 2 and 3 represent a slow and a fast tissue compartment, respectively. Compartment 4 in the 4 compartment model represents a "very slow" tissue compartment. As can be seen, excretion is thought to occur via the feces, the product coming from either compartment 1 or from compartment 2 (the fast tissue compartment). In more simplified models, compartments 1 and 2 would be combined but because of the somewhat different excretory pattern of this agent, they were considered separately since biliary elimination and fecal excretion continued long after urinary excretion became undetectable.

The choice of operational models was made on the basis of the observed fit of the analytical data to the computer projections for plasma concentration, cumulative urinary- and fecal excretion-time curves for the treated monkeys. The four-compartment model was judged to better represent the actual results. The computer-based projections of the fate of mirex in the monkeys revealed that the maximum concentration of agent in the "very slow" compartment did not occur until 3 to 12 months after the single injection. Though continuous excretion of mirex occurred in both urine and feces, the decline in body burden within 5 years was estimated to be 1 to 2% of the amount present at the end of the first year, signifying that the biological half-life of mirex would exceed the life spans of the animals. Some 85 to 87% of the administered mirex was found sequestered in the body fat of the animals within 106 to 388 days of treatment with a single dose. This sequestration and the lack of biotransformation were considered to be major factors responsible for the prolonged biological half-life.

While the toxicity of lipophilic, environmentally persistent pesticides is effectively nullified by the process of sequestration in body fat, it should be realized that any

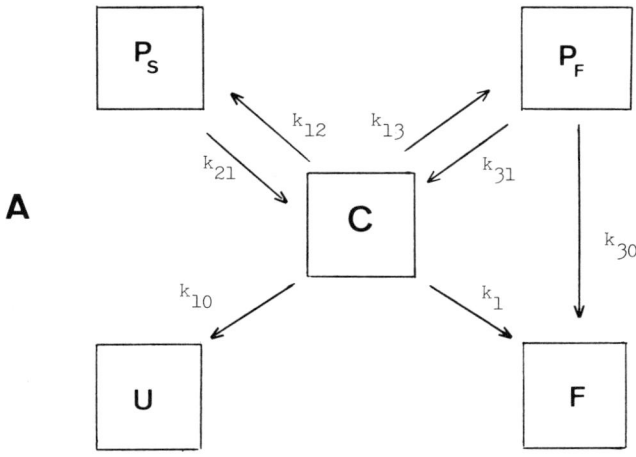

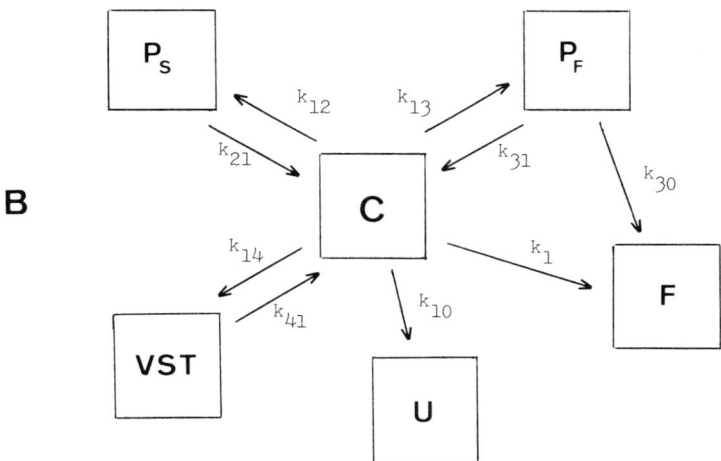

FIGURE 22. Schematic diagrams for A, a three-compartment and B, a four-compartment, open-system mamillary model. The compartments consist of the central compartment (C); a peripheral fast tissue (P_F) and a peripheral slow-tissue (P_S) compartment; urinary (U) and fecal (F) excretory compartments. In the case of the four-compartment model, an additional very slow tissue (VST) storage compartment representing adipose tissue, skin, etc. is also shown. Fecal excretion is treated separately from urinary elimination and is measured either by parameter k_1 originating from the central compartment or by parameter k_{30} which originates from the fast peripheral compartment. Models are redrawn from Pittman et al.[97]

physiological response requiring caloric input, i.e., migration, food deprivation, pregnancy, parturition and lactation, cold acclimatization, etc., will result in a "reactivation" and a redistribution of the stored agent from the peripheral storage compartment back into the central compartment. If the levels mobilized are sufficiently high, then toxic signs will be elicited as target organs accumulate significant concentrations. Early

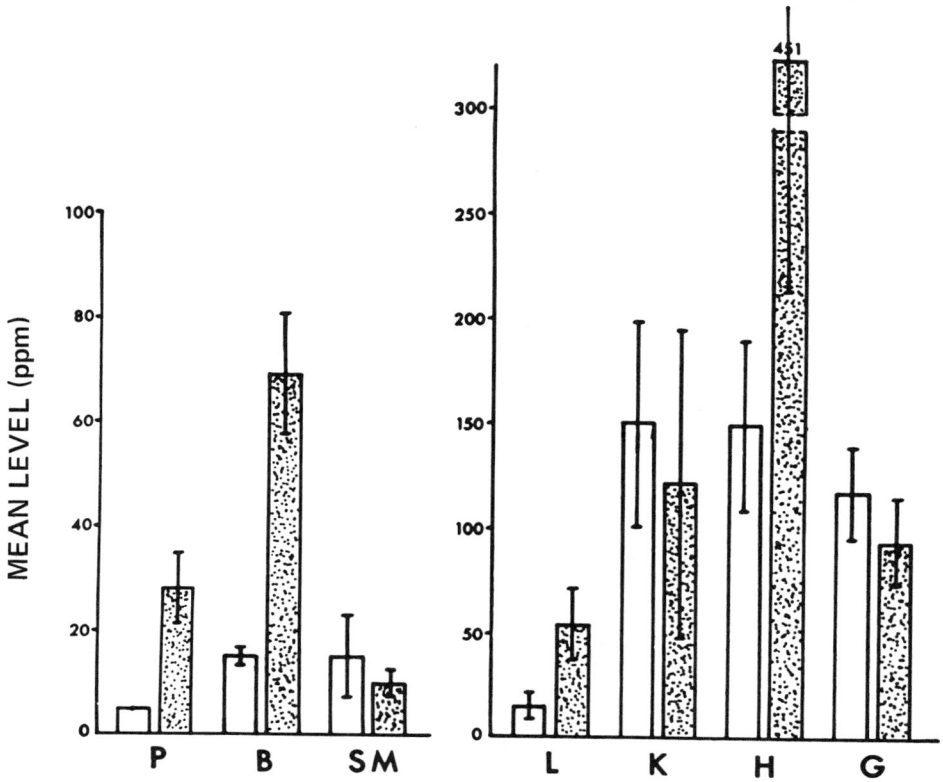

FIGURE 23. The mobilization and tissue redistribution of total toxic residues (o,p'-DDT, p,p'-DDD, p,p'-DDT) from the body fat of White Leghorn cockerels fed 250 µg of technical DDT in their diet for 15 weeks prior to a 50% reduction in the food provided. The control (open bars) and food-deprived (hatched bars) groups each contained 10 birds. The tissues analyzed for residues included blood plasma (P), brain (B), skeletal (breast) muscle (SM), liver (L), kidney (K), heart (H) and intestinal tract (G). The mean concentrations are shown with the vertical lines representing the SEM. No toxicity was observed in any birds during the treatment period but, with food deprivation, the translocation of residues was accompanied by severe neurological signs of poisoning. (From Ecobichon, D. J. and Saschenbrecker, P. W., *Toxicol. Appl. Pharmacol.*, 15, 420, 1969. With permission.)

studies in rats revealed that both DDT and dieldrin could be mobilized by starvation or food deprivation, the biliary secretion and fecal excretion of the parent compounds and metabolites being markedly enhanced.[98,99] When DDT-treated cockerels were deprived of a calorically adequate diet, a dramatic mobilization of the parent compound and metabolites into the blood occurred with subsequent redistribution of the agent(s) into other tissues as is shown in Figure 23.[100] As a consequence of the high brain levels of "toxic" residues (p,p'-DDT, o,p'-DDT and p,p'-DDD), signs of DDT poisoning were observed, the stages progressing from barely perceptible tremors in the tail feathers to curling of the digits, an ataxic gait, the inability to balance upright or rise voluntarily, to severe generalized tremors, tonic convulsions, and death.[100] Exercise, starvation, and cold exposure all caused the elevation of blood DDT levels in male rats treated subacutely with DDT.[101] Similar experiments have been conducted with DDT-fed pigeons.[102] Similar observations have been made for the mobilization and redistribution of PCB congeners, in fed and fasted pigeons, the fractional turnover of stored PCBs not being dependent upon the chemical structure of the congener but being related to the fractional turnover of the lipid pool.[103] It was found that, even with the highly sequestered agent mirex, in rhesus monkeys, food deprivation resulted in an almost two-fold increase in the blood plasma level of the agent.[97]

In summary, the pharmacokinetic disposition of pesticides can be considered in terms of simple or complex models, depending upon the physicochemical properties of the particular agent being studied. The two-compartment open model will suffice for most studies, but if the agent is very lipid-soluble and only slowly degraded in vivo, a three-compartment model including a "deep" storage area with the appropriate clearance constant(s) will be essential to insure a proper consideration of the persistence of such chemicals. Rapid biotransformation and/or elimination of the particular chemical being studied will simplify the kinetic model required to predict the distribution and persistence in vivo.

REFERENCES

1. Matsumura, F., *Toxicology of Insecticides,* Plenum Press, New York, 1975, 4.
2. Brown, A. W. A., *Ecology of Pesticides,* Wiley-Interscience, New York, 1978, 13.
3. Carson, R., *Silent Spring,* Houghton Mifflin, Boston, 1962.
4. Johnson, D. W. and Lew, S., Chlorinated hydrocarbon pesticides in representative fishes of southern Arizona, *Pestic. Monit. J.,* 4, 57, 1970.
5. Symons, P. E. K., Dispersal and toxicology of the insecticide fenitrothion; predicting hazards of forest spraying, *Residue Rev.,* 68, 1, 1977.
6. Pearce, P. A., Brun, G. L., and Witteman, J., Off-target fallout of fenitrothion during 1978 forest spraying operations in New Brunswick, *Bull. Environ. Contam. Toxicol.,* 23, 503, 1979.
7. World Health Organization, Safe use of pesticides, *WHO Tech Rep. Ser.,* 513, 42, 1973.
8. Matsumura, F., *Toxicology of Insecticides,* Plenum Press, New York, 1975, chap. 1.
9. Kearney, P. C., Plimmer, J. R., and Helling, C. S., Soil chemistry of pesticides, in *Encyclopedia of Chemical Technology,* Vol 18, 2nd ed., Kirk, R. E. and Othmer, D. F., Eds., 1969, 515.
10. Anon., A look at world pesticide markets, *Farm Chem.,* 141, 38, 1977.
11. Cremlyn, R., *Pesticides. Preparation and Mode of Action,* John Wiley & Sons, Toronto, 1978, 9.
12. Brown, A. W. A., The ecological implications of insecticide usage in malaria programs, *Am. J. Trop. Med. Hyg.,* 21, 829, 1972.
13. Westlake, W. E. and Gunther, F. A., Occurrence and mode of introduction of pesticides in the environment, in *Organic Pesticides in the Environment,* Advances in Chemistry Series, 60, 110, 1966.
14. Woodwell, G. M., Craig, P. P., and Johnson, H. A., DDT in the biosphere: where does it go?, *Science,* 174, 1101, 1971.
15. Wheatley, G. A., Pesticides in the atmosphere, in *Environmental Pollution by Pesticides,* Edwards, C. A., Ed., Plenum Press, New York, 1973, 365.
16. Spencer, W. F., Movement of DDT and its derivatives into the atmosphere, *Residue Rev.,* 59, 91, 1975.
17. Hinden, E., May, D. S., and Dunstan, G. H., Distribution of insecticides sprayed by airplane on an irrigated corn plot, in *Organic Pesticides in the Environment,* (Advances in Chemistry Series), 60, 132, 1966.
18. Poulsen, E., Rester af pesticider i afgreder foder og levneds midler, *Jord. Pesticider Forureningsradet Sekretariat Publ.,* 17, 143, 1971.
19. Ware, G. W., Cahill, W. P., Gerhardt, P. D., and Witt, J. M., Pesticide drift: IV. On-target deposits from aerial application of pesticides, *J. Econ. Entomol.,* 62, 1982, 1970.
20. Ware, G. W., Apple, E. J., Cahill, W. P., Gerhardt, P. D., and Frost, K. R., Pesticide drift II. Mist blower vs. aerial application of sprays, *J. Econ. Entomol.,* 62, 844, 1969.
21. Fisher, R. W. and Hikichi, A., Orchard sprayers, *Ontario Ministry Agric. Food Publ.,* 373, 44, 1971.
22. Brann, Jr., J. L., Factors affecting the thoroughness of spray application, *N.Y. State Agric. Exp. Stn. J.,* Paper 1429, 1965.
23. Akesson, N. B. and Yates, W. E., Problems relating to application of agricultural chemicals and resulting drift residues, *Annu. Rev. Entomol.,* 9, 285, 1964.
24. Woodwell, G. M., Toxic substances and ecological cycles, *Sci. Am.,* 216, 24, 1967.
25. Decker, G. C., Weinman, C. J., and Bann, J. M., A preliminary report on the rate of insecticide residue loss from treated plants, *J. Econ. Entomol.,* 43, 919, 1950.
26. Lloyd-Jones, C. P., Evaporation of DDT, *Nature (London),* 229, 65, 1971.

27. Orgill, M. M., Schmel, G. A., and Petersen, M. R., Some initial measurement of airborn DDT over Pacific northwest forests, *Atmos. Environ.*, 10, 827, 1976.
28. Crabbe, R., Krzymien, M., Elias, L., and Davie, S., New Brunswick spray operations: measurement of atmospheric fenitrothion concentrations near the spray area, *Natl. Res. Counc. of Can. Rep.*, LTR-UA-56, 1980.
29. Liss, P. S. and Slater, P. G., Flux of gases across air-sea interface, *Nature (London)*, 247, 181, 1974.
30. Cohen, J. M. and Pinkerton, C., Widespread translocation of pesticides by air transport and rain-out, in *Organic Pesticides in the Environment*, (Advances in Chemistry Series), 60, 163, 1966.
31. Risebrough, R. W., Huggett, R. J., Griffin, J. J., and Goldberg, E. D., Pesticides: transatlantic movements in the northeast trades, *Science*, 159, 1233, 1968.
32. Seba, D. B. and Prospero, J. M., Pesticides in the lower atmosphere of the Northern Equatorial Atlantic Ocean, *Atmos. Environ.*, 5, 1043, 1971.
33. Bidleman, T. F. and Olney, C. E., Chlorinated hydrocarbons in the Sargasso Sea atmosphere and surface water, *Science*, 183, 516, 1974.
34. Harvey, G. R. and Steinhauer, W. G., Atmospheric transport of polychlorinated biphenyls to the North Atlantic, *Atmos. Environ.*, 8, 777, 1974.
35. Arthur, R. D., Cain, J. D., and Barrentine, B. F., Atmospheric levels of pesticides in the Mississippi delta, *Bull. Environ. Contam. Toxicol.*, 15, 129, 1976.
36. Tabor, E. C., Contamination of urban air through the use of insecticides, *Trans. N. Y. Acd. Sci.*, Vol. II, 569, 1966.
37. McEwen, F. L. and Stephenson, G. R., *The Use and Significance of Pesticides in the Environment*, J. Wiley & Sons, New York, 1979, chap. 16.
38. Abbott, D. C., Harrison, R. B., Tatton, J. O'G., and Thomson, J., Organochlorine pesticides in the atmospheric environment, *Nature (London)*, 208, 1317, 1965.
39. Rudd, R. L., *Pesticides and the Living Landscape*, University of Wisconsin Press, Madison, 1964, 250.
40. Ecobichon, D. J., Chlorinated hydrocarbon insecticides: Recent animal data of potential significance for man, *Can. Med. Assoc. J.*, 103, 711, 1970.
41. Ecobichon, D. J., Hydrolytic mechanisms of pesticide degradation, in *Advances in Pesticide Science*, Part 3, Geissbuhler, H., Ed., Pergamon Press, New York, 1979, 516.
42. Kulkarni, A. P. and Hodgson, E., Metabolism of insecticides by mixed function oxidase systems, *Pharmacol. and Ther.*, 8, 379, 1980.
43. Edwards, C. A., *Persistent Pesticides in the Environment*, CRC Press, Boca Raton, Fla., 2nd ed., 1973, 170.
44. White, D. H., King, K. A., Mitchell, C. A., Hill, E. F., and Lamont, T. G., Parathion causes secondary poisoning in a laughing gull breeding colony, *Bull. Environ. Contam. Toxicol.*, 23, 281, 1979.
45. Hickey, J. J., Keith, J. A., and Coon, F. B., An exploration of pesticides in a Lake Michigan ecosystem, *J. Appl. Ecol.*, (Suppl 3), *Pesticides in the Environment and Their Effects on Wildlife*, Moore, N. W., Ed., 1966, 141.
46. Mulla, M. S., Keith, J. O., and Gunther, F. A., Persistence and biological effects of parathion residues in waterfowl habitats, *J. Econ. Entomol.*, 59, 1085, 1966.
47. Freed, V. H., Schmedding, D., Kohnert, R., and Haque, R., Physical chemical properties of several organophosphates: some implications in environmental and biological behavior, *Pest. Biochem. Physiol.*, 10, 203, 1979.
48. Kenaga, E. E., Chlorinated hydrocarbon insecticides in the environment. Factors related to bioconcentration of pesticides, in *Environmental Toxicology of Pesticides*, Matsumura, F., Boush, G. M., and Misato, T., Eds., Academic Press, New York, 1972, 193.
49. Decker, G. C., Bruce, W. N., and Bigger, J. H., The accumulation and dissipation of residues resulting from the use of aldrin in soils, *J. Econ. Entomol.*, 58, 266, 1965.
50. Nash, R. G. and Woolson, E. A., Persistence of chlorinated hydrocarbon insecticides in soils, *Science*, 157, 924, 1967.
51. Lichtenstein, E. P., Fuhremann, T. W., and Schulz, K. R., Persistence and vertical distribution of DDT, lindane and aldrin residues, 10 and 15 years after a single soil application, *J. Agric. Food Chem.*, 19, 718, 1971.
52. Lichtenstein, E. P. and Schulz, K. R., The effect of moisture and microorganisms on the persistence and metabolism of some organophosphorus insecticides in soil, *J. Econ. Entomol.*, 57, 618, 1964.
53. Kearney, P. C., Nash, R. G., and Isensee, A. R., Persistence of pesticide residues in soils, in *Chemical Fallout*, Miller, M. W. and Berg, G. G., Eds., Charles C Thomas, Springfield, Ill., 1969, 54.
54. Hurtig, H., Significance of conversion products and metabolites of pesticides in the environment, in *Environmental Quality and Safety*, Vol. 1, Coulston, F. and Korte, F., Eds., Academic Press, New York, 1972, 58.

55. McEwen, F. L. and Stephenson, G. R., *The Use and Significance of Pesticides in the Environment*, John Wiley & Sons, Inc., New York, 1979.
56. Gladtke, E. and von Hattingberg, H. M., *Pharmacokinetics. An Introduction*, Springer-Verlag, New York, 1979.
57. Filov, V. A., Goluben, A. A., Liublina, E. I., and Tolokontsen, N. A., *Quantitative Toxicology (Selected Topics)*, John Wiley & Sons, New York, 1979, 94.
58. Wagner, J. G., Pharmacokinetics, in *Annual Review of Pharmacology*, Vol. 8, Elliott, H. W., Ed., Annual Reviews, Palo Alto, Calif., 1968, 67.
59. Wagner, J. G., *Biopharmaceutics and Relevant Pharmacokinetics*, Drug Intelligence, Hamilton, Ill., 1st. Ed., 1971.
60. Levy, G. and Gibaldi, M., Pharmacokinetics, in *Handbook of Exper. Pharmacology, New Series*, Vol. 8, Eichler, O., Farah, A., Herken, H., and Welch, A. D., Eds., Springer-Verlag, New York, 1975, 1.
61. Gehring, P. J., Watanabe, P. G., and Blau, G. E., Pharmacokinetic studies in evaluation of the toxicological and environmental hazard of chemicals, in *Advances in Modern Toxicology*, Vol. 1, Part 1, Mehlman, M. A., Shapiro, R. E., and Blumenthal, H., Eds., Hemisphere Pub., Washington, D.C., 1976.
62. Tuey, D. B., Toxicokinetics, in *Introduction to Biochemical Toxicology*, Hodgson, E. and Guthrie, F. E., Eds., Elsevier, North Holland, New York, 1980, 40.
63. Levine, R., *Pharmacology. Drug Actions and Reactions*, Little, Brown, Boston, 1973, chap. 5.
64. Black, W. D. and Ecobichon, D. J., A pharmacodynamic study of DDT in rabbits following acute intravenous administration, *Can. J. Physiol. Pharmacol.*, 49, 45, 1971.
65. Robinson, J., The burden of chlorinated hydrocarbon pesticides in man, *Can. Med. Assoc. J.*, 100, 180, 1969.
66. Deichmann, W. B., Keplinger, M., Dressler, I., and Sala, F., Retention of dieldrin and DDT in the tissues of dogs fed aldrin and DDT individually and as a mixture, *Toxicol. Appl. Pharmacol.*, 14, 205, 1969.
67. Bovard, K. P., Fontenot, J. P., and Priode, B. M., Accumulation and dissipation of heptachlor residues in fattening steers, *J. Anim. Sci.*, 33, 127, 1971.
68. Niimi, A. J., Hexachlorobenzene (HCB) levels in Lake Ontario salmonids, *Bull. Environ. Contam. Toxicol.*, 23, 20, 1979.
69. Ecobichon, D. J. and Saschenbrecker, P. W., Pharmacodynamic study of DDT in cockerels, *Can. J. Physiol. Pharmacol.*, 46, 785, 1968.
70. Morgan, D. P. and Roan, C. C., The metabolism of DDT in man, in *Essays in Toxicology*, Hayes, W. J., Jr., Ed., Academic Press, New York, 1974, 39.
71. Bovard, K. P., Priode, B. M., Whitmore, G. E., and Ackerman, A. J., DDT residues in the internal fat of beef cattle fat contaminated apple pomace, *J. Anim. Sci.*, 20, 824, 1961.
72. McCully, K. A., Villeneuve, D. C., McKinley, W. P., Phillips, W. E. J., and Hidiroglou, M., Metabolism and storage of DDT in beef cattle, *J. Assoc. Offic. Agric. Chem.*, 49, 966, 1966.
73. Robinson, J., Roberts, M., Baldwin, M., and Walker, A. I. T., The pharmacokinetics of HEOD (dieldrin) in the rat, *Food Cosmet. Toxicol.*, 7, 317, 1969.
74. Lindstrom, F. T., Gillette, J. W., and Rodecap, S. E., Distribution of HEOD (dieldrin) in mammals I Preliminary model, *Arch. Environ. Contam. Toxicol.*, 2, 9, 1974.
75. Hayes, W. J., Jr., Review of the metabolism of chlorinated hydrocarbon insecticides especially in mammals, in *Annual Review of Pharmacology*, Vol. 5, Cutting, W. C., Ed., Annual Reviews, Palo Alto, Calif., 1965, 27.
76. Braeckman, R. A., Godefroot, M. G., Blondeel, G. M., Belpaire, F. M., and Willems, J. L., Kinetic analysis of the fate of methyl parathion in the dog, *Arch. Toxicol.*, 43, 263, 1980.
77. Higuchi, K., Ed., *PCB Poisoning and Pollution*, Academic Press, New York, 1976.
78. Kay, K., Polybrominated biphenyls (PBB) environmental contamination in Michigan, 1973-1976, *Environ. Res.*, 13, 74, 1977.
79. Ecobichon, D. J. and Comeau, A. M., Comparative effects of commercial Arochlors on rat liver enzyme activities, *Chem. Biol. Interactions*, 9, 341, 1974.
80. Bush, B., Tumasonis, C. F., and Baker, F. D., Toxicity and persistence of PCB homologs and isomers in the avian system, *Arch. Environ. Contam. Toxicol.*, 2, 194, 1974.
81. Johnstone, G. J., Ecobichon, D. J., and Hutzinger, O., The influence of pure polychlorinated biphenyl compounds on hepatic function in the rat, *Toxicol. Appl. Pharmacol.*, 28, 66, 1974.
82. Ecobichon, D. J. and Comeau, A. M., Isomerically pure chlorobiphenyl congeners and hepatic function in the rat: influence of position and degree of chlorination, *Toxicol. Appl. Pharmacol.*, 33, 94, 1975.
83. Hansell, M. M., Ecobichon, D. J., Comeau, A. M., and Cameron, P. H., The relationship between retention of pure chlorobiphenyl congeners and hepatic function in the rat, *Exp. Mol. Pathol.*, 26, 1977.

84. Matthews, H. B. and Anderson, M. W., Effect of chlorination on the distribution and excretion of polychlorinated biphenyls, *Drug Metab. Dispos.*, 3, 371, 1975.
85. Lutz, R. J., Dedrick, R. L., Matthews, H. B., Eling, T. E., and Anderson, M. W., A preliminary pharmacokinetic model for several chlorinated biphenyls in the rat, *Drug Metab. Dispos.*, 5, 386, 1977.
86. Tuey, D. B. and Matthews, H. B., Pharmacokinetics of 3,3′,5,5′-tetrachlorobiphenyl in the male rat, *Drug Metab. Dispos.*, 5, 444, 1977.
87. Matthews, H. B. and Anderson, M. W., The distribution and excretion of 2,4,5,2′,5′-pentachlorobiphenyl in the rat, *Drug Metab. Dispos.*, 3, 211, 1975.
88. Morales, N. M., Tuey, D. B., Colburn, W. A., and Matthews, H. B., Pharmacokinetics of multiple oral doses of selected polychlorinated biphenyls in mice, *Toxicol. Appl. Pharmacol.*, 48, 397, 1979.
89. Tuey, D. B. and Matthews, H. B., Distribution and excretion of 2,2′,4,4′,5,5′-hexabromobiphenyl in rat and man: pharmacokinetic model predictions, *Toxicol. Appl. Pharmacol.*, 53, 420, 1980.
90. Hammer, R. and Bozler, G., Pharmacokinetics as an aid in the interpretation of toxicity tests, *Arzneim. Forsch.*, 27, 555, 1977.
91. Colburn, W. A. and Matthews, H. B., Pharmacokinetics in the interpretation of chronic toxicity tests: the last-in, first-out phenomenon, *Toxicol. Appl. Pharmacol.*, 48, 387, 1979.
92. Waters, E. M., Huff, J. E., and Gerstner, H. B., Mirex: an overview, *Environ. Res.*, 14, 212, 1977.
93. Ivie, G. W., Gibson, J. R., Bryant, H. E., Begin, J. J., Barnett, J. R., and Dorough, H. W., Accumulation, distribution and excretion of mirex-^{14}C in animals exposed for long periods to the insecticide in the diet, *J. Agric. Food Chem.*, 22, 646, 1974.
94. Mehendale, H. M., Fishbein, L., Fields, M. and Matthews, H. B., Fate of mirex-^{14}C in the rat and plants, *Bull. Environ. Contam. Toxicol.*, 8, 200, 1972.
95. Medley, J. G., Bond, C. A., and Woodham, D. W., The cumulation and disappearance of mirex residues. I. The tissues of roosters fed four concentrations of mirex in their feed, *Bull. Environ. Contam. Toxicol.*, 11, 217, 1974.
96. Weiner, M., Pittman, K. A., and Stein, V., Mirex kinetics in the rhesus monkey, I. Disposition and Excretion, *Drug Metab. Dispos.*, 4, 281, 1976.
97. Pittman, K. A., Weiner, M., and Treble, D. H., Mirex kinetics in the rhesus monkey. II. Pharmacokinetic model, *Drug Metab. Dispos.*, 4, 288, 1976.
98. Dale, W. E., Gaines, T. B., Hayes, W. J., Jr., Storage and excretion of DDT in starved rats, *Toxicol. Appl. Pharmacol.*, 4, 89, 1962.
99. Heath, D. F. and Vandekar, M., Toxicity and metabolism of dieldrin in rats, *Brit. J. Ind. Med.*, 21, 269, 1964.
100. Ecobichon, D. J. and Saschenbrecker, P. W., The redistribution of stored DDT in cockerels under the influence of food deprivation, *Toxicol. Appl. Pharmacol.*, 15, 420, 1969.
101. Brown, J. R., The effect of environmental and dietary stress on the concentration of 1,1-bis(4-chlorophenyl)-2,2,2-trichloroethane in rats, *Toxicol. Appl. Pharmacol.*, 17, 504, 1970.
102. Findlay, G. M. and DeFreitas, A. S. W., DDT movement from adipocyte to muscle cell during lipid utilization, *Nature (London)*, 229, 1971.
103. DeFreitas, A. S. W. and Norstrom, R. J., Turnover and metabolism of polychlorinated biphenyls in relation to their chemical structure and the movement of lipids in the pigeon, *Can. J. Physiol. Pharmacol.*, 52, 1080, 1974.

Chapter 3

THE NERVOUS SYSTEM

Robert M. Joy

I. INTRODUCTION

The nervous system is the primary system that provides control for the organism over its external and internal environments. It receives inputs from these environments through various receptors, integrates such information, extracts meaning, and finally generates responses aimed at maximizing the organism's survival. It controls both voluntary and involuntary processes.

The nervous system of even simple organisms is capable of tremendous complexities and varieties of response. In higher vertebrates, and especially in man, this capability is truly astonishing and remains primarily beyond understanding. Although the structure and function of simpler parts of the nervous system, e.g., the peripheral nerves, autonomic nervous system, and first and second order sensory and motor neurons, have been partly worked out, the bases underlying higher level integration and phenomena such as consciousness, learning, and memory remain poorly understood.

Because of the unique properties of the nervous system, it is frequently and importantly involved in toxic responses to drugs and to chemicals in the environment. The participation of the nervous system in a toxic reaction may result from direct or indirect causes. Direct involvement occurs when chemicals stimulate or depress, or sometimes destroy, nervous tissue. The actions of chemicals may involve all or a part of the nervous system. Both specific and nonspecific interactions have been described. Chemicals such as dieldrin and endrin appear to enhance the excitability of nearly all neurons, and thus are relatively nonspecific in their actions. On the other hand, organophosphate insecticides act specifically upon cholinergic systems by inhibiting acetylcholinesterase. At the time of maximal involvement of cholinergic neurons, the activity of neurons utilizing other transmitter substances remains relatively unaffected.

Indirect involvement of the nervous system in toxic reactions is also commonly observed. This should be clearly differentiated from direct effects. Indirect involvement occurs when the organism attempts to escape from or compensate for toxic actions directed at other loci. These include such behaviors as escape reactions directed towards removing the animal from the toxic environment, licking, vocalization, biting, chewing, etc. Responses of nervous origin arising indirectly from irritation of skin or mucous membranes, electrolyte imbalances, or toxic actions occurring at voluntary or smooth muscle sites are usually quite easy to distinguish from direct neural toxicities. Indirect reactions to toxic effects developing in the pulmonary or cardiovascular systems or following exposure to metabolic poisons are frequently more difficult to distinguish. Convulsions, which are usually considered to be a direct neurotoxic effect, commonly occur secondarily to anoxia. In the latter case they are more properly classified as an indirect, not a direct, neurological consequence. Such distinctions are of more than casual interest. They are critical in establishing both the sites and mechanisms underlying toxic responses.

It is the intent of this chapter to provide a general review of aspects of nervous system structure and function. Emphasis is placed on material most useful to the understanding of changes produced by the insecticides, particularly in the mammalian organism. It is assumed that the reader has a general knowledge of the nervous system. For those wishing a more comprehensive coverage, one of the general references given at the end of this chapter should be consulted.

which is absent from the axon, represents the nodal points of the endoplasmic reticulum which exists throughout the soma and dendrites. The endoplasmic reticulum is the major protein-synthesizing organelle and manufactures in 1 to 3 days an amount of protein equal to the total protein content in the cell. Much of this is related to the formation and storage of neurotransmitter substances and is transported down the axon to the terminal by axoplasmic transport. Microtubules and neurofilaments are also found in the cell body and in other parts of the neuron. These are filamentous protein structures oriented parallel to the long axis of the cell or its processes. The microtubules are straight, unbranched structures, 200 to 260 Å in diameter, with a central core of 100 to 140 Å. They appear to be involved in fast axoplasmic transport but may also play some role in providing strength and rigidity to the neuron. In addition to these organelles, the Golgi apparatus, lysosomes, and various pigmented inclusion bodies are usually present.

The dendrites represent all the processes of the cell body except for the specialized axonal process (axon). They are usually numerous and serve to increase the surface area of the neuron available for receiving synaptic input. Neurons will have one or more main dendrites which successively branch and branch to form many smaller processes. Small protrusions of various sizes and shapes, called dendritic spines, thorns, or gemmules, are present on many dendrites. These are specializations of the dendrite upon which synaptic contacts occur.

The axon of the typical neuron arises from a cone shaped region of the cell body, the axon hillock. The initial segment of the axon is both the smallest region in diameter and the region with the lowest threshold to electrical activation. Distal to the initial segment the axonal diameter enlarges and the diameter remains constant out to the terminal ending or until the axon branches. Myelinated neurons have their axons ensheathed by myelin segments, each segment provided by an oligodendrocyte or neurolemmal cell. Between segments the axonal membrane is exposed. These regions are termed the "nodes of Ranvier". Unmyelinated axons do not possess such a segmented sheath. Rather they tend to collect in bundles which are loosely enclosed into troughs formed by neurolemmal cells or oligodendrocytes. The terminal end of an axon most often is profusely branching, each branch ultimately terminating in a synaptic ending.

Mechanisms are available within the neuron which allow the passage of substances from the soma to axon terminal and vice verse. Two types of axoplasmic transport from soma to axon terminal have been demonstrated. There is a slow transport involving bulk flow of materials and axoplasm, mitochondria, lysosomes, and vesicles, at a rate of 1 to 3 mm/day. A rapid transport system also exists that is capable of transport at velocities of up to 100 mm/day. The most rapid rate of transport so far established is for the axons in the hypothalamohypophyseal tract where velocities of 2800 mm/day have been reported. Because rapid transport can be eliminated by treating the axon with colchicine, microtubules are probably involved. Retrograde transport, from axon terminal to cell body, is also established. The purpose of axoplasmic transport includes the transport of enzymes involved in neurotransmitter synthesis and of neurosecretory granules to the axon terminal from their source of synthesis in the cell body.

Neurons assume a vast array of forms (Figure 2) in accordance with the functions they subserve. All parts of the neuron, that is cell body, dendritic arborizaton and axonal aborization, can possess different shapes. Common types include the unipolar neurons, in which the cell body has a single extension giving rise to both dendritic and axonal branches. Such neurons are found in dorsal root ganglia and in the olfactory bulb. Bipolar neurons possess extensions at both ends of the cell body, one dendritic and the other axonal. This morphological type is characteristic of retinal bipolar cells, some cochlear, and vestibular ganglion cells. Multipolar neurons have several extensions of the cell body, usually one axonal and the rest dendritic. Most neurons are of this type.

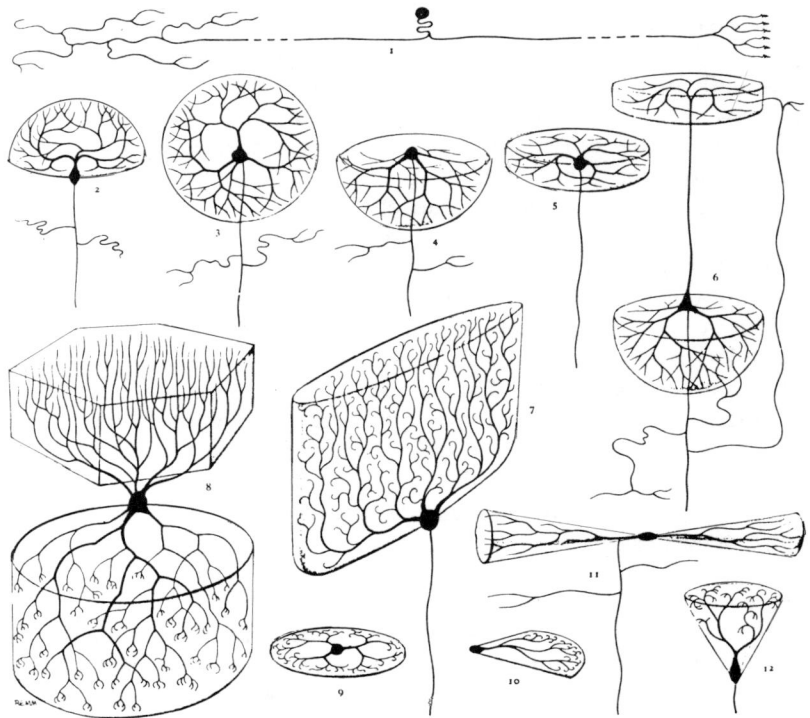

FIGURE 2. Patterns of variations of neuronal geometry: (1) unipolar, sensory ganglionic neuron; (2) bipolar neuron; (3-8); multipolar neurons: (3) stellate neuron, with (4), (5), and (11) which are modifications of this pattern; (6) pyramidal neuron with an apical and a series of basal dendrites, and recurrent axon collaterals, from the cerebral cortex; (7) Purkinje neuron from the cerebellar cortex; (8) Golgi neuron from the cerebellar cortex; (9) and (10) amacrine cells lacking axons; (12) glomerular neuron (mitral cell) from the olfactory bulb, showing recurved dendritic tips. (From Warwick, R. and Williams, P. L., *Gray's Anatomy,* 35th British ed., W. B. Saunders, Philadelphia, 1973, 766. With permission.)

In addition to the division of neurons into simple morphological categories based upon cell shape, other schemes are frequently employed. Thus neurons may be classified as Golgi type I neurons, which have long axons connecting different parts of the CNS or as Golgi type II neurons, with short axons that terminate entirely in the area immediately around the cell. Some neurons even lack an axon altogether, such as the amacrine cells found in the retina. It appears that axonless cells are characteristic elements in all sensory pathways. They, along with other inhibitory neurons in sensory ganglia and thalamic nuclei, are essential to the sharpening of the sensory image through lateral inhibition.

Certain common cell types are named on the basis of their dendritic morphology. Neurons with a roughly spherical dendritic field (surrounding the whole cell body) are termed stellate cells. Such cells are common throughout the cortex, reticular formation, and spinal cord. Pyramidal cells are so named because the cell body is pyramidal or conical in shape. Two sets of dendrites, an apical and a basal set, are also characteristic of the pyramidal cell. These types are common in the cortex and in the hippocampus. Fusiform cells possess conical dendrites emerging from one or both ends of the cell body. A final commonly described neuronal type with a remarkable dendritic field is the cerebellar Purkinje cell. In these neurons a primary dendrite emerges apically and branches and rebranches to form a two-dimensional, fan-like array.

C. The Synapse

When it was first established that the central nervous system was composed of large numbers of individual neurons rather than a continuous syncytial network, the question was raised as to how such neurons were joined. Light microscopy suggested, and the electron microscope subsequently confirmed, that neurons did not enter into intimate contact with each other, but rather were always separated by a 200 to 500 Å space. Specializations of the membranes of both cells at such close encounters suggested that they served the function of interneuronal communication. This concept has been completely confirmed. These loci are now termed "synapses". Because synapses conduct only in one direction as a rule, asymmetry is found in the structural elements. The most common type of synaptic junction is that between an axon (presynaptic element) and a dendrite (postsynaptic element). The presynaptic axon terminal is usually expanded into a bulb or bouton. These boutons may synapse with dendritic spines or the flat surface of a dendrite, forming an axodendritic synapse. If the axonal bouton synapses directly onto the soma, the synapse is termed axosomatic. It is also possible for the bouton to synapse with another axon forming an axoaxonal synapse. In addition to these synaptic types, which are by far the most numerous in the CNS, synapses between two dendrites (dendrodendritic), two cell bodies (somatosomatic), a cell body and an axon (somatoaxonic), and a cell body and a dendrite (somatodendritic) have been observed.

With few exceptions, all synapses in the mammalian CNS are chemical in nature. These possess three components: a presynaptic element, a postsynaptic element, and a synaptic cleft of approximately 200 Å between the two elements. Within the presynaptic boutons are small, spherical, synaptic vesicles containing the neurotransmitter material. During activity it is thought that these vesicles attach to specific sites on the presynaptic membrane and subsequently release their contents into the synaptic cleft. The postsynaptic membrane is specialized in that there are receptors capable of responding to transmitter molecules present within it. When a molecule of nuerotransmittercombines with a receptor, the membrane permeability is altered. Thus, "information" is transferred across the synapse. Chemical synapses are asymmetrical and unidirectional in that transmitter is present and can only be released from one element, not both.

A few electrical synapses have been reported to exist in the mammalian CNS. They are much more commonly observed in lower vertebrates. Such synapses also are comprised of three elements: a presynaptic element, a postsynaptic element, and a narrow gap of about 20 Å in between. Because the two membranes of the involved elements are so closely apposed, the contact point is usually termed a "gap junction" or a "tight junction". Electrical synapses are symmetrical because no synaptic vesicles exist, and the pre- and postsynaptic membranes are indistinguishable. Many, if not most, are bidirectional. Activity can pass through the synapse in either direction. The significance of an electrical synapse lies in the fact that it acts to effectively couple the pre- and postsynaptic elements into one unit. Thus, any activity that arises in one or the other element is immediately and completely coupled into the other. In a sense the electrical synapse acts to convert two or more elements into one. In cardiac muscle, where tight junctions are common between muscle fibers, activation of one fiber spreads immediately and inevitably to all other fibers, making their composite behavior one of unity.

When first formed, synapses are recognizable only as dense zones on each side of the synaptic cleft. With maturation, presynaptic and postsynaptic characteristics become recognizable. The presence of nondifferentiated zones separated by a synaptic cleft has also been observed in adult nervous systems, possibly representing new synapses being formed or old synapses being destroyed. A plasticity in synapse production and removal is an essential element in most theories of memory and learning in which the number of synapses along a specific conduction pathway must increase or decrease according to use. However, evidence supporting this concept is still equivocal. What has been established is that synapses may become either more or less easy to excite with repeated use or may change in size. An alternative way in which synaptic efficacy could be altered would be to increase or decrease the number of receptor sites on the postsynaptic membrane. Either a pre- or a postsynaptic mechanism would require increased protein synthesis. In this regard the capability of synthesizing new protein appears necessary to the laying down of long-term memory.

III. COMPARATIVE ANATOMY OF THE NERVOUS SYSTEM

A. Phylogenetic Development

In unicellular organisms, specialized regions of the cell function as sensory receptor and effector. The development of a true nervous system is obviously only possible with multicellular development. The simplest neural elements occur in simple animals such as the primitive coelenterates. These are epithelial sensori-motor cells in which the functions of sensory reception and motor activity are both represented in the same cell. The nervous system subsequently evolved by the separation of receptor and effector components into discrete separate sensory, and motor units. As the sensory and motor elements became further and further separated in space, additional cells became interposed between them and served to "connect" the sensory and motor cells to one another. By the incorporation of more and more of the additional cells into the system to facilitate communication, a simple nerve net of the type found in some coelenterates was created. The next evolutionary step resulted in the aggregation of neurons into ganglia.

The general increase in body size that has occurred during the evolution of higher organisms results frequently in a large spatial separation of central neurons from sensory receptors and motor effectors. Communication in such circumstances was solved by developing neurons possessing long processes. These subsequently developed the properties of modern axons and became capable of rapidly transmitting activity between other widely separated cells. In the more complex organisms, afferent or sensory axons connect receptors with a central nervous system while efferent or motor axons connect the central nervous system with various effectors. The afferent and efferent fibers, receptors, and those nerve cells lying outside the central nervous system, taken together, make up the peripheral nervous system.

Although neurons, per se, have become somewhat more differentiated and specialized during evolution, this factor alone does not explain the tremendous increase in capacity seen in man as compared to simpler organisms such as worms and insects. In the main, nervous complexity has occurred as a result of a tremendous increase in the number of nerve cells and in their interconnections.

B. Insect Nervous System

The insect nervous system represents a midpoint along the evolutionary pathway that culminates in man. In most insects, the nervous system is a ventrally located, double ganglionated cord. In the embryo, each segment possesses its own ganglia, but in the adult some of the ganglia fuse so that there may be only two in the head, three in the thorax, and five in the abdomen (Figure 3). In some insects, such as the common housefly, additional fusion occurs, obscuring the segmental design. The most obvious differences between the insect and mammalian nervous system are the much greater degree of encephalization and overall complexity of the latter. The various segmental ganglia of the insect are the analogue of the mammalian brain and spinal cord.

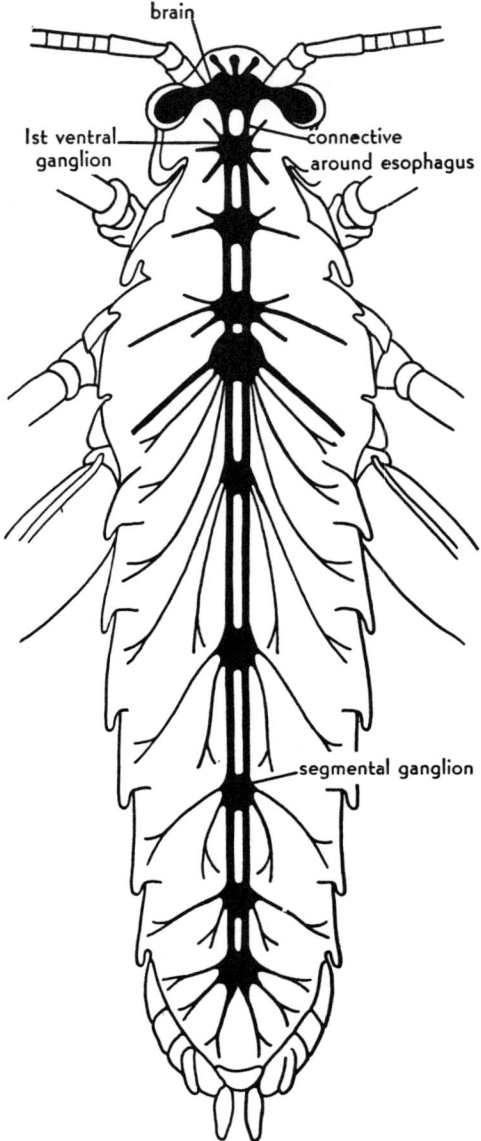

FIGURE 3. The insect nervous system. (From Buchsbaum, R., *Animals Without Backbones*, University of Chicago Press, Chicago, 1958, 284. With permission.)

In insects, the brain lies above the esophagus and between the eyes. It is joined to the first ventral ganglia by a pair of connectives which encircle the gut. These connectives consist of axons with their associated glial cells and connective tissue sheaths. The brain has no centers for coordinating muscular activity, and it serves primarily as a sensory relay which receives inputs from the sense organs on the head. Its outflow is largely directed to the first ventral ganglion (subesophageal ganglion) of the same side. Peripheral nerves arise from the first ventral ganglia to innervate the muscles of the head and mouth. The remainder of the ganglia represent a rudimentary spinal cord.

Typically, each somite has its own pair of ganglia. These give rise to the various sensory and motor nerves that innervate that somite. The segmental ganglia are joined together by connectives containing nerve processes. All synaptic connections in the insect central nervous system (CNS) occur in the ganglia, none in the connectives. In a typical ganglia the ganglion cells are located at the periphery. Their processes and the axonal endings entering the ganglia from without form a profuse central neuropil. The ganglia and connectives are sheathed in a neural lamella of varying thickness. This sheath is formed by connective tissue elements and is similar in structure to collagen. Many of the properties of this sheath are similar to that of the myelin sheaths in vertebrate nerves.

The peripheral nerves of insects arise from the various segmental ganglia. They contain both sensory and motor fibers, all of which are nonmyelinated. Both excitatory and inhibitory motor fibers exist, in contrast to mammals where only excitatory motoneurons occur. The insect motor fibers make multiple synaptic connections with muscle cells and do not form the typical motor end plate region characteristic of mammalian muscle. The autonomic nervous system of insects is restricted to a single system, the sympathetic or visceral system, which innervates the heart, gut, spiracles, and other organs. There are no peripheral autonomic ganglia in insects. Autonomic fibers arising from the CNS go directly to the effector organs before synapsing.

C. The Mammalian Nervous System

The mammalian brain is shown schematically in Table 1. During development, the nervous system begins as a plate of epidermal cells which eventually coalesces into a tube-like structure. The neural tube develops at one end three distinct areas of accelerated cellular growth which eventually form the brain. These regions are termed the prosencephalon, mesocephalon, and rhombencephalon. With maturation each of these regions differentiates to form the characteristic structure of the mature brain. The prosencephalon gives rise to the telencephalon and diencephalon, the mesocephalon becomes the mesencephalon, and the rhombencephalon gives rise to the metencephalon and myelencephalon.

The telencephalon contains the striatal nuclei, which are most fully developed in birds, and the cortex, which is by far most well developed in the mammals. The corpus callosum, the broad sheet of myelinated cell processess which connects the two cerebral hemispheres together, is also part of the telencephalon. Within the diencephalon lie the thalamus, with its associated specific and nonspecific nuclei, the subthalamus, and the hypothalamus. The latter is importantly involved in autonomic functions and is intimately associated with the pituitary gland. The mesencephalon is characterized by the four colliculi dorsally and by the tegmentum ventrally. Lying in between these regions is the mesencephalic portion of the reticular formation. The reticular formation actually forms the central core of the brain from the anterior edge of the mesencephalon through the myelencephalon. The pons and the cerebellum are both easily distinguished structures lying within the metencephalon while the myelencephalon retains its tube-like appearance with minimal morphologic changes. The posterior two-thirds of the myelencephalon is called the medulla oblongata, which merges without further demarcation with the spinal cord.

The spinal cord represents the posterior part of the CNS. It is entirely enclosed within the vertebral canal, is sementally arranged, and gives rise to the various peripheral nerves. Sensory and motor function are concentrated within the dorsal and ventral halves of the spinal cord, respectively. Afferent fibers enter the cord by way of the dorsal roots (in the periphery these same afferent fibers may end as bare nerve endings or may terminate upon specialized sensory receptor cells). Some of these fibers synapse within the same cord segment, others synapse at adjacent segments, while some course

Table 1
DEVELOPMENT OF THE MODERN VERTEBRATE NERVOUS SYSTEM

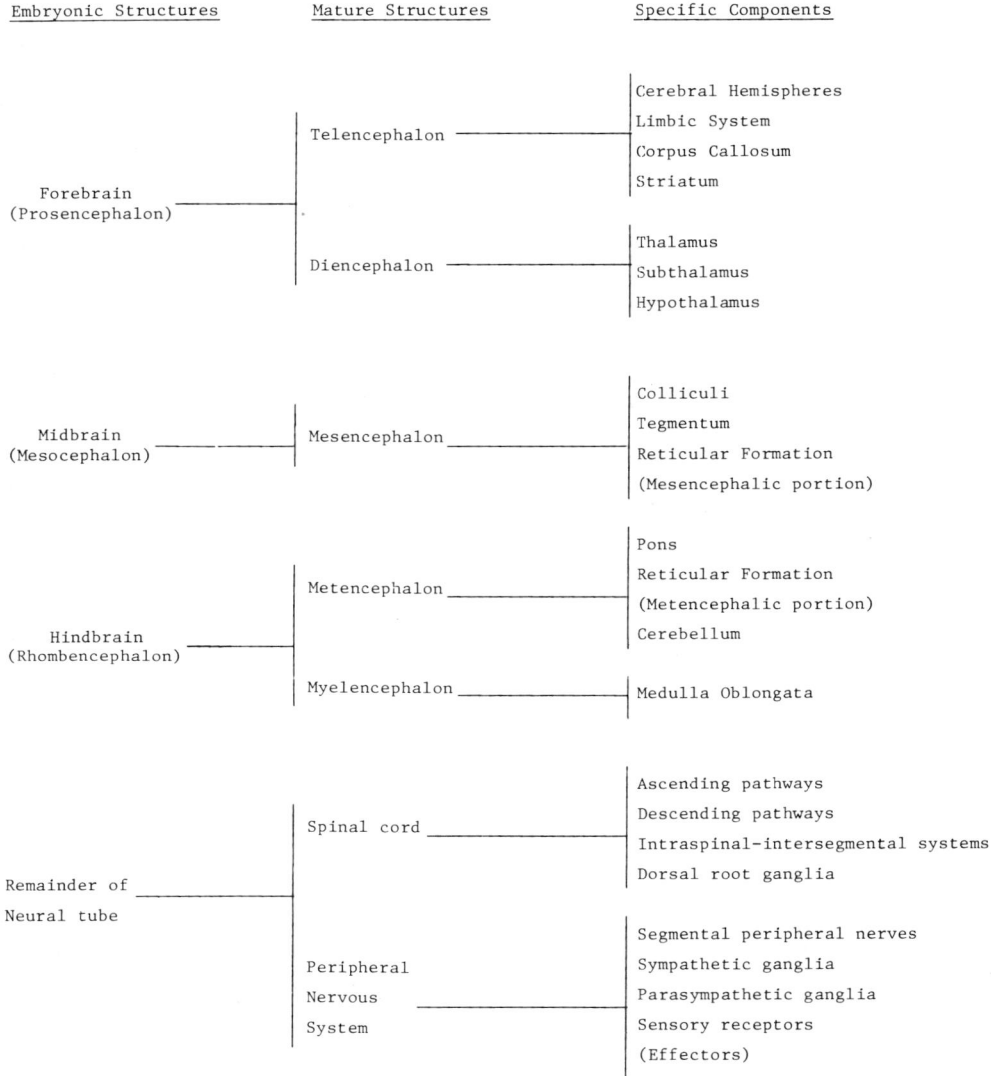

along sensory pathways to synapse within the brain. Efferent fibers originate from cells located within the ventral part of cord. These course out through the ventral roots. Those fibers destined to innervate skeletal muscle and muscle spindles continue without synapsing to the muscle fibers they innervate. Those destined to innervate smooth or cardiac muscle and glands (the preganglionic autonomic fibers) synapse once before reaching their effectors. The parasympathetic fibers synapse primarily in intramural ganglia, closely apposed with the effector. The sympathetic fibers synapse in various ganglia, some of which comprise the vertebral sympathetic chain lying alongside of the spinal cord just outside the vertebral canal, others of which exist at various locations within the abdominal cavity. Postganglionic fibers arise from these parasympathetic or sympathetic ganglia and go to innervate the effector organs.

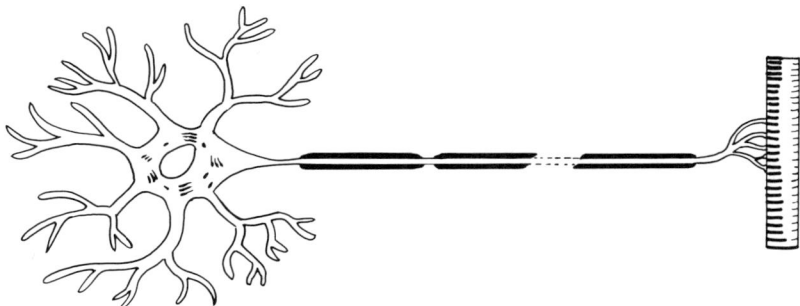

FIGURE 4. The neuron as an information system. The **Information processing section**—consists of the cell soma and dendrites. These comprise a unit which receives information from other cells by way of its synaptic inputs. This section of the neuron converts its various inputs into small changes in electrical potential across its membrane. **Threshold detecting section**—consists of the axon hillock and the initial segment of the axon. These regions monitor the net change in electrical potential occurring on the soma and dendritic portions, and if appropriate they initiate the action potential. **Transmitting section**—consists of the axon, whose only purpose is the faithful transmission of the action potential along the length of the axon to its terminals. **Transducer section**—is the axon terminals which respond to the action potential by releasing a chemical substance (transmitter) which diffuses across the synaptic space to the next cell. There the transmitter reacts with receptors on the soma or dendritic membranes (information processing section) of the next cell.

IV. ELECTROCHEMICAL PROPERTIES OF NEURONS

The neuron is a cell specially designed to respond to, evaluate, and transmit information. It is helpful to think of the neuron as differentially specialized to perform these various functions (Figure 4). It possesses an information processing section consisting of the cell soma and dendrites. These regions act as chemical to electrical transducers and convert afferent input information into small changes in electrical potential. The axon hillock and the initial segment of the axon can be viewed as a threshold detecting section. These regions effectively monitor the net change in electrical potential occurring on the somatic and dendritic surfaces and, if appropriate, initiate an action potential. The axon is the transmitting section of the neuron and has as its major purpose the faithful propagation of the action potential from its origin in the hillock to the terminal endings. The various axon terminals behave as electrical to chemical transducers. The electrical signal (action potential) causes the release of a chemical substance (neurotransmitter) which diffuses across the synaptic space to the next cell. There the transmitter acts upon the soma and dendrites of the next cell to start the process anew.

This view of the neuron emphasizes that function depends largely on a combination of chemical and electrical activities. Intracellular communication relies primarily upon electrical events whereas intercellular communication is predominately chemical in nature.

A. Background Concepts
1. Unicellular Measurements

Neurons are typically 10 to 50 μm in diameter and from 1 to 10 cm in length. Most of the length of a neuron is accounted for by the long axonal process. Neurons weigh only nanograms. The majority of events associated with the generation and alteration of membrane potentials occur within a few milliseconds. Longer term, often metabolically related, changes in excitability have been observed with periods of seconds, minutes, or hours. Membrane potentials range from 0 to 100 mV. Concentration of various intracellular substances may range from picomolar to millimolar.

2. Ion

An ion is a molecule or atom possessing a net electric charge. Positive ions are formed by the loss of an electron while negative ions are formed by the gain of an electron. In nerve cells the primary positive ions are Na^+, K^+, and Ca^{++}. The primary negative ions are Cl^- and intracellular proteins, which tend to carry a net negative charge at physiological pH.

3. Coulomb (C)

A coulomb is the quantity of electric charge produced by 6.2×10^{18} ions of the same charge. One drop (0.05 mℓ) of saline solution would contain about 2 C of electrical charge — 1 C of Na^+, and 1 C of Cl^-. The electric charges contained in one drop of saline equal the number of electrons flowing through a 100 W light bulb in about 2 sec. From this perspective the amount of charge contained in a drop of saline seems pretty small. However, from the neuron's point of view it is gigantic. A nerve fiber 10 µm in diameter and 10 cm long (average size) would have to produce 500 million action potentials to move the amount of charge in a drop of saline across its membrane. At an average rate of 10 impuses/sec this would take about 580 days.

4. Ampere

An electric current is produced when electrically charged particles move from one location to another. This could be electrons moving along a wire or it could be ions flowing through the cell membrane. A current of 1 A flows when 1 C of charge moves per second. For the 100 W light bulb about 1 A of current flows through the filament each second. For an average nerve cell a current of a few µA may flow through the membrane at the peak of the action potential. Because the nerve fiber is so small, however, the current density (current flow per unit area of membrane) may be several mA/cm² of membrane surface.

5. Volt

The volt is a measure of the imbalance in net charge that exists between two regions. In a biological system the existence of a voltage (potential difference) always indicates that there are relatively more positive or negative charges in one region as compared to another. An imbalance of just a few ions can create very large potentials or voltages. For example, if we were to place a drop of saline into each of two separate tubes, then by some process exchange all the Na^+ from one of the tubes for all of the Cl^- in the other (all the positive ions are in one tube and all the negative charges are in the other), the tubes would have a potential difference of about 40 trillion (4×10^{13}) V. To create a potential difference of 100 mV across the cell membrane of a nerve, an imbalance of only a few thousand ions is necessary. It is worth emphasizing that large changes in potential across a neuronal membrane may result from the movement of only a few ions across that membrane.

6. Ohm

The ohm is a measure of the resistance a material offers to the flow of current. By definition, if 1 A of current flows through a substance when a potential of 1 V exists across it, then the substance has a resistance of 1 Ω. Because ions move only through pores, which make up a very small percentage of the total membrane surface, the membrane serves as a barrier to ionic movement. In other words, the membrane partially resists the free flow of ions. A square centimeter of membrane has a resistance at rest of approximately 500 Ω (5 KΩ).

7. Permeability

The permeability of a membrane to an ion is defined as the velocity at which that ion can move through it down an electrochemical gradient. Its value depends upon the mobility of the ion in solution, the size of the electrochemical gradient, and the size and number of channels in the membrane. Because velocity of movement is measured, permeability is expressed in centimeters per second. The permeabilities of a typical spinal motoneuron soma during the resting state have been shown to be approximately:

Ion	Permeability (10^{-6} cm/sec)	Relative permeability
K^+	2	1
Na^+	0.04	0.02
Cl^-	2	1

True permeabilities are difficult to measure. In most instances data for nerve cells are presented in terms of relative permeabilities and in changes in relative permeabilities.

B. The Generation of a Membrane Potential

It has been indicated that a potential difference invariably results when the distribution of positive and negative charges between two regions is not equal. For a nerve cell the two regions are the intracellular and extracellular volumes. These are separated from each other by the cell membrane which acts as a barrier to free, rapid exchange of substances from one volume to the other. When a membrane potential is demonstrable it means that the intracellular volume contains a few more net positive or net negative charges per unit volume than does the extracellular volume. When the intracellular compartment has more net positive charges than the extracellular compartment, the cell is said to possess a positive membrane potential. Conversely, if the intracellular volume contains a net negative charge, a negative membrane potential will exist. It is now universally accepted that a neuron invariably possesses a negative membrane potential (inside compared to outside) when at rest which may range from cell to cell from 40 to 90 mV. From this it can be deduced that the interior of the cell must contain a surplus of negative charges, or a deficit of positive charges.

The neuron creates and maintains the membrane potential by two important means. First it expends metabolic energy to create and maintain concentration gradients for various ions. Second it controls the permeability of its membrane to the various ions. This permeability is different for the various ions and, in addition, can be changed over a wide range. These two factors account primarily for the various potentials that the neuron employs in its function: the resting potential, the action potential, and the various synaptic potentials.

The establishment of a concentration gradient coupled with selective permeability induces the membrane potential in the following way. For simplicity, first consider a nerve cell which has pumped K^+ ions into the intracellular fluid in exchange for Na^+ ions, which the cell has pumped out into the extracellular fluid. Again, for simplicity, assume that the nerve membrane is totally impermeable to any ionic movement during the period in which the ionic gradients are being established. The pumping process continues until sizable concentration gradients have been established both for K^+ and Na^+. Throughout this period of establishing a gradient, there would be no potential difference because positive ions are being equally exchanged. There is no change in the net positive charge on either side of the membrane.

At some point in time the neuron brings into play the second mechanism required to induce a potential difference across its membrane, the development of selective permeability. In this example let the neuron develop a slight permeability to K^+ ions while remaining impermeable to the movement of Na⁺ ions. Because of the concentration gradient that the neuron has created for K⁺, these ions will begin to flow outwards through the membrane. In the process of exiting from the intracellular space, the K⁺ ions take with them a positive charge. This results in a charge separation across the membrane such that there are more positive charges outside the cell. The consequence of this movement is to produce a potential difference across the membrane directed so that the interior of the neuron is negative in potential to the extracellular fluid. In this example, the movement of K⁺ will continue down its concentration gradient until the force represented by the imbalance in concentration is just balanced by an equal, but oppositely directed, electrical force created by the separation of charges, e.g., until the chemical and electrical forces reach an equilibrium. The value of the electrical potential at equilibrium can be determined for a single ion with the Nernst equation:

$$V = \frac{RT}{Fn} \ln \frac{(C_1)}{(C_2)}$$

Where: R = universal gas constant, T = absolute temperature, F = Faraday number, n = valence of ion, C_1, O_2 = ionic concentrations on either side of membrane.

At body temperature and for univalent ions the Nernst Equation simplifies to:

$$V \cong 60 \text{ mV} \cdot \log \frac{C_1}{C_2}$$

It is interesting to note that in this hypothetical neuron, (which is permeable to only one ionic species) the membrane potential, which has developed because of the charge separation, will be maintained indefinitely. It is therefore a capacitor as well. In effect, in proposing a semipermeable membrane it has been necessary to assume a structure which behaves like a mosaic of conductances (the ionic channels and capacitors (the areas which have no channels) which are effectively arranged in parallel. It may easily be calculated that for a membrane like that of a nerve cell, with a capacitance of a few microfarads per square centimeter, the charge separation associated with a transmembrane potential of about 90 mV is equivalent to an excess charge on either side of the membrane corresponding to only a very small proportion of the ions present in a region a few angstroms thick. Although any alteration of membrane potential will involve movement of ions, the corresponding change in ionic concentrations will be negligible.

The Nernst equation is a valid means of determining a membrane potential contribution of a single ion. For conditions where more than one ion can penetrate a membrane, a more comprehensive equation is necessary. The Goldman, Hodgkin, Katz (GHK) equation is used to provide an approximate solution for membrane potentials when more than one ion is moving at the same time, a situation more consistant with real neuron behavior. When dealing with more than one ion, (relative) membrane permeabilities must also be considered to determine the contribution of each ion to the overall potential at equilibrium. The GHK equation in simple form is:

$$V = -60 \text{ mV} \cdot \log \frac{P_K (K_i) + P_{Na} (Na_i) + P_{Cl} (Cl_o)}{P_K (K_o) + P_{Na} (Na_o) + P_{Cl} (Cl_i)}$$

Where:
    V = potential difference: inside referenced to outside of the membrane
    P_K, P_{Na}, P_{Cl} = membrane peremeabilities for K⁺, Na⁺, and Cl⁻ respectively.

(K_o), (Na_o), (Cl_o) = extracellular K^+, Na^+ and Cl^- concentrations.
(K_i), (Na_i), (Cl_i) = intracellular K^+, Na^+ and Cl^- concentrations.

In most nerve cells, the three ions, K^+, Na^+ and Cl^- are the only major determinants of the membrane potential. In some, however, Ca^{++} may also play a role in generating potential differences. In that case, Ca^{++} would also have to be added into the GHK equation as given above.

C. The Resting Potential

The resting potential in the typical neuron is the result of the active development of concentration gradients for the ions, K^+ and Na^+. The value of the resting potential is determined both by the size of the concentration gradients and by the relative permeabilities of the membrane to these ions. The major contribution is provided by K^+. Na^+ contributes relatively little to the resting potential. Chloride ions are typically passively distributed between the intracellular and extracellular compartments in accordance with the potential difference across the membrane. In the typical neuron the resting potential lies between the potassium and the sodium equilibrium potentials, each of which can be determined with the Nernst equation. These are near -90 mV for potassium and $+45$ mV for sodium, respectively. Resting potentials are usually within the range of -50 mV to -80 mV. If the membrane permeabilities for both Na^+ and K^+ were equal, the resting potential would be exactly in between, that is, -20 to -25 mV. The fact that the resting potential is much nearer to the potassium equilibrium potential than to the sodium equilibrium potential indicates that the permeability of the membrane to potassium at rest is much greater than its permeability to sodium. The relative permeability for K^+ is usually 20 to 50 times that for Na^+.

Because the nerve membrane is permeable to both K^+ and Na^+, a constant slow leak for K^+ occurs from inside to outside of the membrane which is exactly balanced by a correspondingly slow leak of Na^+ from outside to inside. Because of these leaks, the concentration gradients established by the neuron would slowly disappear unless the neuron had a mechanism for maintaining them. These gradients are actively maintained by the neuron, which expends energy to pump the sodium, which has leaked into the cell, back out. In conjunction with this, potassium is pumped back into the cell. Energy is required for this process since the ionic movement is actually uphill, that is, against the existing ionic gradients. The energy for this process is provided by ATP. If oxidative metabolism is interrupted, the concentration gradients slowly disappear. In certain situations the pumping appears to occur on a one-to-one basis in which one sodium ion is exchanged for one potassium ion. In these situations, no potential is produced by the pump since no separation of charge occurs. In other situations, more of one ion, usually sodium, is pumped than is the other. In these circumstances the pump can actually contribute directly to the membrane potential. Such pumps are termed "electrogenic". In addition to the pumps described for Na^+ and K^+, other pumps exist within the membrane. For example, Ca^{++} is also actively pumped out of neurons to maintain a calcium gradient, but the specifics of this pumping process appear complex. Normally the movement of calcium is more important in the kinetics of transmitter release than in the development and maintenance of a resting potential.

D. The Action Potential

A common feature of viable nerve cells is the capacity to develop an abrupt, transient change in membrane potential termed the action potential (Figure 5). This event usually is initiated in the initial segment of the axon and is conducted down the axon to the axon terminals. The action potential is an all-or-none event in that, if it occurs at all, it always occurs with the same amplitude and velocity of conduction.

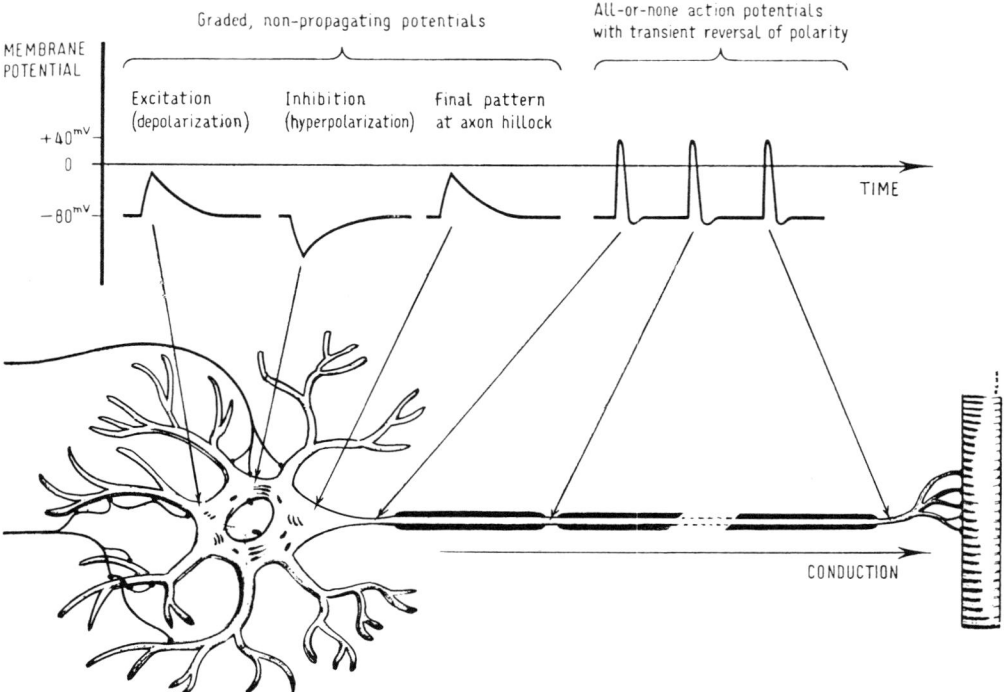

FIGURE 5. A diagram showing the types of change in electrical potential which can be recorded across the cell membrane of a motor neuron at the points indicated by the arrows. Excitatory and inhibitory synapses on the surfaces of the dendrites and soma cause local graded changes of potential which summate at the axon hillock. These may initiate a series of all-or-none action potentials which are conducted along the axon to the effector terminals. (From Warwick, R. and Williams, P. L., *Gray's Anatomy*, 35th British ed., W. B. Saunders, Philadelphia, 1973, 770. With permission.)

The action potential can be initiated by any of a number of factors, e.g., by neurotransmitter action, receptor responses, mechanical deformation or even injury, all of which induce a decrease (depolarization) in the resting memmbrane potential. Increases (hyperpolarizations) in the esting membrane potential have no such effect. When the potential of the initiating segmnt is depolarized beyond a certain point, the displacement of the potetial increases out of proportion with the initiating stimulus, and the recovery of the resting state is delayed. This response has been demonstrated to be due to a progressive increase in the membrane permeability to Na^+ (Figure 6). This process is characterizd by positive feedback, as additional permeability to Na^+ leads to further dddepolarization of the membrane, which leads to additionl permeability, and so on. In consequence, the membrane potential L (+ + 45 mV).

The sudden depolarization stemming fr and progressive increase in sodium ermeability is followed by events which just as rapidly return the membrane potential to its resting level. One of these events is the rapid shutting off of sodium permeability. The other is a slower, more prolonged increase in potassium permeability. The effects of these two actions will be that no more Na^+ enters the cell while K^+ begins to leave. There is at this time a large electrochemicl gradient forcing K^+ from the cell. As the K^+ leaves, positive charges are removed from the interior of the cell, causing the membrane potential to become more negative. This process will continue until enough K^+ has left the cell to match the Na^+ that originally entered. With the balancing of charges, the membrane potential will be returned to the original resting condition. The cost of this process to the cell has been the entry of a small number of

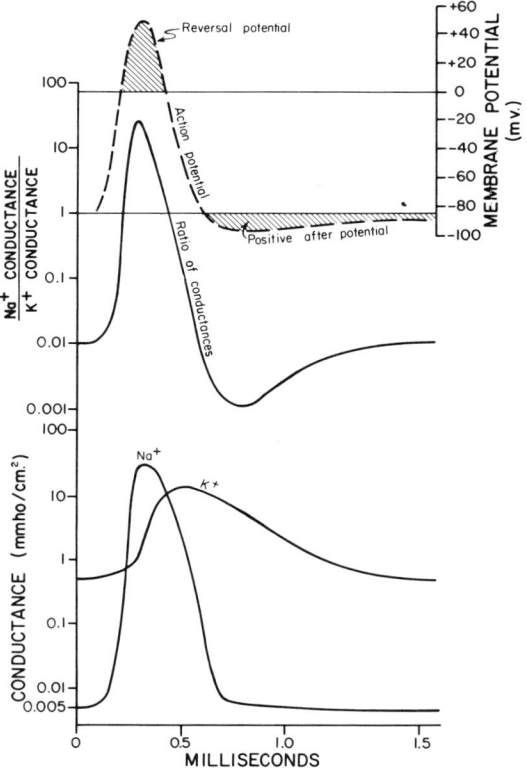

FIGURE 6. Changes in sodium and potassium conductances during the course of the action potential. Note that sodium conductance increases several thousand-fold during the early stages of the action potential, while potassium conductance increases only about thirty-fold during the latter stages of the action potential and for a short time thereafter. (From Guyton, A. C., *Structure and Function of the Nervous System*, W. B. Saunders, Philadelphia, 1972, 16. With permission.)

Na^+ ions and the loss of an equal number of K^+ ions. These must be pumped back across the membrane in order to maintain the needed ionic gradients. The action potential actually produces only a small displacement of ions. The average cell can produce thousands of action potentials before the concentration gradients for Na^+ and K^+ are seriously altered. Under pathologic or toxicologic conditions, however, the rate of pumping may not keep pace with the rate of Na^+ entry. In such situations, the erosion of the ionic gradients can have serious consequences upon neuronal function.

E. Electrotonic Potentials

Electrotonic potentials are passively distributed changes in membrane potentials occurring at locations removed from a locus of active depolarization or hyperpolarization. They are absolutely essential to the initiation and propagation of the action potential. They also provide a means by which a neuron can differentially "weigh" the influences of synaptic inputs arriving upon its somatodendritic surfaces. These potentials arise because of the cable-like properties that a neuron possesses.

In a perfect cable (one that consists of a central conductor having no resistance to

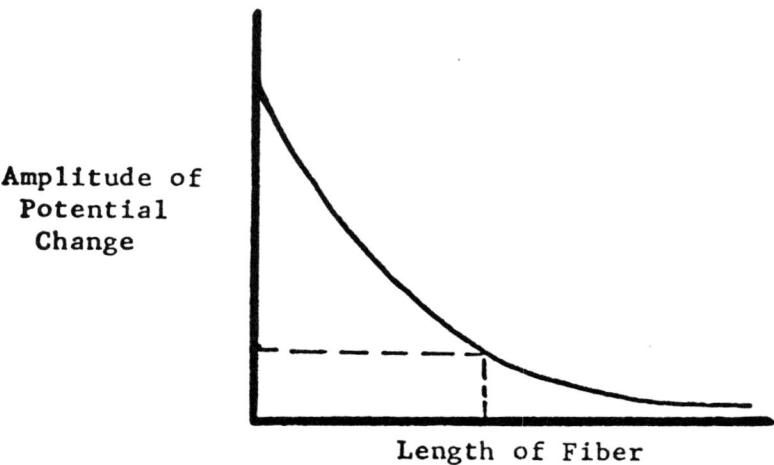

FIGURE 7. The loss of signal amplitude in a cable or its analogue, the nerve fiber, as a function of distance. The loss is exponential with distance. The length constant, shown by the dotted lines, is the distance along the cable at which the signal has decreased in amplitude to 0.39 of its original value.

electron or ion flow surrounded by a perfect insulator having infinite resistance to electron or ion flow) a change in electrical potential induced at any point along the cable would be reflected without loss along the entire cable. There would be no attenuation of a signal along the cable. No cable is perfect, however, and some loss in signal inevitably occurs due to core resistance and leakage through the insulator. The loss will be exponential in amplitude along the length of the cable or its analogue, the nerve fiber. (Figure 7). The rate of loss is defined in terms of a length constant, which is the length of cable over which a signal will travel before it is attenuated to 0.39 (1/e) of its original size. The length constant in turn is determined from:

$$\text{Length Constant} = \left[\frac{R_{insulator} \times \text{Radius}}{2 \times R_{conductor}} \right]^{1/2}$$

In a good telephone cable the length constant is measured in miles. In a nerve cell, however, the resistance of the axoplasm is sufficiently high (about 50 Ω-cm) and the resistance of the membrane sufficiently low (about 5000 Ω-cm^2) that the length constant is measured in millimeters.

Electrotonic conduction will not transmit a depolarization or a hyperpolarization very far along a nerve membrane without severe attenuation. Certainly the neuron cannot rely upon such a mechanism to convey the consequence of an action potential developing at the initial segment of the axon to the axon terminals, which may be several centimeters removed. This problem of transmission was solved by the neuron by making the entire axonal membrane capable of initiating a complete action potential when depolarized beyond a certain point. The membrane, however, is effectively coupled along its length due to electrotonic effects, and the length constant, at least in unmyelinated fibers, directly determines the rate of propagation of the action potential along the axon. The reason for this can be seen by considering the following example. Suppose a neuron possesses an axon with a radius of 5 μm. Using the values of resistance for the axoplasm and membrane given above, the length constant for that axon would be approximately 1.6 mm. Further assume a resting potential of -70 mV and a threshold potential for initiating an action potential at any point along the axon of

−55 mV. The potential induced by electrotonic spread along the axon from the point of initiation of the action potential would be

At the site of initiation	+45
At 1 length constant (1.6 mm)	−25mV
At 2 length constants (3.2 mm)	−53mV
At 3 length constants (4.8 mm)	−63mV
At 4 length constants (6.4 mm)	−67mV

The electrotonic potential produced at any point along the axon by an action potential converting the local potential to +45 mV would be of sufficient intensity to drive the adjacent membrane below its threshold for initiating its own action potential for a distance of about 3 mm. This front of depolarization preceeding the action potential itself provides the stimulus for the progressive movement of the potential down the axon. The rate of propagation is determined by the length of membrane driven below threshold in front of the advancing action potential.

In myelinated axons, the myelin formed by Schwann cells or by the oligodendroglia is closely applied to the nerve membrane except at the nodes of Ranvier. This myelin effectively increases the internodal membrane thickness and, secondarily, greatly increases the apparent membrane resistance. From the length constant equation it can be seen that any increase in the effective membrane resistance will increase the length constant. Since membrane resistance may be increased by 100-fold by myelin, the effective length constant may be increased to 10 times the value for a nonmyelinated fiber of the same diameter. Thus a myelinated fiber of the same diameter of 5 μm as used in the previous example could effectively depolarize a length of axon 32 mm long below its threshold for initiating an action potential. On account of this long depolarizing front and the fact that current can only flow out of a myelinated fiber at the nodes, the action potential appears to skip from one node to another, and may skip two or three nodes at a time. This type of conduction of the action potential has been given the name "saltatory conduction".

F. Synaptic Potentials

Almost all communication between neurons and other neurons, muscles fibers, glands, etc. is brought about in mammalian systems by the release of chemical substances called neurotransmitters. The nature, synthesis, and storage of these chemicals will be considered later. At present only the actions of these transmitters upon the receptive surfaces (soma and dendrites) of a neuron or effector will be considered.

Molecules of neurotransmitter, when released from a presynaptic axon terminal, diffuse across the synaptic cleft to complex with specific receptors on the postsynaptic membrane. In most instances, this process results in a rapid and transient change in the permeability of the postsynaptic membrane to one or more ionic species. The permeability changes result in ion flow across the membrane. Frequently this is associated with changes in the membrane potential in the region adjacent to the receptor.

Certain neurotransmitters produce an excitatory postsynaptic potential (EPSP) (Figure 5). These substances produce changes which result in an increased permeability of the postsynaptic membrane to both Na^+ and K^+ ions. The permeability changes drive the membrane potential towards a value intermediate between the sodium and potassium equilibrium potentials at about 0 to −10 mV. The thresholds of the soma and dendritic membranes for producing an action potential are very high, and it is rare for the EPSP to induce an action potential in these structures. Instead the depolarization is conveyed electrotonically to the axon hillock and is summated there with all other synaptic potentials. For synaptic potentials, the length constant is important in determining the influence that a synapse will have upon subsequent action potential generation. Synapses located close to the axon hillock will produce potential changes that will undergo less attenuation than will potentials produced at synaptic loci farther removed. For a great many neurons, all of the EPSPs will be attenuated to such a degree that no single synapse is capable of causing the initiation of an action potential by itself. In these cases, many synapses must summate with one another, either spatially or temporally, to drive the hillock below its threshold potential.

Other transmitter substances can provoke changes in ionic permeability that result in the hyperpolarization of the underlying postsynaptic membrane (Figure 5). In these situations the permeability for K^+ and/or Cl^- is increased. The resulting hyperpolarization is termed an inhibitory postsynaptic potential (IPSP) because it results in a movement of the potential at the axon hillock away from the threshold for generating an action potential. The influence of such IPSPs is also conducted to the axon hillock electrotonically where it is summated with all other synaptic influences. As is the case with EPSPs, the relative ability of an IPSP to influence the hillock will be determined largely by the distance between the synapse and the hillock. In many neurons, the inhibitory synapses are predominately axosomatic and relatively closer to the hillock than are the excitatory synapses, which are predominately axodendritic.

In axo-axonal synapses, synaptic action results in a different set of consequences than in axosomatic or axodendritic synapses. Axo-axonal synapses tend to modify the amount of transmitter released by the postsynaptic element. Such actions are termed presynaptic inhibition or presynaptic facilitation, depending upon whether the amount of transmitter release is decreased or increased. The mechanisms underlying such changes are not completely clear. However, it is known that presynaptic inhibition is accompanied by a depolarization of the membrane of the recipient axon and by a decrease in the amount of transmitter released from that axon subsequent to the arrival of an action potential at its terminal. Presynaptic facilitation is accompanied by a hyperpolarization of the membrane of the recipient axon and by an increase in the amount of transmitter released from that axon subsequent to the arrival of an action potential at its terminals.

All of the synaptic effects described above can be considered as fast, transient phenomena which involve the momentary alteration of membrane conductances through the engagement of receptors at specialized synaptic junctions. This type of synaptic action is primarily designed to change the excitability of the recipient neuron for a brief period of time, a few milliseconds to perhaps half a second.

A slower, more sustained synaptic action has also been reported. In these cases the neurotransmitter acts as a first messenger and interacts with a specifie receptor on the postsynaptic cell. The interaction of the neurotransmitter with its receptor results in the activation or inactivation of the enzymatic function of the receptor or a closely applied macromolecule. An example is the activation of adenyl cyclase by norepinephrine, leading to an increase in intracellular cyclic-AMP levels which in turn leads to a variety of metabolic and regulatory adjustments. These adjustments may change membrane resistance, ionic gradients, affinity of membrane bound receptors, etc. and in this way alter the excitability of the responding cell for relatively long periods of times, many seconds to perhaps minutes.

V. NEUROTRANSMISSION AND NEUROTRANSMITTERS

Tremendous changes have occurred during the last decade in the recognition of substances which function as neurotransmitter or neuromodulators in the nervous system. The number of substances that have (nearly) completely met the criteria required of neurotransmitters has increased from 3 or 4 10 years ago to 20 to 30 today (Table 2).

Table 2
SUBSTANCES PROPOSED AS NEUROTRANSMITTERS

Monoamines
  Norepinephrine
  Acetylcholine
  Dopamine
  Serotonin
  Epinephrine
  Histamine

Amino Acids
  Gamma-aminobutyric acid
  Glutamic acid
  Glycine
  Taurine

Neuropeptides
  Methionine-enkephalin
  Leucine-enkephalin
  Substance P
  Neurotensin
  Beta-endorphin
  ACTH
  Angiotensin II
  Oxytocin
  Vasopressin
  Vasoactive intestinal polypeptide
  Somatostatin
  Thyrotropin releasing hormone
  Luteinizing-hormone releasing hormone
  Bombesin
  Carnosine
  Cholecystokinin-like peptide

Much of the credit for this tremendous increase in understanding must go to the development of better instrumentation and analytical techniques which have made it possible to identify and quantify incredibly small (nanogram) amounts of material. Many of the most recently identified compounds are neuropeptides containing from 5 to 39 amino acids. The primary structure of most of these has been determined.

Chemical substances must meet certain criteria before they can reasonably be accepted as being neurotransmitters. The compound must be present in the presynaptic terminal. Enzymes for its synthesis and degradation must exist at or near the synapse or plausable mechanisms must be demonstrated which can account for its uptake and removal from the synaptic region. The substance must be released from the presynaptic fiber by the normal stimuli produced by the presynaptic neuron. This criterion has been the most difficult to demonstrate for most putative neurotransmitters because of the difficulty of collecting and measuring the material extruded from a presynaptic terminal. Finally the substance, when experimentally administered, must produce the same effect on the postsynaptic membrane as does the material released by nervous activity.

At a typical synapse, the process of neurotransmission can be viewed as being made up of four phases: (1) synthesis and storage, (2) transmitter release, (3) receptor activation, and (4) transmitter inactivation. Most neurotransmitters are synthesized within the presynaptic terminal by enzymes made in the cell body and transported to the ending by axoplasmic transport. The synthesized transmitter substance is stored in vesicles in the terminal. In the case of norepinephrine, storage "granules" are synthesized within the cell soma, and the granules are transported to the terminal. Synthesis of norepinephrine can occur during the transport of the granules as well as in the terminal.

The release of neurotransmitter is triggered by a sequence of events which normally begins with the propagation of an action potential into the terminal region. The depolarization that accompanies the action potential induces an increase in the permeability to Ca^{++} ions in the terminal membrane. A large electrochemical gradient exists for Ca^{++} which is inwardly directed, and, as a result, Ca^{++} moves into the fiber. The Ca^{++} which enters the terminal triggers a sequence of events which results in the movement of some synaptic vesicles to the terminal membrane, the fusion of the vesicle and terminal membrane, and the exocytosis of the vesicular contents out into the synaptic cleft. This process appears to be terminated by the removal of Ca^{++} from the terminal cytoplasm, either back to the extracellular fluids or perhaps by uptake into mitochondria.

Once the neurotransmitter is released from the presynaptic terminal it diffuses across the synaptic cleft, a distance of approximately 200 Å. On the postsynaptic side the neurotransmitter complexes with a membrane bound macromolecule, its receptor. The complexation can lead to either fast, transient changes in the ionic permeabilities of the postsynaptic membrane or to slower, more sustained changes in postsynaptic cell excitability. The complexation process is rapidly reversible with a half-life for occupation of the receptor of just a few milliseconds.

The consequences of transmitter release are terminated by various means. Catabolizing enzymes may be present on the postsynaptic side of the cleft which rapidly break the transmitter down into inactive substances. Part of the transmitter may be taken back up into the presynaptic terminal to be reused once again. Some of the neurotransmitter simply diffuses away from the synaptic region and is processed elsewhere.

Certain of the transmitters have been clearly identified as being involved in specific systems whose functions are well known. Many of these appear to be importantly involved in neurotoxic actions. As such these transmitters deserve more specific consideration.

A. Acetylcholine

Acetylcholine has been identified as the neurotransmitter for many important systems. It is the neurotransmitter released by all alpha and gamma motoneurons, all preganglionic autonomic neurons, all postganglionic parasympathetic neurons, some postganglionic sympathetic neurons, and some neurons within the central nervous system. Two distinct receptors have been identified for acetylcholine, the nicotinic and the muscarinic receptor. An additional receptor may exist for acetylcholine within the central nervous system as the effects of acetylcholine there can not be readily explained on the basis of muscarinic and nicotinic receptors, alone. The nicotinic receptor typically evokes fast, transient changes in the membrane potential of the recipient cell whereas the muscarinic receptor, when activated, evokes slower, more sustained changes in membrane potential. The latter may be mediated by cyclic-GMP.

Acetylcholine is synthesized from acetyl-CoA and choline within the presynaptic terminal by the enzyme choline acetylase (Figure 8). The acetylcholine formed is stored in small, lightly staining synaptic vesicles which are concentrated around the synaptic contact area. The release of acetylcholine is Ca^{++} dependent. The entire content of a synaptic vesicle is released into the cleft in an all or none manner. This makes the release quantal in nature. Once in the synaptic cleft, the acetylcholine is rapidly destroyed by acetylcholinesterase which is found predominately upon the postsynaptic membrane in close proximity to the receptors for acetylcholine. The acetylcholine is hydrolyzed to choline and acetate, and the choline is actively pumped back into the presynaptic terminal to be used to synthesize more acetylcholine.

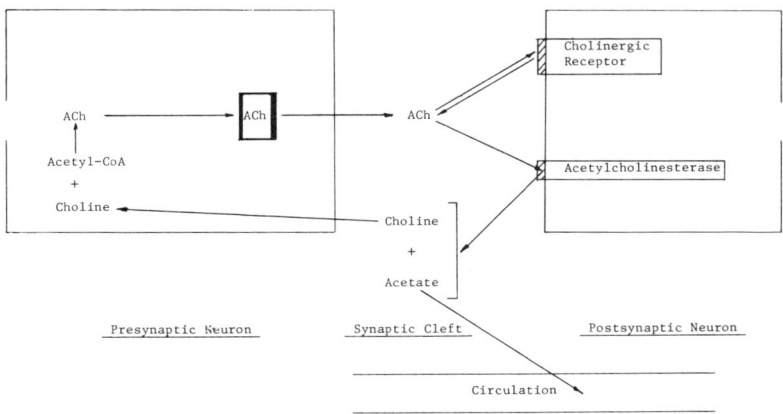

FIGURE 8. The cholinergic synapse. Presynaptic components include synthesis, storage and release of acetylcholine. Postsynaptic components include the interaction of acetylcholine with its receptors and with acetylcholinesterase.

The termination of the action of acetylcholine, once released from a nerve terminal, is critically dependent upon the presence of a functioning acetylcholinesterase. Substances such as the organophosphate insecticides and the carbamate insecticides inhibit cholinesterases. This leads to an intensification and prolongation of the action of acetylcholine at all loci in the body where it is released. The impact of this will be dealt with in a later chapter.

B. Norepinephrine

Norepinephrine has been identified as the neurotransmitter released by most postganglionic sympathetic neurons. It is probably also the neurotransmitter released by many central neurons. Cell bodies containing norepinephrine are found in clusters through the medulla, pons, and midbrain, mostly within the reticular formation. One nucleus, the locus coeruleus, is primarily made up of norepinephrine containing cells. Two major types of receptors for norepinephrine have been identified, each of which appears to have two or more subtypes. These have been named "alpha" and "beta" receptors. Alpha receptors are found distributed throughout the body and, when activated, usually induce active, excitatory responses. Beta receptors are also widely distributed and, when activated, usually induce passive, depressor responses. There are exceptions to this rule, however. The most notable exception occurs in the heart. The receptors there are beta receptors. Activation of these receptors leads to increased heart rate and increased contractility of the myocardium.

Norephinephrine is manufactured from the amino acid tyrosine in three enzymatically controlled steps (Figure 9). The transmitter is then stored within membrane-bound vesicles in association with certain storage proteins. These vesicles or granules are concentrated over a fairly long length of the nerve terminal in varicosities. An action potential arriving at the terminal typically evokes the release of norepinephrine from many of the varicosities. The release of norepinephrine is Ca^{++} dependent and is probably quantal in nature. When released the norepinephrine crosses the synaptic cleft to bind to specific receptor proteins embedded in the postsynaptic membrane, triggering a series of reactions that culminate in short-term (electrical) and long-term effects on the recipient neuron. The action of norepinephrine is terminated in a number of ways. Part of it is taken back up into the presynaptic terminal and degraded within the terminal or repacked into the vesicles. A part of it is degraded by an enzyme, catechol-O-methyltransferase, which is present within the synaptic region. Part of it

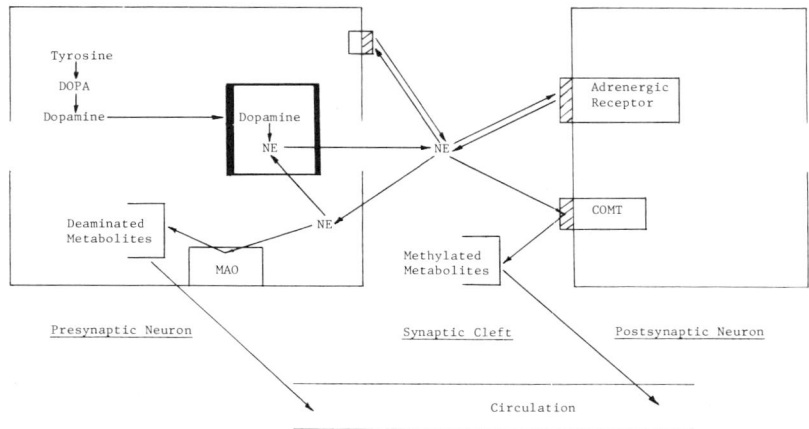

FIGURE 9. The adrenergic synapse. Presynaptic components include the synthesis, storage and release of norepinephrine. A presynaptic receptor also exists which modulates the amount of transmitter release. Postsynaptic components include the interaction of norepinephrine with its receptors. Termination of action is performed both by presynaptic (reuptake and degradation by MAO) and postsynaptic (degradation by COMT) processes. MAO:monoamine oxidase, COMT: catechol-O-methyltransferase.

simply diffuses away. Presynaptic receptors for norepinephrine have been identified on neurons which secrete it. These receptors appear to be involved in a control system which regulates the amount of norepinephrine synthesis by the terminal.

C. Dopamine

Dopamine is the immediate synthetic precursor of norepinephrine. In addition it appears to function in its own right as a neurotransmitter. The largest dopaminergic tract originates in the substantia nigra and projects to the striatum. In Parkinson's disease, this nigrostriatal pathway degenerates with a correlative decrease in brain dopamine. Other dopaminergic tracts have been demonstrated in the central nervous system. Dopamine may also function as a neurotransmitter at selected sites in the periphery that receive sympathetic innervation.

D. Serotonin

Cell bodies containing serotonin have been observed in a series of nuclei in the lower midbrain and pons, the raphe nuclei. The axons from these cells pass in the median forebrain bundle to terminate largely in and around the hypothalamus. The raphe nuclei, and therefore serotonin, have been shown to have important effects upon sleep and wakefulness patterns. Destruction of these nuclei induces insomnia while electrical stimulation at a physiological frequency produces somnolence.

E. Gamma-Aminobutyric Acid (GABA)

The earliest demonstration of a neurotransmitter function for GABA was in crayfish where GABA was found to be released from the inhibitory fibers going to muscles and stretch receptors. There is presently strong evidence supporting the role of GABA as an inhibitory transmitter in the vertebrate central nervous system.

In mammals, GABA has been detected in brain and spinal cord but not in peripheral nerve tissue. The highest concentrations of GABA are found within the diencephalon and the colliculi. There is good evidence that GABA is released by certain neurons within the cerebellum, cortex, retina, and spinal cord. At all these loci the action of GABA seems universally inhibitory.

F. Glycine

There is growing evidence supporting the role of glycine as an inhibitory transmitter in the spinal cord. Strychnine reversibly antagonizes the effects of iontophoretically applied glycine on spinal neurons. This effect of strychnine is probably responsible for the convulsant activities strychnine produces.

G. Glutamate

Glutamate has been proposed as an excitatory neurotransmitter. It almost certainly is the neurotransmitter released by excited nerves that innervate crustacean muscles. It has been proposed that glutamate, and perhaps aspartate, are the neurotransmitters released by primary sensory endings in the spinal cord. It may function similarly at a number of central synapses.

VI. FUNCTIONAL ORGANIZATION OF THE NERVOUS SYSTEM

The nervous system has evolved as a system that provides for an organism some degree of control over its environment. To do this the nervous system must be supplied information about the state of that environment. This is provided through a number of different types of sensory receptors. The neurons which are primarily concerned with conveying the information transduced by these receptors and with processing it for its content comprise the sensory part of the nervous system. A large part of the brain is concerned with the integration of sensory data, with comparing it to past experiences, identifying various environmental components, and developing useable judgements as to the most appropriate actions that need be undertaken. The neurons involved in these functions form the integrative part of the nervous system. A third part of the system, the motor part, is directly concerned with the translation of desired intents for movement into patterns of neuronal discharge to be provided to the various muscles of the body. All three functional components, sensory, integrative, and motor, are fairly distinct and separable within the older parts of the nervous system. These functions, however, become more and more overlapping in newer sections of the brain until, at the level of the cortex, it is difficult to separate purely sensory, motor and integrative systems. Only a brief consideration of the major functional divisions of the nervous system can be included here.

A. Sensory Systems

1. Receptors

All animals have cells, termed receptors, that have developed to sample the environment. Receptors are either specialized cells connected to afferent fibers, are modified nerve fibers, or are actual nerve fibers freely ending in the skin or deep tissues. During evolution receptors have evolved to become more sensitive and more adapted to the needs for survival.

All receptors function by changing their membrane potential when energy from the environment reaches them. The intensity of the stimuli impinging on the receptor is usually signalled by the rate of impulse production in the adjacent nerve fiber. Receptors are suitably specialized so that they respond to only certain types of energy. This provides for separation of sensory information into specific channels which can be fine-tuned for processing only that type of information.

2. Cutaneous, Deep and Visceral Sensation

A number of receptor types exist in the skin and deeper tissues which are responsive to stimuli such as touch, pressure, pain, warmth and cold. Their activation induces action potentials in the associated nerve fibers which convey the information into the spinal cord. These fibers may synapse in the cord or be conveyed by a number of pathways to the brain. The various pathways conveying somatic sensation converge onto specific sensory nuclei in the thalamus and eventually project to the somatosensory areas of the cerebral cortex. These stimuli can provoke simple spinal cord reflexes and induce a nonspecific arousal of the entire nervous system when sufficiently noxious. They also give rise to conscious perception of sensation. Certain of these afferents pass to the cerebellum and related motor areas to assist in unconscious adaptation of posture and muscle tone. Indirectly some of these afferents reach autonomic centers and may induce changes in autonomic function, particularly those associated with temperature regulation and the maintenance of blood pressure.

3. Visual

Vision is the most complex and dominant sensory system in man. It is the sensory source that normally provides more than half of all the conscious perception man has of his environment. The receptor cells lie in the retina of the eye and are of two types. The rods are extremely sensitive to light and are largely concentrated in the periphery of the retina. They are not responsive to the color of light. Rods are the receptor type that dominate in vision during periods of low light intensity (scotopic vision). The cones possess much higher thresholds for activation. Because of their spacing and location in the retina, they provide for greater visual acuity than do the rods. Cones dominate during day vision (photopic vision) and are the color-sensitive elements in the retina.

Within the retina, visual input is partly processed, than transmitted into the central nervous system by way of the optic nerves. Most of the visual afferents synapse in the thalamus, and are subsequently distributed to the cortical and subcortical vision centers. The visual cortex processes the input and provides largely for the conscious "picture" of the environment we perceive. Subcortical centers may contribute to conscious vision, but they also provide visual information necessary for the great many reflexive adjustments needed to focus and orient the eyes. These same centers interact with motor areas and provide a visual perspective of space to aid in orientation and posture.

The separation of function between cortical and subcortical regions is not absolute. These areas are heavily interconnected and provide coordinated processing of incoming data. Much of cortical processing is conveyed to subcortical areas and used in reflexive activities. Subcortical efferents passing to cortex provide some information, for example brightness, that eventually reaches consciousness. This division between cortical and subcortical holds true for all sensory systems.

4. Auditory

Receptors responsive to sound waves are located in the cochlea. Fluctuations in air pressure are actually converted by the ear into traveling waves within the fluid bathing the receptors. The movement of the fluid bends sensitive hairs that are part of the receptors. Action potentials generated by the nerve fibers attached to these receptors are transmitted via the statoacoustic nerve to subcortical areas and to the auditory cortex. As indicated for the visual system, the auditory cortex is primarily involved in converting the information reaching it into a conscious perception of the stimulus. Subcortical centers process the information and provide it for use by motor areas for behavioral responses.

5. Vestibular

The vestibular apparatus contains receptors capable of detecting the acceleration of the body relative to the environment. Gravitational forces, which are also accelerative in nature, are equally perceived. Vestibular output is conveyed to the vestibular nuclei in the brain stem. These provide information to many brain stem, cerebellar, and spinal areas involved in the maintenance of equilibrium.

6. Olfactory

The olfactory receptors are located in a specialized portion of the nasal mucosa. Each receptor is a true neuron, and the olfactory mucosa is the region in which neurons come into the closest contact with the environment. The neurons possess short, thick dendrites containing cilia that project to the mucosal surface. These cilia are chemoreceptors which are responsive to molecules dissolving in the mucosal layer around them. The axons of these neuronal receptors pass to the olfactory bulb, and from there pass into the limbic region of the brain. There are no direct neocortical pathways for olfactory input.

The sensitivity of olfaction in certain species is astounding. For example man can detect the odor of methyl mercaptan in air at a concentration of less than 1 ng (10^{-9} G)/ℓ. Man can distinguish between 2000 and 4000 different odors. The sense of smell is generally more acute in women than in men, and in women it is most acute at the time of ovulation.

7. Gustatory

The taste buds contain chemoreceptors which respond to substances dissolved in the saliva bathing them. These receptors exist in certain parts of the tongue and pharyngeal cavity. The receptors give rise to sensations identified as sweet, salt, sour, and bitter. In most animals, but not in man, there are taste receptors that respond to pure water also.

The nerve fibers activated by the gustatory receptors reach the brain stem by passing through the cranial nerves. Taste sensation eventually joins up with the various somatosensory pathways and projects via the thalamus to the cortex. Taste does not have a separate cortical projection area but is represented in the area of the cortex that subserves cutaneous sensation of the face.

B. Motor Systems

The initiation and control of posture and movement are among the most important functions of the nervous system. The involved structures, the motor centers, are distributed throughout the central nervous system from the spinal cord to the cerebral cortex. With increasing complexity of the nervous system, motor function was improved not so much by the remodeling of existing structures as by the addition of more proficient supplementary control systems. Figure 10 illustrates in block diagram form the major motor centers and their interconnections. Motor centers within the spinal cord and brain stem function primarily to maintain posture and coordinate, at the final stages, all voluntary and involuntary movement. The motor cortex, basal ganglia, and cerebellum are primarily oriented to providing the programs to lower motor centers that allow for coordinated, sequential and productive movements. The cortex in a broader sense is also largely responsible for planning sequences and for setting motor activities into motion.

1. Spinal Cord

The spinal cord gives rise to the final common pathways for all motor activity, the alpha and gamma motoneurons, going to the body musculature. Motor centers in the brain stem subserve similar functions for the muscles of the head. The motoneurons exit at all segments of the spinal cord. They receive inputs from three major sources: (1) segmental afferent input from muscle spindles, Golgi tendon organs and various sensory receptors in the skin and deep structures; (2) intraspinal interneurons; and (3) supraspinal pathways.

The muscle spindles and Golgi tendon organs are responsive to the length of muscles and the tension upon muscles, respectively. Input from these sources evokes reflexes

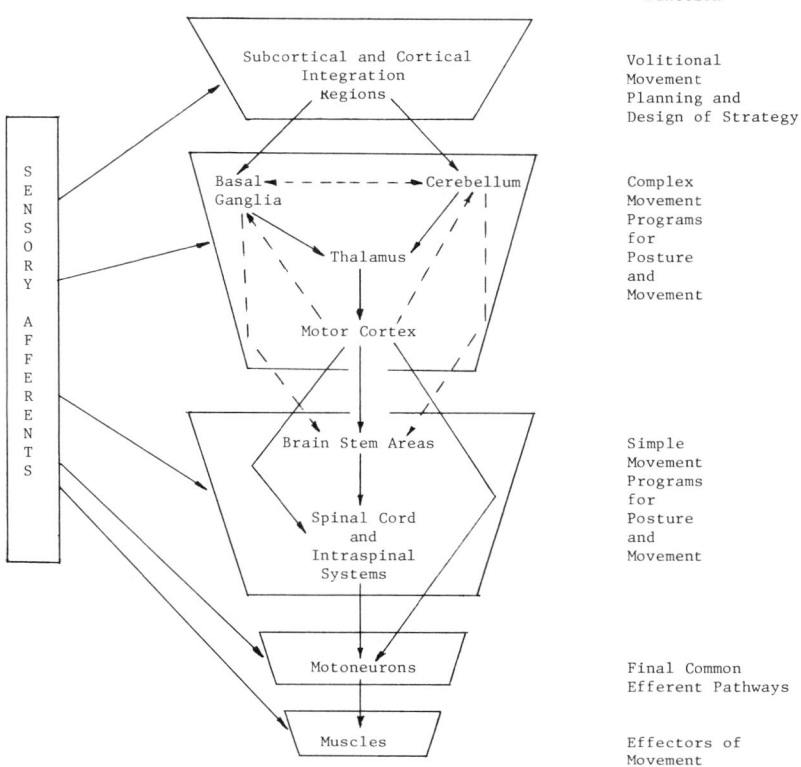

FIGURE 10. Major motor centers and their interconnections. Five levels of motor integration are shown. The types of function each level is most involved in is shown on the right.

which tend to oppose any change in the position of joints and limbs in space. The monosynaptic stretch reflex (knee-jerk reflex) is largely evoked by muscle spindle activity. The jack-knife reflex is largely mediated by Golgi tendon organ input.

Afferent input arising from other receptors, such as those in the skin, can also trigger motor reflexes. In these situations the reflex arc is always polysynaptic with one or more interneurons interposed between the afferent ending and the motor neuron. The most prominent reflex of this type is the flexor reflex, which is demonstrable as a rapid pulling away of a limb to a strong aversive stimulus. Additional reflex arcs are always activated simultaneously with the flexion reflex so as to minimally disrupt equilibrium. The extensor muscles of the same limb are inhibited during the time that the flexors are stimulated. Exactly opposite effects are induced in the contralateral limb. There, the flexors are inhibited and the extensors facilitated. This is often termed the crossed extensor reflex.

Intersegmental reflexes also occur in the spinal cord. These are mediated via propriospinal tracts that interconnect various sections of the cord. These are largely responsible for the reflexive coordination of the function of the fore- and hindlimbs. For example, aversive stimulation to one hind limb will produce a flexion of the affected limb and an extension of the other hind limb by segmental means. In addition both forelimbs will be stimulated by propriospinal neurons to extend. These various spinal reflexes are frequently obscured by the activity produced by supraspinal efferents.

In the spinal cord reflexes exist which allow an animal to retain a stable position in space and to carry out simple movements directed towards removing the animal from aversive stimuli. The connectivity of the spinal cord is such that it can generate rather complex motor acts when an appropriate signal is provided. The various spinal reflexes can be viewed as a set of elementary posture and movement subroutines which can be used as required. The beauty of this is that the higher level motor centers are freed from the requirement to orchestrate every single muscle when producing a purposeful movement.

2. Brainstem

The various motor centers of the brain stem lie in and around the reticular formation from the posterior edge of the medulla to the rostral edge of the midbrain (mesencephalon). These centers are chiefly responsible for reflex control of posture and spatial orientation of the body. A number of sensory inputs reach this area, the most important of which are the afferents arising from the vestibular apparatus and from stretch and joint receptors of the muscles of the neck. The output from these centers to the spinal cord is tonic in nature and, as a consequence, posture is adopted, corrected and maintained without conscious control.

In experimental animals the removal of all brain rostral to the pons (decerebrate animal) results in a marked increase in the tone of the entire extensor musculature, a condition termed "decerebrate rigidity." Such animals can be set upon their feet and will remain standing indefinitely, with all four legs rigidly straight and the head and tail bent backwards. The muscle tone in decerebrate rigidity can be altered by movement of the head. If the head is bent further backwards the extensor tonus of the hindlimbs decreases. The extensor tonus of the forelimbs increases. Bending the head downwards evokes the opposite effect. Turning the head to the right produces an increase in extensor tone of the two right limbs. The new positions reached will be maintained as long as the head is kept in the altered position. These effects are commonly referred to as the tonic neck reflexes.

In the decerebrate animal just described only the medulla and the pons remain attached to the spinal cord. If the midbrain is also retained the motor performance of the animal is greatly improved in two ways. The muscle tone is normal, not excessive, and the animal can set itself into normal postural positions if displaced from that position. The animal can right itself from any position back to a normal posture. This righting process always occurs in a specified sequence. First the head is brought to a normal, chin down position, largely through vestibular activation. Second the neck and trunk are brought into alignment with the head.

The animal with intact brainstem and spinal cord differs little from a normal, fully intact, animal with respect to posture, positioning and simple walking and jumping responses. However, spontaneous movement is lacking. External stimuli must be provided to initiate any motion.

3. Basal Ganglia

The basal ganglia form an important link between the motor cortex and the remainder of the cortex. Association cortical areas project heavily to the basal ganglia which in turn project heavily to the motor cortex by way of certain thalamic nuclei. The basal ganglia, appear to function in the role of converting plans for movement, generated largely in the associational cortical regions, into programs for movement. They appear to provide a spatial and temporal pattern of nerve impulses which are further processed at the motor cortex or at brainstem centers. These programs seem primarily oriented for the carrying out of slow and sustained components of coordinated movements.

Some of the functions of the basal ganglia can be inferred from the motor activity of people with Parkinson's disease. In this condition there is a deficiency in the concentration of the neurotransmitter within the nigrostriatal pathway which results in the disruption of the function of this area. The affected individuals exhibit a relative absence or an impairment of slow, sustained movements (akinesia), an increased muscle tonus (rigor), and a passive tremor, which occurs only at rest and is suppressed during movements.

4. Cerebellum

The cerebellum in many ways performs analogous functions with the basal ganglia, and its interconnections to motor cortex and brainstem motor nuclei are quite similar. It also appears involved in the programming of movement patterns. Functionally, however, the systems are complementary rather than equivalent. Whereas the basal ganglia are primarily involved in programming slow, steady movements, the cerebellum seems chiefly responsible for the programming of rapid, transient movements. It also functions to provide rapid correction for movements in progress and serves additionally in coordinating posture and movement into cohesive, efficient patterns.

Removal of the cerebellum results in a number of motor abnormalities. One of these is an inability to acheive the proper timing and coordination of the activities of the various muscles participating in a movement (asynergy). Motion tends to become decomposed into a series of successive movements with large, uncompensated error. Movements such as those occurring during walking, speaking, writing, etc. may become impossible. In addition, an intention tremor is seen during movement. It does not occur during rest. Finally muscle tone is frequently inappropriately low (hypotonus).

5. Motor Cortex

The motor cortex occupies a unique position among motor centers. It functions both as the last supraspinal station involved in providing programs for movement and as the first of a chain of structures responsible for the execution of movement. The latter function is easiest to observe after damage to the motor cortex or its efferent projections. In man, complete interruption of all cortical outflow results ultimately in the development of a spastic paralysis which mimics in many ways the state of decerebrate rigidity. It is of great interest that the selective lesioning of the corticospinal (pyramidal) tract, which eliminates all direct connections between the motor cortex and spinal cord, produces minimal effects upon movement. There is a certain loss of dexterity, but no other important consequences.

C. The Autonomic Nervous System

The autonomic nervous system provides the innervation to the heart, blood vessels, glands, viscera, and other smooth muscle of the body. The innervation of smooth muscle resembles that of striated muscle in that the autonomic systems are organized into afferent (sensory), integrative, and efferent (motor) parts. Most organs and glands receive a dual innervation from the two divisions of the autonomic nervous system, the sympathetic and the parasympathetic divisions. The efferent autonomic fibers are usually tonically active and maintain a constant level of tone in the effectors. The consequences of autonomic innervation will depend upon the relative activities of the sympathetic and parasympathetic divisions. This will vary with the behavioral state of the animal.

1. Afferent Components

The majority of afferent input to autonomic centers arises from receptors located in and around the viscera, glands, blood vessels, and heart. Part of these course with the somatic afferent fibers and enter the spinal cord through the dorsal roots. Perhaps the greatest number, however, are carried by the vagus nerve into the brain. Visceral afferent fibers synapse at various locations in the brain stem, midbrain and the hypothalamus, as well as in the spinal cord. The hypothalamus has deservedly acquired the title of "head ganglion" of the autonomic nervous system. It serves as the major integrative center for most involuntary functions and participates in stimulating behavior designed for maintaining body integrity.

2. Efferent Components

The efferent limb of the autonomic nervous system differs from the somatic in having two peripheral fibers between the spinal cord and the effector, rather than just one. These nerve elements are termed preganglionic and postganglionic dependent upon their position relative to the autonomic ganglia. The cell bodies of the preganglionic neurons are located in the intermediolateral cell columns of the spinal cord or in homologous motor nuclei of the cranial nerves. These fibers are predominately myelinated and small in diameter. The axons synapse onto the postganglionic cells whose cell bodies make up the autonomic ganglia. These ganglia exist at various locations outside the central nervous system. The postganglionic fibers are usually unmyelinated.

a. Parasympathetic Division

Preganglionic parasympathetic fibers exit from the central nervous system in two regions (Figure 11). A cranial outflow occurs via cranial nerves III, VII, IX, and X which goes eventually to ganglia located throughout the head and neck and viscera. From there, postganglionic fibers innervate smooth muscle and glands of the eye and the pharyngeal, thoracic, and abdominal cavities. A second parasympathetic outflow occurs from sacral segments of the spinal cord. These fibers course in the pelvic nerves and provide parasympathetic innervation to the viscera in the lower abdominal cavity.

b. Sympathetic Division

Sympathetic nerves exit from the central nervous system at each spinal segment from upper thoracic to lower lumbar regions of the cord. The preganglionic sympathetic fibers pass out through the ventral roots to the vertebral ganglionic chains (Figure 12). A given fiber may synapse there, pass upward or downward along the chain to synapse at a different segmental level or continue through the chain ganglia forming the splanchnic nerves. The latter synapse at additional ganglia located along the ventral aspect of the aorta. Postganglionic fibers arising from the vertebral ganglionic chain pass to the spinal nerves and are distributed peripherally to the sweat glands, pilomotor muscles and blood vessels. Postganglionic fibers arising from the aortic ganglia go to innervate the glands and smooth muscle of the abdominal and pelvic viscera. The sympathetic distribution to the head and neck comes entirely from the cervical sympathetic ganglia.

3. Integrative Components

The influences that act to regulate the pattern of nervous activity to the autonomic effectors are analogous to those in the somatic nervous system. The autonomic efferents (preganglionic and postganglionic fibers) represent the final common pathway to all autonomic effectors, and their activity controls effector behavior. These pathways tend to be tonically active, providing constant input to the effectors. The level of activity in these pathways is capable of great fluctuation. Effector behavior will be determined largely by the relative level of activity of both the parasympathetic and sympathetic innervation to it. These two systems produce activities that are commonly opposite in type, but the combined effect is better considered to be complementary rather than antagonistic in function.

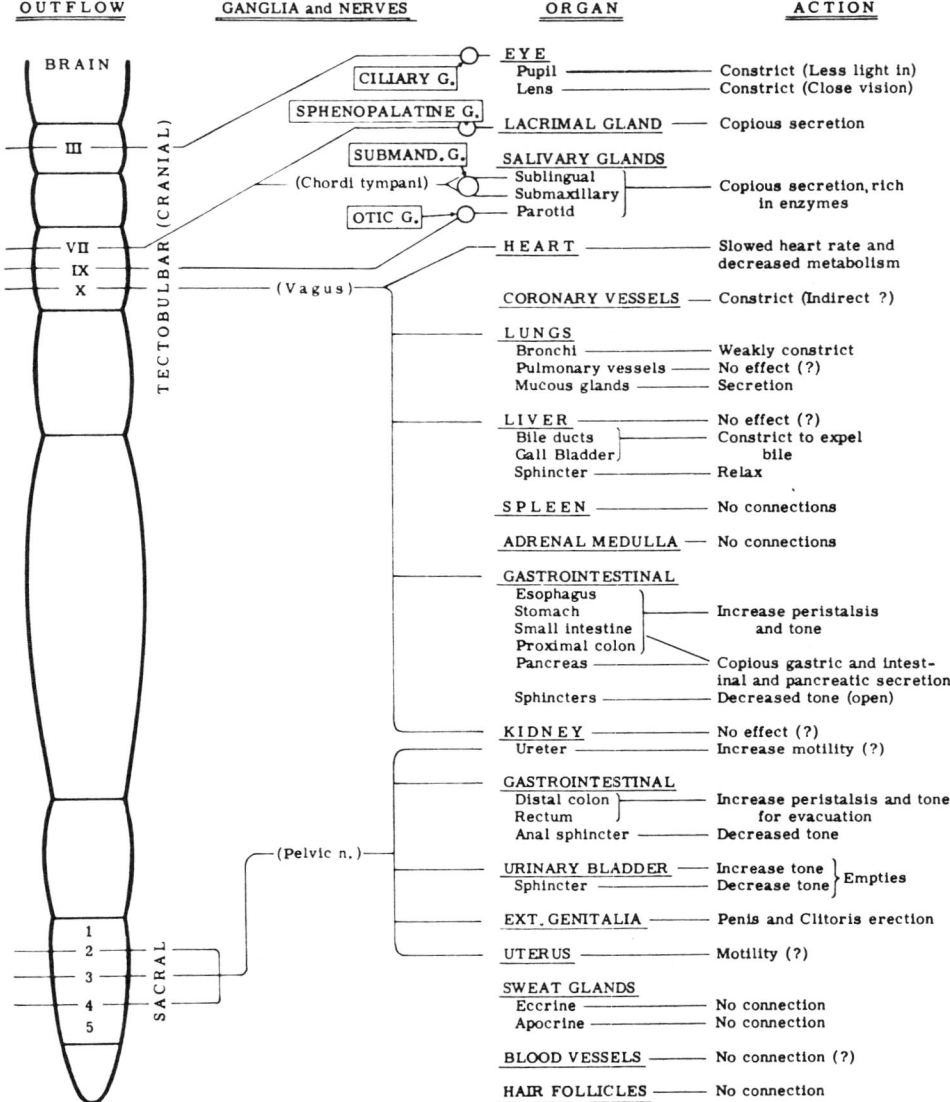

FIGURE 11. The parasympathetic division of the autonomic nervous system. Characteristic effects induced in various effectors by activation of parasympathetic innervation are indicated on the right. (From Selkurt, E. E., *Physiology*, 4th ed., Little, Brown, Boston, 1976, 189. With permission.)

The efferent autonomic nerves are regulated by diverse influences. The simplest input to them comes from afferent fibers within the same spinal cord segment. Segmental reflex arcs of this type are always polysynaptic and are characterized by a high degree of divergence and overlap. Intersegmental reflexes also exist which provide for the coordination of autonomic activity occurring over more than one segment. Such reflexes aid in producing coordinated function within organs that are innervated by more than one cord segment. They also are demonstrable in conditions such as "referred pain".

Supraspinal control is largely provided by autonomic centers that exist throughout the brain stem, midbrain, hypothalamus, and limbic system. Many of these centers are not well defined anatomically and are not recognizable as specific nuclei. By way

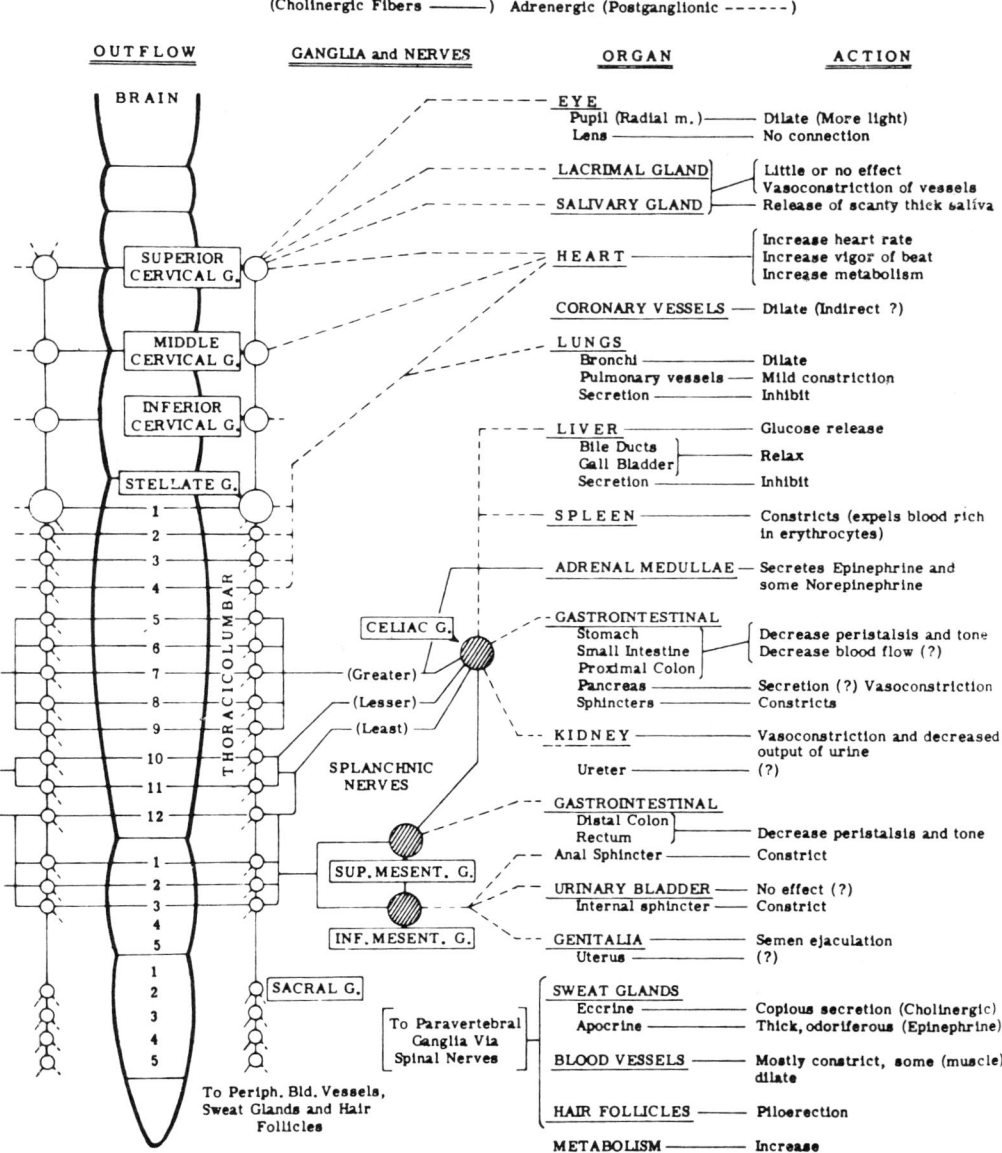

FIGURE 12. The sympathetic division of the autonomic nervous system. Characteristic effects induced in various effectors by activation of sympathetic innervation are indicated on the right. (From Selkurt, E. E., *Physiology*, 4th ed., Little, Brown, Boston, 1976, 188. With permission.)

of these centers integrated autonomic control can occur. Supraspinal inputs are largely responsible for the regulation of functions that are not segmental in nature. These include such functions as the maintenance of blood pressure, regional blood flow, body temperature, hormonal release, etc. These centers, particularly those in the hypothalamus and limbic areas also give rise to a wide variety of behaviors designed to maintain the body in optimal condition. Many components of directed behaviors, such as food and water seeking, sleeping and reproductive behaviors, are initiated or facilitated by autonomic components.

The hypothalamus can legitimately be considered as the most important of the autonomic integrative centers. Its position in the brain, directly above the pituitary gland, which it controls, and its connections through the brain stem allow it to control both neuronal and hormonal activity. It provides the necessary programming responsible for the overall homeostasis, or state of dynamic equilibrium, of the autonomic effectors. It is largely responsible for the coordination of activities between different effectors that maintain a condition of maximal adaptability to meet moment to moment needs. The hypothalamus also participates in defining appropriate levels of input and elimination of necessary body constituents. It determines the "desirable" blood pressure and largely decides the rate of flow of blood through various tissues under different conditions of need. It defines "normal" temperature and institutes heat-saving or heat-losing activities as necessary. It monitors and controls the levels of fluid compartments and their contents. It can affect these by altering the desire for water intake and elimination as well as by controlling levels of salt, sugar, or total caloric intake.

In experimental animals, the removal of the entire cerebrum, leaving an intact hypothalamus, reduces the diversity of behavioral patterns, but does not eliminate them. Behaviors oriented towards thermoregulation, eating, drinking, defense and reproduction can be elicited, but they lack adaptability to environmental fluctuations and are not maintained after the immediate stimulus is removed. Electrical stimulation of selected regions of the hypothalamus may initiate or stop behavior in progress. Stimulation at selected sites can result in continuous eating or drinking until the stimulation is stopped. Conversely, stimulation at other sites can completely block eating or drinking. Such findings suggest that the hypothalamus contains a large number of different neuronal populations that are capable of activating somatomotor, autonomic, and endocrine responses to a variety of different situations and in a variety of different patterns.

The full expression of behavior that can be elicited by hypothalamic activation requires the present of the cortex, particularly the limbic cortex. With an intact limbic system the behaviors described above become adaptive rather than stereotypic in type. Rapid changes in patterns of behavior can occur to the fluctuations in the state of the environment the animals meets. The limbic system in turn is influenced by neocortical areas. A current view of the limbic system is that it is a substrate for species-specific behavior. Also it is importantly involved in emotions, affective behavior, feelings, etc. Not only can the limbic system facilitate the adaptation of autonomic oriented behaviors, it can also start them. Emotional stimuli inevitably lead to anticipatory changes in autonomic function. These are initiated via the limbic system and processed through the hypothalamus to the autonomic centers elsewhere.

4. Comparative Roles of the Sympathetic and Parasympathetic Divisions

The sympathetic and parasympathetic systems have contrasting but complementary functions in regulating the internal environment. In normal animals the sympathetic system is active at all times. Only the degree of activity changes to maintain optimal conditions to meet a constantly changing environment. During rage or fright, the sympathoadrenal axis can discharge as a unit. In these situations, sympathetically innervated structures over the entire body are simultaneously affected. Heart rate is accelerated, blood pressure rises, the blood is shifted from the skin and viscera to the skeletal muscles, blood sugar levels increase, bronchioles dilate, all in anticipation of the increased metabolic activity that will be required to combat an adverse environment. It contrast, the parasympahtetic system is never activated as a unit. It is organized more for discrete and localized discharges on an organ-to-organ basis. The parasympathetic innervation is more concerned with optimizing the individual for functions involving the conservation and restoration of energy rather than with the expenditure of energy. When activated, the heart is slowed, the blood pressure lowered, and the digestive tract is stimulated in ways which aid in the digestion of food and the absorption of nutrients. The parasympathetic innervation also controls the emptying of the bowel and bladder.

Neither system is essential to life, although extensive loss of the parasympathetic innervation, with the associated loss of bowel and bladder control, is more difficult to overcome than would be a comparable loss in sympathetic innervation. Animals lacking sympathetic nerves and lacking the adrenal medulla can lead fairly normal lives within the sheltered confines of a laboratory. They can grow, become pregnant, and give birth to young. Certain reproductive deficiencies, however, can be demonstrated. Gestation may be abnormal, with a high incidence of stillbirths. Sympathectomized mothers will not lactate. Such animals also have a slightly reduced basal metabolic rate. There is surprisingly little alteration in most autonomic processes. Digestion continues normally, heart rate and blood pressure are normal and vasomotor tone, at rest, is well maintained. If such an animal is stressed, however, deficiencies in adaptation become evident. For example, body temperature cannot be regulated when environmental temperature varies, there is no compensatory rise in blood sugar to meet increased metabolic needs, and other compensatory responses, usually associated with flight or fright, are missing. Compensatory vascular responses to hemorrhage, oxygen lack, excitement and work are lacking.

VII. THE HIGHER FUNCTIONS OF THE BRAIN

In man, as well as most other animal species, the most unique nervous functions are those we typically associate with the cerebral cortex. Such processes as consciousness, memory, spontaneous activity, and anticipatory capacity are shared by many animals. Certain more complex functions such as language and abstraction, are only well developed in man. The neural substrates underlying these complex behaviors is poorly understood. Moreover, experimentation is difficult, particularly when dealing with functions only fully developed in man.

The cortex did not develop as a replacement for more primitive structures. Rather, it evolved as an extension of the nervous system, capable of more complicated processing than older centers. As a consequence, the cortex is heavily interconnected to many subcortical structures by great bundles of fibers. The heirarchical development of the brain has resulted in there being specific regions of cortex that are coupled to subcortical structures primarily associated with sensory or motor functions. In addition, large regions of cortex are not connected at all to subcortical areas. These serve as the cortical association or integrations areas (Figure 13).

The cortical regions in which sensory pathways terminate are termed sensory cortical areas. Different sensory areas exist for somatic, visual, auditory, and olfactory sensory processing. Taste is processed in the somatosensory region. Damage to these primary sensory areas leads to severe losses in perception and discrimination of these sensations. The perceptual changes show up mainly as a loss in acuity of sensation or as a loss in sensation from a particular region of the sensory field. The major motor area of the cortex lies in front of the somatosensory area. From it the majority of efferent fibers leave the cortex to innervate subcortical centers. Large areas of cortex are neither sensory nor motor, but exist to process and integrate the data from the many sensory sources. These areas have been termed "nonspecific" or "associative" cortex. They have undergone a tremendous increase in size during evolution and represent the greatest volume of cortex in man The increase in size of the associative regions has paralleled quite well the increase in the complexity of behavior that a species can evoke. Damage to these areas frequently results in the loss of skills requiring the integration of more

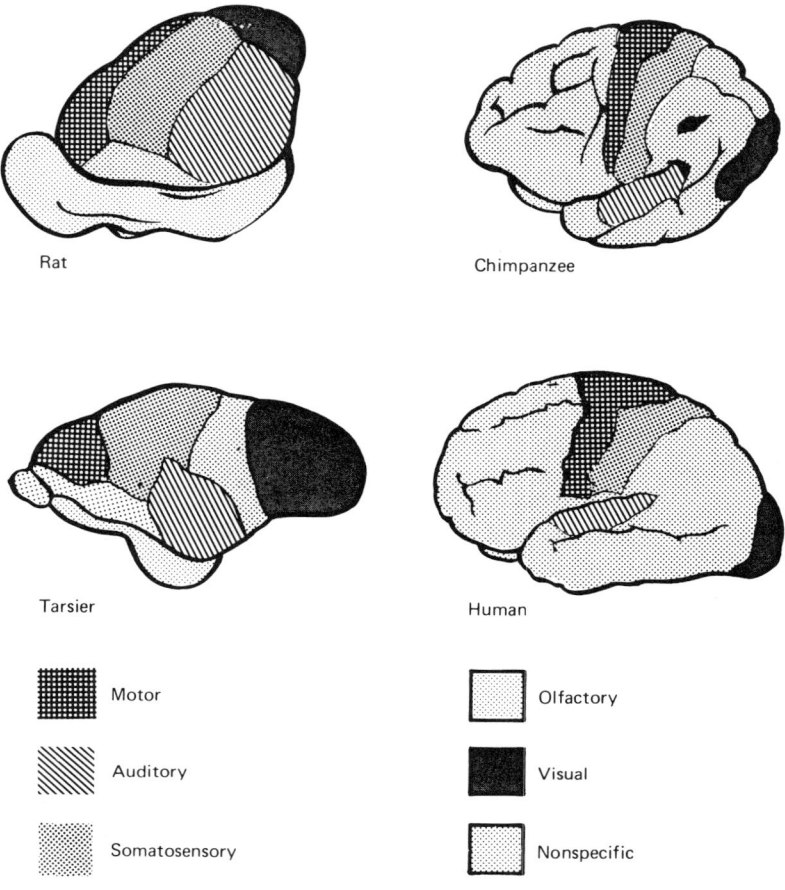

FIGURE 13. Side views of the brains of several mammals, illustrating the relative areas occupied by motor, sensory and association cortices. Note the massive increase in association cortex in the anthropoid ape and even more in the human brain. The considerable difference in size of the different brains should also be kept in mind. (From Schmidt, R. F., *Fundamentals of Neurophysiology*, 2nd ed., Springer-Verlag, New York, 1978, 271. With permission.)

than one sensory system. Deficiencies in abstraction capabilities, spatial perceptions, speech, and object recognition are also commonly associated with lesions within these areas.

A. Consciousness

Consciousness is a condition of awareness, the return of which is experienced when one awakens from sleep. The state of consciousness is most easily described in terms of its attributes. These include the capacity to purposefully focus attention upon objects, the capacity to create and use abstract ideas and to express them in symbols, the capacity to weight information and judge its significance, the capacity to predict future events on the basis of past experiences, the capacity to recognize self, and the capacity to create and appreciate esthetic and ethical concepts. Different degrees of consciousness, at least as measured by the richness of conscious capabilities, exist between animal species. Brain dysfunction or injury can bring about a loss, often selective, of certain conscious capabilities in man.

The state of consciousness does not appear to reside specifically in just one part of the brain. It is possible only when coordinated functioning of cortical and subcortical areas occurs. A perception of self seems to be totally dependent upon the status of the interconnections between parts of the brain. Consciousness is not a singular thing, either, and evidence for distinctive and separate consciousness within one individual has been observed in certain people whose intercortical connections have been totally severed. In such "split-brain" subjects, the functions and memories of the two sides of the brain are separable and unique. The left half literally does not know that the right half exists and possesses a separate and unique consciousness.

B. Speech

From studies of neurologically damaged patients, it has been determined that the brain areas involved in speech are usually restricted to the left cortical hemisphere. A number of speech areas exist there with different roles. Damage to one of these areas (Broca's area) produces motor aphasia, a condition in which the individual understands speech and can formulate thoughts consciously into words but has difficulty in producing the motor patterns necessary to verbalize them. The muscles involved can be used readily and fully for other actions. Damage to a second area (Wernicke's area) results in a sensory aphasia. In this condition individuals have difficulty understanding speech. It is often impossible for them to differentiate between words having meaning and nonsense syllables. These patients have no difficulty in speaking themselves. Damage to the corresponding brain areas in the right cortical hemisphere do not usually affect speech. Damage there produces impairments in spatial orientation with variable symptomology. An individual so affected may be unable to recognize familiar surroundings, to make three-dimensional drawings, or to estimate the distance between two objects.

C. Learning and Memory

The capacity to receive, integrate, store, retrieve, and use past experiences represents learning and memory. These processes are elementary and absolutely essential to all conscious functioning. Without memory or the capacity to learn, an organism would never rise above the developmental stage represented by hard-wired neuronal functions such as the reflex arc and totally sterotypic behavioral routines.

The formation of memory involves the storage of concepts and relationships more than specific details. Retrieval regenerates the concept which is usable directly as thought or may be verbalized by transforming the concept into appropriate symbols. Storage in memory seems to occur in steps. The immediate storage occurs into a "short-term" store which has a retention time of only seconds or minutes. In order for material to be remembered over a longer time it must be transferred into "long-term" stores. This transfer is not automatic but is keyed by practice and by the significance placed upon the need to remember a particular event. Extremely complex trains of input can be memorized, particularly by repetition. These trains tend to be recalled sequentially and completely, such as the memory of a complex piece of music or a poem. Often such material is difficult or impossible to recall except by starting at the beginning of the sequence.

The form in which memories are stored remains completely unknown. Short-term memory must be maintained in some type of dynamic network capable of re-exciting an appropriate neural pattern representing the input. Longer term storage may involve structural changes that alter neuronal connectivity between certain cells or groups of cells or may be coded in some type of intracellular storage form. Storage onto RNA and proteins have both been suggested, but definitive evidence is lacking for this at present.

D. Esthetics, Ethics, and Related Behaviors

Very little can be said about the means by which the nervous system generates judgements of value, beauty, rightness, wrongness, or generates reactions based upon these judgements. Certainly the frontal lobes participate in these processes, probably in close interaction with the limbic system, to which the frontal lobes are extensively connected. This belief is supported by observations on patients having frontal lobe damage. Such individuals commonly show unusually impulsive, uninhibited, irritable, euphoric, childlike, or psychologically labile behavior.

Frontal lobe damage does not seem to modify intelligence, as measured by performance on standarized tests. However, individuals with such damage frequently exhibit such behavioral deficits as a lack of drive or motivation, an absence of intention or desire, and a lack of planned, purposeful, long-term activities. The same subjects may show perseveration, the tendency to persist at a particular task long after common sense would suggest that it should stop. It is difficult for such people to adjust and adapt to changes in their environments. Often these same patients lose their previous abilities to judge good from bad, beautiful from ugly, etc. Certainly the loss of frontal lobe function severely reduces the human attributes normally considered to develop with maturation. This loss results in a rather child-like approach to situations and to making judgments. Patience is limited and frustration rapidly leads to upset.

REFERENCES

1. Noback, C. R. and Demarest, R. J., *The Human Nervous System,* 2nd ed., McGraw-Hill, New York, 1975.
2. Warwick, R. and Williams, P. L., *Gray's Anatomy,* 35th British ed., W. B. Saunders, Philadelphia, 1973.
3. Schmidt, R. F. *Fundamentals of Neurophysiology* 2nd ed., Springer-Verlag, New York, 1978, 339.
4. Selkurt, E. E. *Physiology,* 4th ed., Little, Brown, Boston, 1976.
5. Guyton, A. C. *Structure and Function of the Nervous System,* W. B. Saunders Philadelphia, 1972.
6. Buchsbaum, R. *Animals Without Backbones,* University of Chicago Press, Chicago, 1958.
7. Ganong, W. F. *Review of Medical Physiology.* Lange, Los Altos, Calif., 1969.
8. House, E. L. and Pansky, B. *A Functional Approach to Neuroanatomy,* McGraw-Hill, New York, 1960.
9. Truex, R. C. and Carpenter, M. B., *Strong and Elwin's Human Neuroanatomy,* 5th ed., Williams & Wilkins, Baltimore, 1964.
10. Cooper, J. R., Bloom, F. E., and Roth, R. H., *The Biochemical Basis of Neuropharmacology,* 3rd ed., Oxford University Press, New York, 1978.
11. Kuffler, S. W. and Nicholls, J. G. *From Neuron to Brain: A Cellular Approach to the Function of the Nervous System,* Sinauer Associates, Sunderland, Maine, 1976.
12. Jenkins, T. W. *Functional Mammalian Neuroanatomy,* Lea & Febiger, Philadelphia, 1972.
13. Goodman, L. S. and Gilman, A. *The Pharmacological Basis of Therapeutics,* 5th ed., Macmillan, New York, 1975.
14. Chusid, J. G., *Correlative Neuroanatomy & Functional Neurology,* Lange, 15th ed., Los Altos, Calif., 1973.
15. Katz, B., *Nerve, Muscle, and Synapse,* McGraw-Hill, New York, 1966.
16. McGeer, P. L., Eccles, J. C. and McGeer, E. G., *Molecular Neurobiology of the Mammalian Brain,* Plenum Press, New York, 1978.
17. Spencer, P. S. and Schaumburg, H. H., *Experimental and Clinical Neurotoxicology,* Williams & Wilkins, Baltimore, 1980.
18. *Scientific American,* 241 (3), 1979.

Chapter 4

CHLORINATED HYDROCARBON INSECTICIDES

Robert M. Joy

I. INTRODUCTION

The synthesis of dichlorodiphenyltrichloroethane (DDT) by Zeidler in 1874 marked the beginning of a new era characterized by the large scale utilization of synthetic chemicals to control various pests and pest-borne diseases. Zeidler was never aware of the insecticidal properties of DDT. It remained for Paul Müller to "rediscover" DDT in 1939 while he was searching for an effective contact poison against clothes moths and carpet beetles. Müller demonstrated the remarkable effectiveness of DDT as an insecticide and was awarded the Nobel Prize for his discoveries in 1948.

The discovery of the insecticidal properties of DDT, and subsequently those of other chlorinated hydrocarbon insecticides (CHIs), including benzene hexachloride, aldrin, dieldrin, endrin, and chlordane (Table 1) between 1939 and 1945 had immediate consequences. These chemicals were used in massive amounts during World War II for the control of mosquito-borne diseases such as malaria, dengue fever, and filariasis. Metcalf[1] has summarized the critical role that the CHIs have played both in world health and in world food production. Their impact has been truly remarkable. It has been estimated that by 1953, DDT alone had saved approximately 50 million lives and averted 1 billion cases of human disease.[2] DDT has been suggested to be the single most important factor in the world population explosion between 1950 and 1970.[1] The CHIs, prior to 1970, were the standard agents used in insect control. They were registered for use on over 300 agricultural commodities and were universally employed to control such pests as the codling moth, pink bollworm, gypsy moth, and the spruce budworm. In retrospect, it is obvious that much abuse of these substances occurred during this same time period. This was due to an overly optimistic view of the selectivity of action of the CHIs in destroying pests without severely affecting beneficial organisms and, in part, to a lack of knowledge about insect resistance and long-term adverse effects.

The hazards inherent in the large scale use of the CHIs began to emerge during the same period. Hayes[3] published a review article in 1959 focusing upon the pharmacology and toxicology of DDT which contains 685 references, most dealing with some aspect of acute or chronic toxicity to man, domestic animals, or wildlife. The environmental impact of massive CHI application was presented publicly and dramatically by Carson[4] in *Silent Spring*. More careful and balanced assessments soon became available.[5,6] At present, strict controls exist upon the use of CHIs in most countries to limit their adverse impact on the environment. This use limitation followed logically from their implied hazards. CHIs tend to persist unchanged in the environment for long periods of time, they tend to accumulate in soils, and they may be translocated from their locus of application into rivers, lakes, and oceans. Because of their persistence and high lipid-solubility, they enter into all organisms and tend to accumulate along various food chains in all ecosystems. This accumulation can lead to undesirable kills of many invertebrates, who are particularly sensitive to their toxic actions. The ingestion of CHI-containing organisms by various predators results in high body burdens of the CHIs and their metabolites in the predator. This mechanism has been implicated in certain widescale loss of fish and birds.[7] Problems such as these have led to severe restrictions in the use of the more persistent CHIs. In addition to the already apparent

Table 1
ACUTE TOXICITY OF CHLORINATED HYDROCARBON INSECTICIDES[a]

Insecticide	Rat-oral LD$_{50}$ mg/kg	Rat-dermal LD$_{50}$ mg/kg	Human-lowest[b] LD$_{50}$ mg/kg
Dichlorodiphenylethanes			
DDT	113	2500	50
DDD (Rhothane)	113	—	500
DMC (Dimite)	500	—	500
Dicofol (Kelthane)	575	1000[c]	500
Chlorobenzylate	700	—	500
Methoxychlor	5000	—	6430
Hexachlorocyclohexanes			
Lindane (gamma-BHC)	88	500	180
Benzene hexachloride (BHC)	100	—	50
Cyclodienes			
Endrin	3	15	5
Telodrin	5	5	5
Isodrin	7	23	5
Endosulfan	18	74	50
Heptachlor	40	195	50
Dieldrin	46	60	5
Toxaphene	60	780[c]	40
Aldrin	67	98	1.25
Strobane	200	—	50
Chlordane	283	700	40

[a] Toxicity values taken from Fairchild[15] unless otherwise indicated.
[b] The lowest reported dose introduced by any route other than inhalation over any given period of time which resulted in death.
[c] Toxicity values taken from Hayes.[16]

impact of the CHIs, there are real concerns remaining about the consequences of the continuous presence of pesticide residues in man and domestic animals. There is concern, too, that continued exposure may result in oncogenic, mutagenic, and teratogenic effects which might not become apparent until many years of chronic exposure have occurred. These latter concerns are not restricted to the CHIs and apply universally to any chemical substance. However, the long persistence of CHIs in the environment and in the tissues of animals and man makes their removal very difficult if and when unacceptable toxicity should occur.

The content of this chapter is limited to the assessment of the neurological consequences of CHI exposure. Primary consideration will be given to the symptoms that follow acute or chronic exposure in man and animals. The mechanisms of action responsible for the evolving symptomology will be examined with emphasis upon the functional changes occurring at the level of the nervous system. Underlying biochemical and biophysical mechanisms will only be briefly reviewed. The aim is not to be exhaustive but rather to consolidate views regarding neurological effects, their loci, and their mechanisms. Consequently, distinctions such as dose-effect relationships or variations in responses associated with pharmacokinetic parameters in published data will be ignored except where they contribute to an understanding of effects. For more general reviews of CHI toxicology the reader is referred to the articles by Hayes,[3,8] O'Brien,[9] Jager,[10] Matsumura,[11] and Hrdina et al.[12]

II. CLASSIFICATION AND CHEMISTRY

The CHIs are usually classified on the basis of their chemical structures and methods of synthesis. The dichlorodiphenylethane derivatives include DDT and related compounds. The hexachlorocyclohexanes include benzene hexachloride, which chemically is a mixture of six distinct isomers, the most active of which is the gamma-isomer. In purified form, the gamma-isomer is lindane. The cyclodiene derivatives are a collective group of synthetic polycyclic hydrocarbons. Many are produced by a Diels-Alder condensation reaction including aldrin, dieldrin, endrin, and isodrin. Toxaphene and strobane are prepared by chlorinating naturally occurring camphenes and terpenes. Regardless of their specific chemical structures, all of the CHIs share many similar chemical and physical properties. They consist primarily of carbon, hydrogen, and chlorine atoms; some contain oxygen and/or sulfur atoms. All are characterized by possessing numerous carbon-chlorine bonds. All possess cyclic carbon chains. They are also characterized by their lack of any particular active intramolecular sites, by their apolarity and lipophilicity, and by their chemical unreactivity.[11] The following discussion of the chemistry of the CHIs follows closely that of Matsumura.[11]

A. Dichlorodiphenylethane Derivatives

DDT:

$Cl-\text{C}_6\text{H}_4-\text{CH}(CCl_3)-\text{C}_6\text{H}_4-Cl$

1,1,1-Trichloro-2,2-bis (p-chlorophenyl) ethane
p,p'-DDT
Dichlorodiphenyltrichloroethane

DDT is a crystalline substance with a melting point of 109°C and a vapor pressure of 1.5×10^{-7} mm Hg at 20°C. It is one of the most nonpolar compounds known having a water solubility of less than 2 ppb. It is soluble in most organic solvents such as hexane. Its tremendously high lipid/water partition coefficient accounts for its capacity to accumulate in fat and fat-containing tissue compartments and to remain there for long periods of time. Technical DDT, the commercial product, consists of about 70% of the p,p'-isomer along with small amounts of the o,p-isomer. DDT is chemically stable. Only small amounts are broken down in the environment via dechlorination into less active products. The primary degradation product is DDE [1,1-Dichloro-2,2-bis (p-chlorophenyl) ethene].

$Cl-\text{C}_6\text{H}_4-\text{CH}(CCl_3)-\text{C}_6\text{H}_4-Cl \longrightarrow Cl-\text{C}_6\text{H}_4-\text{C}(=CCl_2)-\text{C}_6\text{H}_4-Cl$

DDT → DDE

The loss of DDT from the environment by decomposition amounts to only a few percent per year.

DDD:

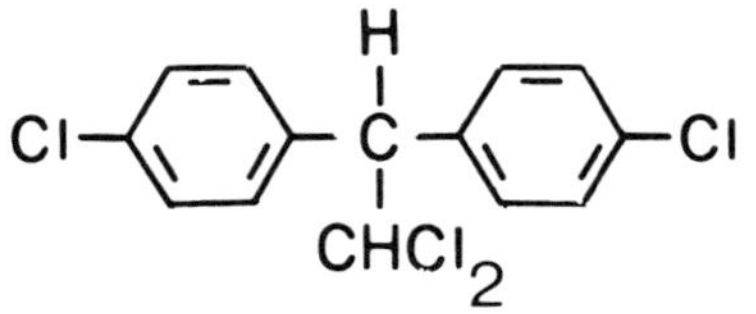

Cl—⟨C₆H₄⟩—CH(CHCl₂)—⟨C₆H₄⟩—Cl

1,1-Dichloro-2,2-*bis* (p-chlorophenyl) ethane
Dichlorodiphenyldichloroethane
Tetrachlorodiphenylethane
TDE
Rhothane

DDD is a white crystalline product with a melting point of 109°C. Its chemical properties are similar to those of DDT. Like DDT it tends to persist in the environment and is only slowly converted into less active compounds.

DMC:

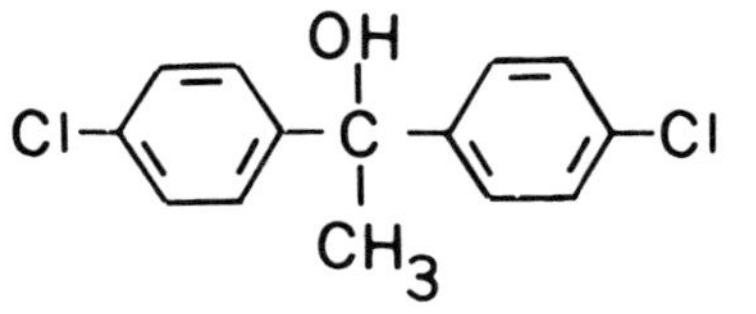

Cl—⟨C₆H₄⟩—C(OH)(CH₃)—⟨C₆H₄⟩—Cl

4,4-Dichloro-α-methylbenzhydrol
DCPC
Dimite

DMC is a colorless crystalline solid with a melting point of 70°C. It is less persistent in the environment than either DDT or DDD.

Dicofol:

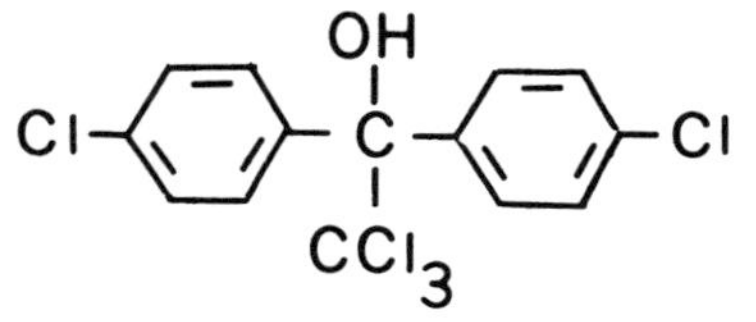

Cl—⟨C₆H₄⟩—C(OH)(CCl₃)—⟨C₆H₄⟩—Cl

4,4-Dichloro-α-(trichloromethyl)-benzhydrol
Kelthane

This compound was available commercially. It is also a metabolic product of DDT in certain insects. It is a brown, viscous oil. It is more water soluble and less persistent than DDT.

Chlorobenzylate:

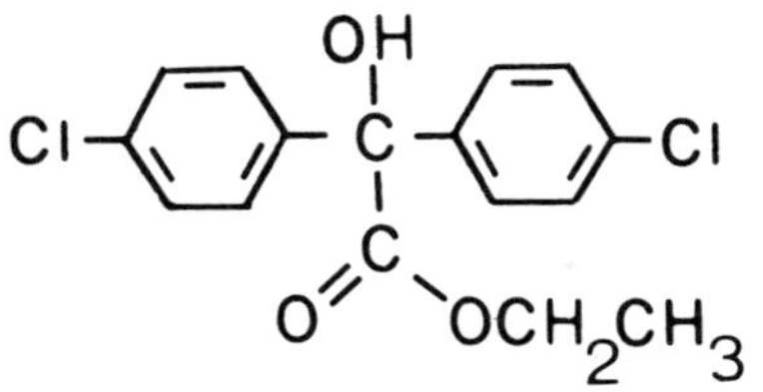

Cl—⟨C₆H₄⟩—C(OH)(COOCH₂CH₃)—⟨C₆H₄⟩—Cl

Ethyl 4,4-dichlorobenzylate

This compound was used primarily as an acaricide. It is not generally toxic to insects. It is a yellow crystalline solid with a melting point of 35 to 37°C.

Methoxychlor:

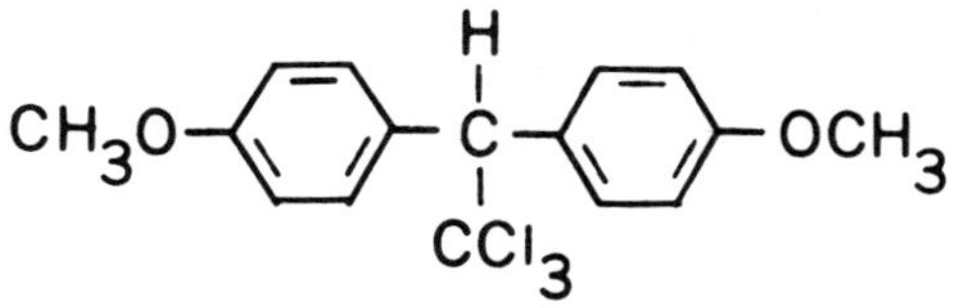

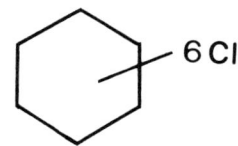

1,1,1-Trichloro-2,2-*bis*(*p*-methoxyphenyl) ethane
DMDT
Methoxy-DDT

Methoxychlor was an important insecticide. It is a white crystalline solid with a melting point of 69°C. Its water solubility is approximately 100 ppb. Methoxychlor is less persistent in the environment than DDT, undergoing dechlorination more rapidly. It does not tend to accumulate to nearly the degree that DDT does in animal tissues or milk.

B. Hexachlorocyclohexane Derivatives

Benezene hexachloride:

1,2,3,4,5,6-Hexachlorocyclohexane
BHC

Benzene hexachloride is formed by chlorinating benzene in the presence of ultraviolet light. This process yields six chemically distinct isomers of hexachlorobenzene. The toxicity of benzene hexachloride is proportional to its content of the gamma-isomer, which is some 50 to 10,000 times more insecticidal than the other isomers. Benzene hexachloride is a grayish or brown amorphous solid with a characteristic odor. It melts around 65°C. It has a water solubility of 10 to 32 ppm.

Lindane:

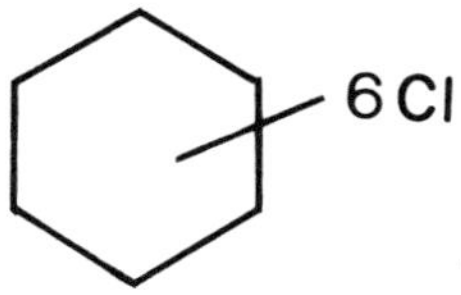

Gamma-1,2,3,4,5,6-Hexachlorocyclohexane
Gamma-BHC
Gammexane

Lindane is the gamma-isomer of benzene hexachloride. It is a crystalline solid with a melting point of 112°C and a vapor pressure of 0.14 mm Hg at 40°C. The high vapor pressure led to its use at one time in vaporizers for insect control. It is practically water insoluble, but is is soluble in most organic solvents.

C. Cyclodiene Derivatives

Aldrin:

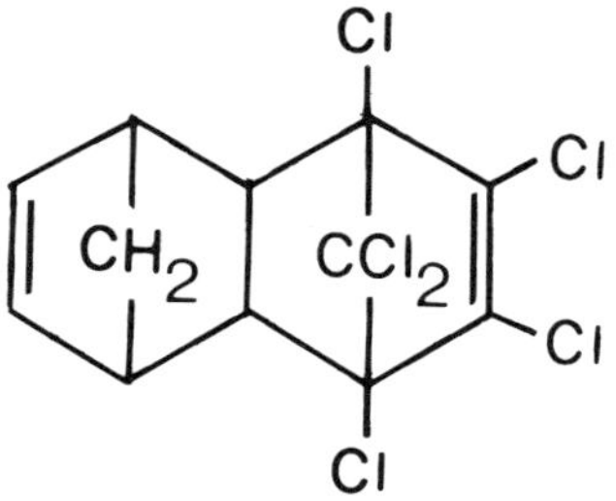

1,2,3,4,10,10-Hexachloro-1,4,4a,5,8,8a-hexahydro-endo,exo-1,4:5,8-dimethanonaphthalene

Aldrin is a white crystalline material with a melting point of 104°C and a vapor pressure of 6×10^{-6} mm Hg at 25°C. Aldrin is stable to alkali and acid. It is readily converted to its epoxide, dieldrin, in the environment and in biological systems. Because of this rapid conversion, dieldrin is generally considered as the primary toxic substance of aldrin.

Dieldrin:

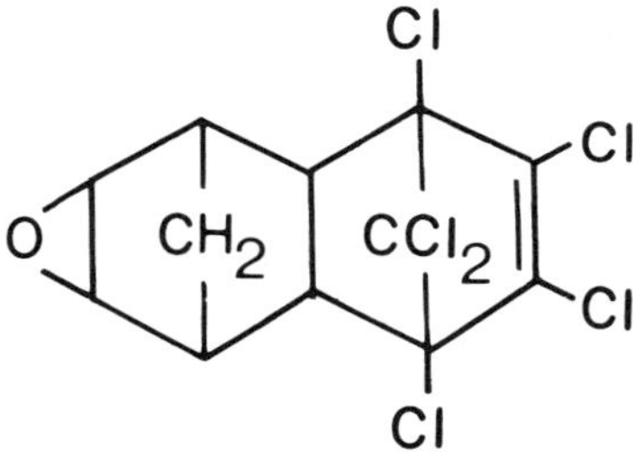

1,2,3,4,10,10-Hexachloro-6,7-epoxy-1,4,4a,5,6,7,8,8a-octahydro-endo,exo-1,4:5,8-dimethanonaphthalene
HEOD

Dieldrin is the epoxide of aldrin. It is a white crystalline substance with a melting point of 173°C and a vapor pressure of 1.8×10^{-7} mm Hg at 25°C. Its water solubility is approximately 0.25 ppm. Dieldrin is extremely stable chemically. It is one of the most persistent insecticides known.

Isodrin is the endo-endo isomer of aldrin. Like aldrin it is readily converted to its epoxide. It has seen only restricted commercial use.

Endrin is the epoxide of isodrin and is the endo-endo isomer of dieldrin. Chemically it is very similar to dieldrin. It is more readily degraded in the environment and more rapidly metabolized by most organisms than dieldrin. As a consequence it is less persistent.

Chlordane:

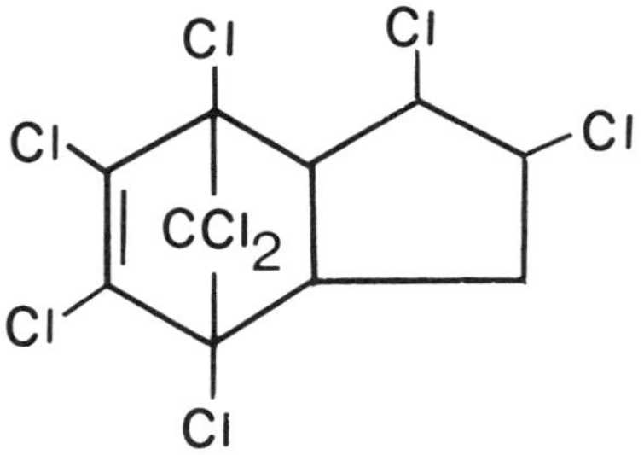

1,2,4,5,6,7,8,8a-Octachloro-2,3-3a,4,7,7a-hexahydro-4,7-methano-1H-indene
Chlordan

Technical chlordane is a dark amber viscous liquid containing approximately 60% chlordane. It is nearly insoluble in water but is miscible with most organic solvents.

Heptachlor:

1,4,5,6,7,8,8-Heptachloro-3a,4,7,7a-tetrahydro-4,7-methanoindene

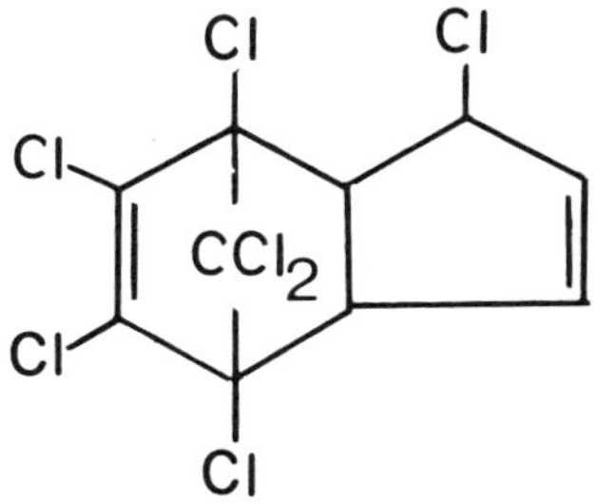

Heptachlor is a white crystalline solid with a melting point of 95°C and a vapor pressure of 3×10^{-4} mm Hg at 25°C. Although heptachlor is quite stable in the environment, it is readily converted to its epoxide in many biological systems. The epoxide is more toxic than heptachlor.

Endosulfan:

6,7,8,9,10,10-Hexachloro-1,5,5a,6,9,9a-hexahydro-6,9-methano-2,4,3-benzo-dioxathiepin 3-oxide
Thiodan

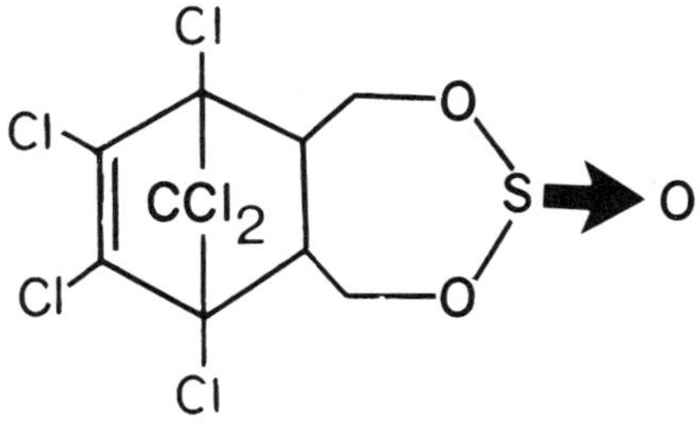

Endosulfan is a mixture of two isomers: the alpha-isomer (a crystalline material with a melting point of 108 to 110°C) and the beta-isomer (a crystalline material with a melting point of 208 to 210° C). The commercial product consists of impure brown crystals with a melting point of 70 to 100° C. Endosulfan is practically insoluble in water, but is soluble in most organic solvents.

Telodrin:

1,3,4,5,6,7,8,8-Octachloro-3a,4,7-7a-tetrahydro-4,7-methanophthalan

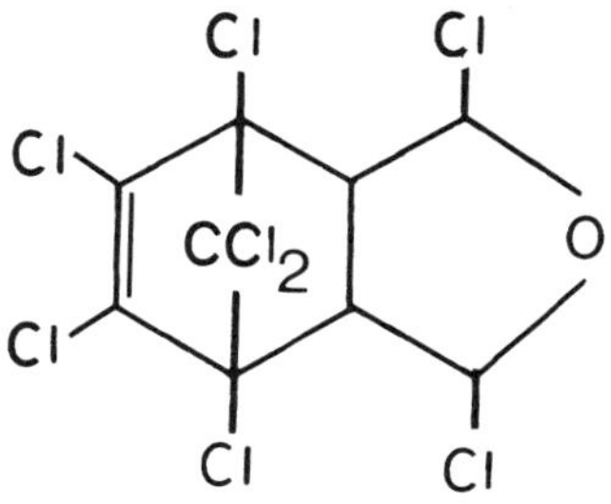

Telodrin is a crystalline solid with a melting point of 120 to 122°C. It is less stable than aldrin. Otherwise its chemistry is similar.

Toxaphene is a complex but reproducible mixture of over 175 chlorinated camphene derivatives. It contains 67 to 69% chlorine by weight. It is a yellow waxy solid with a melting point of 65 to 90°C. It is practically insoluble in water but is soluble in most organic solvents.

Strobane is also a complex mixture of chlorinated camphene and terpene isomers. It contains about 65% chlorine by weight. It is sometimes referred to as terpene polychlorinate. It is a viscous straw-colored liquid with low water solubility.

III. PHARMACOKINETICS IN BIOLOGICAL SYSTEMS

The final impact of CHI exposure upon an organism is determined primarily by the final tissue levels realized. These in turn will depend upon the amount absorbed, the rate of loss by metabolism and excretion, and the capacity of the organism to sequestrate the CHI into relatively inactive stores, such as fat.

The absorptive surfaces of the body for the CHIs are diverse. Direct absorption can occur through the skin and lungs or result from accidental ingestion. Animals may acquire large amounts on their fur which subsequently can be ingested during grooming.

The rate and nature of biotransformation vary with the particular compound and the species of animal. Also important are factors such as age, sex, diet, temperature, disease, and the simultaneous presence of other toxic substances. The metabolic conversion of a CHI typically involves many steps. The general trend in biotransformation is the conversion of a substance to more polar, hydrophilic metabolites. The major organ for metabolism of CHIs is the liver, the hepatic microsomes constituting the primary site of transformation. Biotransformation may have one of two consequences. Some CHIs such as aldrin and heptachlor can be biotransformed into more toxic epoxide derivatives. Other metabolites have little toxicity. Their formation effectively removes the CHI, and its consequences, from the system.

The high lipid/water partition coefficients of the CHIs results in their storage in fat and organs with a high lipid content. Accumulation is affected both by the amount and the rate of exposure. With chronic exposure, accumulation in body fat eventually plateaus in equilibrium with plasma and other body fluids. The CHIs thus stored are effectively sequestered and do not contribute to toxic symptomology. However, they serve as a repository from which the CHI can move back into body fluids as its level there is depleted by metabolism or excretion.

A consideration of the metabolism of each CHI is beyond the scope of this chapter. The reader is referred to DeBruin[13] or Matsumura[11] for more specific information on metabolism.

IV. COMPARATIVE TOXICITY OF THE CHLORINATED HYDROCARBON INSECTICIDES

Table 1 indicates the lowest doses of CHIs reported in the literature to produce acute lethal effects in animals and man. These data should be viewed as approximations since actual toxicity is affected by many factors, such as those just mentioned above. Insecticides are most often used with a solvent which may possess its own toxicity or enhance the absorption of one or more insecticides presented in a mixture. For man, most acute toxicity results from ingestion or skin contact whereas chronic toxicity may result from ingestion, skin contact, or inhalation. As a rule, the cyclodienes are the most toxic, and they possess significant toxicological potential from any route of exposure. The various hexachlorocyclohexanes possess intermediate acute toxicities whereas the DDT analogues possess the lowest acute toxicities. DDT analogues are relatively nontoxic upon dermal or inhalational exposure. The different potential of these substances to produce toxicity in man is amply demonstrated in reports describing toxicity development in spraymen constantly exposed to these substances. Little or no toxic symptoms have been noted in spraymen using DDT[3] whereas symptoms developed in 20 to 30% of spraymen using dieldrin.[14]

In general, a good correlation exists between the oral LD_{50} doses in rats and the lowest doses found to result in human death. The exceptionally low human lethal doses reported for chlordane, aldrin, and dieldrin, as compared to their oral LD_{50} values in rats, may simply reflect their more extensive use and greater availability which would increase the probability of exposing highly sensitive individuals.

The evidence which has accumulated suggests that the CHIs should be separated into two distinctive groups on the basis of their neurotoxic actions. DDT and its analogues produce similar effects on nerve and muscle. They can be considered as one group, with DDT acting as its representative. The remainder of the CHIs, which includes the hexachlorocyclohexanes and the cyclodiene derivatives, share sufficient similarities with each other in regards to their actions and proposed mechanisms on nerve and muscle to be considered as a second group. Certainly, differences exist between members of these two groups and to some extent such separation is arbitrary. However, the similarities in symptom development and evidence from mechanistic studies support such a separation. It is worth emphasizing that this separation is based solely upon an assessment of neurotoxicity.

V. NEUROTOXICITY OF THE DICHLORODIPHENYLETHANE DERIVATIVES

The neurotoxicity of the various DDT analogues is similar with regard to actions on nerve and muscle. DDT has been much more widely studied than the rest of the group combined, and it will be the focus for consideration in this section.

Wide individual variations in susceptibility to poisoning occur with various animal species. These variations are often independent of the route of administration employed and of the solvent or suspending agent used. They may be related in part to differences in body fat, absorptive ability, nutritional state, and diet. Animals with large amounts of body fat are less susceptible to acute poisoning than animals which lack body fat. Fasted animals are more susceptible than animals on an adequate diet. Young animals are usually more susceptible than older ones. Susceptibility generally decreases with increase in the complexity of the living organism. Fish are nearly as sensitive as insects, followed in decreasing order by crustaceans, amphibians, and small animals. Among the mammals, susceptibility decreases roughly in the order: mice, rats, cats, dogs, monkeys, swine, sheep, and goats. Birds are especially resistant to the nervous system effects of DDT. Man is also comparatively resistant.[17]

A. Signs and Symptoms of Poisoning in Animals
1. Acute Exposure
a. Insects

The major symptoms of acute exposure to DDT and its analogues are of nervous origin. This is true for insects, animals, and man. In insects DDT evokes incoordination, ataxia, and tremorous movements of the legs. These symptoms may be followed by immobility and death, or the insect may recover completely depending upon the total level of exposure. The account of DDT's actions in the cockroach by Tobias and Kollros[18] is generally applicable to all insects. They observed a sequence of actions which included: (1) hyperextension of the legs and incoordinated movements; (2) general tremulousness; (3) ataxia; (4) hyperreactivity to external stimulation; (5) loss of righting capabilities; (6) development of two types of leg movements including a high-frequency tremor and a slower flexion and extension movement sequence; (7) the eventual disappearance of tremors leaving only sporadic, isolated movements of the head, tarsi, palpi, cerci, and antennae, and; (8) immobility followed shortly by death.

b. Vertebrates

Vertebrates show a qualitatively similar progression of responses. As expected, differences in the overall development of symptoms have been described by various authors. These can be explained in part by the different routes of administration employed and by the various solvents and/or emulsifiers used. Other differences can be accounted for by the different expectations and intents of the studies undertaken. There appear to be, however, real species differences in responses which warrant consideration.

In rats and rabbits, the first perceptible effect is hyperexcitability, consisting of excessive emotional and physical reactions to stimuli. Spontaneous activity is increased, and much of it is nonpurposeful. As intoxication becomes more severe, blepharospasm, hyperreflexia, and twitching of the ears and vibrissae occur. Massive myoclonic jerks may appear in response to sensory stimulation. A most prominent feature is tremor, which develops as an intention tremor during voluntary movement. Both a fine and a coarse tremor evolve, usually in coexistence. The two components are separable and can be demonstrated by spectral analysis of the tremor.[19] The tremor may become so intense that purposeful movements can no longer be initiated. Episodes of clonic convulsions may occur, but tremors in the rat are frequently so pronounced that convulsions are hard to discern. Tremoring may continue for hours or days depending upon the severity of exposure, and it is the primary cause for the persisting hyperthermia seen in these species. In later stages, death may occur following a convulsive seizure, or it may follow a terminal period characterized by progressive exhaustion, flaccid immobility, and unconsciousness. The cause of death is usually ventricular fibrillation in the former case and respiratory failure in the latter. Recovery is possible from any of these stages, and, when it occurs, all symptomology disappears without signs of irreversible damage.[3,11,12,19-25] Similar findings have been reported for methoxychlor.[26]

Cats, dogs, and monkeys become restless and apprehensive. The first symptoms of motor involvement are blepharospasm and twitching of the ears and vibrissae. Tremors develop at the level of the head and neck and spread over the entire body. The tremors are often severe enough to make purposeful movements difficult. In contrast to the rat or rabbit, convulsive episodes occur early and frequently. Brief myoclonus of all four limbs may accompany sensory stimulation or attempts to move. Convulsions develop abruptly and are generalized tonic-clonic convulsions. They are frequently accompanied by marked pilomotor activity in the cat and by signs of autonomic stimulation in all species. In the postictal period following a convulsion, the tremor is initially absent, and the animal is capable of regaining an upright position. Attempts to move expose a marked dysmetria and the reemergence of an intention tremor. Spontaneous tremoring eventually reappears, dysmetria becomes exaggerated, and additional convulsions may follow. The periods between convulsions become progressively shorter until a status epilepticus-like condition prevails. Eventually the animals become severely depressed and lapse into a comatose condition. Death may ensue. When death occurs suddenly during a convulsive episode, it is generally the result of ventricular fibrillation. When death occurs during the period of exhaustion following many convulsive seizures, it typically occurs as a result of respiratory failure.[3,21,27-29]

The acute toxicity of DDT in man will be covered in detail in a later section of this chapter, and only a brief description of the symptomology will be given here. Upon exposure, the earliest effects involve paresthesias of the mouth and face. Altered motor function leading to ataxia and abnormal stepping reactions also tend to occur early. These symptoms are typically followed by dizziness, confusion, general malaise, headache, and fatigue. Tremor, particularly of the hands is a common symptom of DDT poisoning. Convulsions have resulted from ingesting large quantities of DDT. Very few deaths have been reported.[3]

2. Chronic Exposure

Chronic exposure to DDT produces many of the same symptoms seen with acute exposure. However, their onset, intensity, and progression will vary depending upon the kinetics of exposure. Chronic intoxication may be associated also with loss of weight, anorexia, mild anemia, muscular weakness, and tremors. At sufficiently high exposure levels, convulsive episodes with subsequent muscular weakness, paralysis, coma, and death may occur.[28,30,31]

Associated with these observations are pathological changes which include centrolobular necrosis of the liver, fatty degeneration of the tubular epithelium of the kidney, focal necrosis of cardiac and skeletal muscle, and various alterations in the central nervous system (CNS). The latter vary from study to study but are characterized by diffuse changes in cellular morphology or degeneration at one or more loci.[31-33] The most clear cut histopathological changes of CNS origin were reported by Haymaker et al.[34] In dogs fed 180 to 250 mg/kg/day of DDT in peanut oil by stomach tube, progressively severe symptomology and associated histopathology were noted. Severe tremoring and ataxia developed and were followed by numerous abnormalities in motor function reminiscent of those observed in cerebellar deficiency. Anorexia, listlessness, and weakness set in, weight loss occurred, and four of eight subjects died after 13 to 98 days of such exposure. However, no convulsive seizures were observed.

Post-mortem examination of the brain and spinal cord indicated moderate congestion and some petechial hemorrhaging. Ventricles were normal. No microscopic lesions were observed in the cerebrum, brain stem, peripheral nerves, or eyes. Some chromatolysis was seen in anterior horn cells in the spinal cord. The cerebellum showed the greatest changes ranging from swollen and hyperchromatic cytoplasm and pyknotic nuclei of Purkinje cells to frank degeneration in the dentate and roof nuclei. The extent of pathology mirrored the severity of symptoms observed in the subjects prior to death.

This striking pathology has not been verified by others, however. In the study of Fitzhugh and Nelson[30] and in a detailed, multiple specie study by Globus,[35,36] in which subjects were dosed for well over a year, no specific CNS pathology was found. The reasons for these discrepancies are not clear. However, Haymaker et al.[34] employed a very severe dosing schedule and purposefully attempted to produce the maximally tolerable levels possible in their subjects. The majority of other studies have employed lower doses and have generally reported much less severe symptomology. Part of the variations may represent different intensities of exposure. It is also possible that in the study of Haymaker et al. some of the damage observed may have been secondary to other phenomena. It is interesting, as Globus[36] has observed, that convulsions which are so common a part of the DDT symptomology, particularly in dogs at high exposures, were never observed by Haymaker et al.[34]

B. Sites of Action

DDT analogues are poisons which act primarily at the level of the CNS in man and higher animals. Important peripheral nerve effects also occur, particularly in insects. The combination of central and peripheral actions is responsible for the hyperexcita-

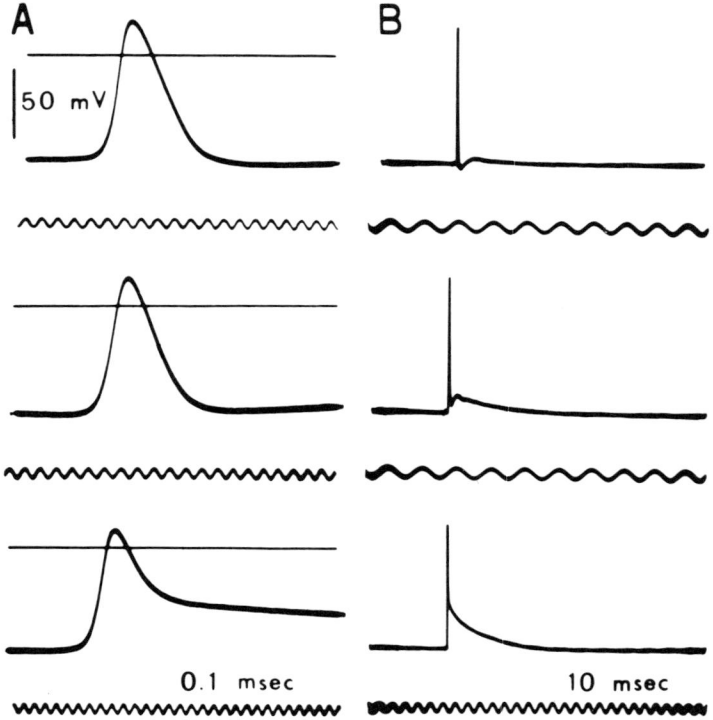

FIGURE 1. Changes in the intracellularly recorded action potential of the cockroach giant axon after treatment with DDT 10^{-4} M. A: from top to bottom, before, 38 min after, and 90 min after treatment with DDT. The horizontal lines indicate zero potential level. B: as in A, but with slower sweep. (From Narahashi, T. and Yamasaki, T., *J. Physiol. (London)*, 152, 122, 1960. With permission.)

bility, generalized tremors, spasticity, and convulsions that characterize toxicity. An important additional effect of these compounds is the sensitization of the myocardium by mechanisms similar to those leading to enhanced neuronal activity. The myocardium is particularly sensitized to the actions of the sympathomimetic amines. Fibrillation and other arrhythmias commonly occur during acute stressful situations and frequently accompany the massive autonomic discharge occurring during convulsive episodes. This is the primary basis for the acute death which can occur during a convulsion.

1. Insects

In insects, DDT and its analogues have an important, and perhaps primary, action on peripheral nerve fibers and their associated sensory receptors. DDT causes a marked increase in and prolongation of the negative afterpotential in individual insect nerve axons[37,38] (Figure 1). DDT induces repetitive discharge in all nerve fibers. Afferent fibers and/or their associated sensory receptors are more susceptible than motor fibers.[39-41] This has been well demonstrated by Roder and Weiant[40] who injected a suspension of DDT in insect saline into the amputated tibial stump of the cockroach, *Periplaneta americana*. The single spikes which characterize the spontaneous discharge of the crural (tibial) nerve are converted to trains of discharges by DDT (Figure 2). This is preceded and accompanied by the development of an afterdischarge to movement of the leg. By amputating sections of the leg and by selective recording from branches of the crural nerve, they demonstrated that the repetitive discharge emanated

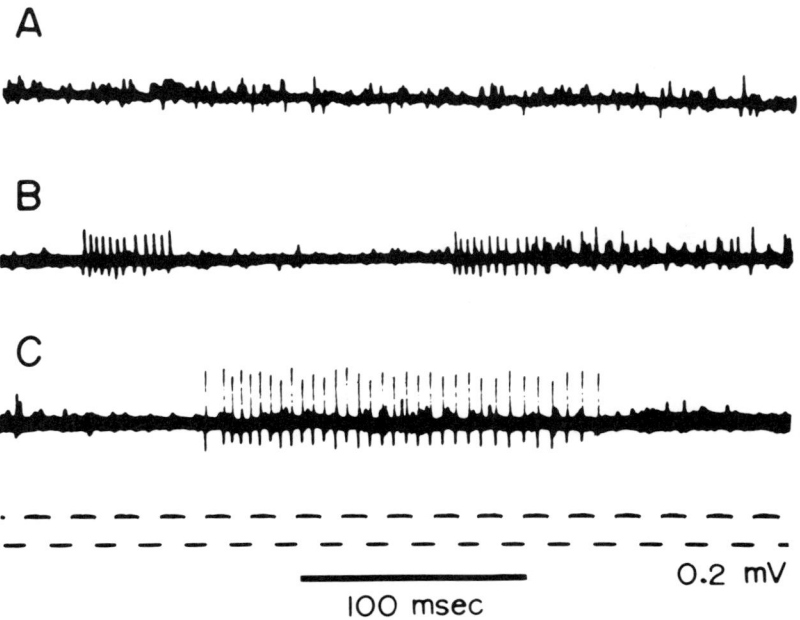

FIGURE 2. Trains of impulses from the sensory cells of the cockroach leg after injection of DDT into the leg. A: before injection of DDT, B and C: after injection. (From Narahashi, T., *Kagaku to Seibutsu*, 4, 134, 1966. With permission.)

from afferent fibers. The most sensitive afferent fibers were those originating from campaniform sensillae, specialized somatosensory receptors in the cockroach. Repetitive discharge could be evoked from other sensory and motor fibers, but only at DDT concentrations 100 to 1000 times greater. In various insect preparations, this relative susceptibility of sensory as compared to motor fibers has been generally confirmed.[42,43]

Similar modes of repetitive discharge generation have been demonstrated for DDD and methoxychlor by Lalonde and Brown.[44] The repetitive discharge patterns of the DDT analogues differ from those produced by CHIs of the hexachlorocyclohexane or the cyclodiene groups and from those evoked by pyrethrins and organophosphates. For DDT and its analogues, a good correlation has been found to exist for insecticidal potencies and for the activity of the various analogues in inducing repetitive discharge.[42]

CNS actions seem to be importantly involved in poisoning in insects, also. The spontaneous discharge in the central nerve cord of the cockroach is increased in frequency, and synaptic transmission is facilitated.[18] Narahashi[43] describes an experiment which suggests that CNS actions play a very important role in symptom expression. DDT has a negative temperature coefficient of action. When just the right amount is given to an insect, the symptoms of poisoning appear at low temperatures but disappear when the temperature is raised. In such a case, the emergence of symptoms can be correlated with changes in nervous activity in both central and peripheral parts of the nervous system. At the low temperature, the poisoned insect exhibits ataxia and convulsions, and both the sensory nerve of the leg and the abdominal nerve cord discharge impulses at high frequencies. When the temperature is raised sufficiently to terminate overt symptoms of toxicity, the impulse discharge rate in the abdominal nerve cord is reduced in frequency while the sensory nerve continues to discharge at high rates. Thus, the expression of toxic symptoms appears to be strongly dependent upon the level of activity within the central as well as the peripheral nervous system.

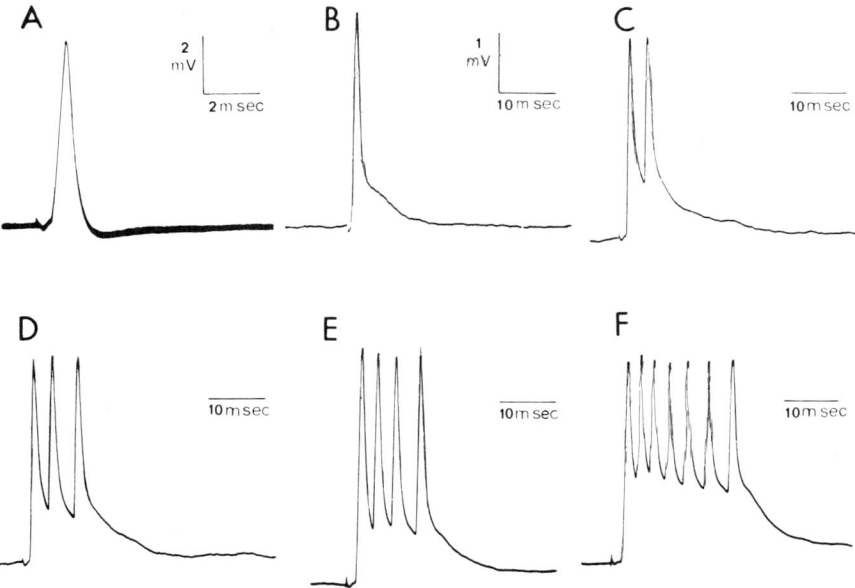

FIGURE 3. Repetitive activity in DDT-treated single (amphibian) nerve fibers, recorded with a suction-pipette electrode. A: Normal action potential of a single nerve fiber, stimulated at a frequency of 10 pulses per second. Several action potentials have been superimposed. B: Action potential with negative afterpotential, 65 min after start of treatment with 5×10^{-4} M DDT. Stimulus frequency in this and the following records, 1 pulse every 2 sec. C—F: Repetitive action potentials superimposed on the negative afterpotential, after 85, 90, 105, and 135 min of treatment with 10^{-4} M DDT, respectively. All records were obtained from the same nerve fiber. Calibration in B applies also to records C to F. (van den Bercken, J., *Eur. J. Pharmacol.*, 20, 205, 1972. With permission.)

2. Vertebrates

In vertebrates, direct effects of DDT and analogues on peripheral systems is demonstrable but overshadowed by effects developing at the level of the CNS. DDT, like pentylenetetrazol, evokes a myotonic response in isolated nerve-muscle preparation of the rat.[45] This appears as an increase in tension and duration of the twitch evoked by a brief stimulus of either the muscle membrane or of the innervating motor nerve fibers. The normal diphasic muscle action potential is followed by a train of spikes after DDT. The isolated sciatic nerve may also respond to stimulation with a burst of spikes following exposure to DDT. Similar peripheral effects have been reported in both frogs and rats.[45-48]

van den Bercken and colleagues[46,48-50] have made detailed studies of the effects of DDT on vertebrate peripheral nerve and have demonstrated that sensory systems are more sensitive to its effects than are motor systems (Figure 3). Repetitive discharges were observed in the lateral line organ and its nerve and in cutaneous touch receptors of various amphibians. Additional components of afferent origin were also observed in the compound sciatic nerve action potential. In experiments where sections of spinal cord were left attached to dorsal and ventral roots, repetitive discharge could be recorded from the dorsal, but not the ventral, root. In rats, repetitive discharge was restricted to afferent fibers and dorsal roots. No repetitive activity was observed from motor fibers or ventral roots.

These various peripheral actions of DDT should enhance and modify motor activity coursing out from the spinal cord to the periphery, but it is doubtful that they play a causative role in the genesis of either tremor or convulsive activity in intact subjects.

A minimal amount of CNS must also be present to elicit spontaneous motor signs similar to those seen in intact subjects. In most cases, the more advanced the species, the more important an intact neuraxis becomes to the evolution of toxicity. In frogs, a few segments of spinal cord and the efferent motor nerve suffice to produce DDT-evoked movements.[51] Dorsal roots and afferent fibers are not required.[52] In rats with low thoracic spinal transections, tremoring can be evoked from the hindlimbs only if the reflex status of those limbs is good.[47] Even then, however, the tremor is quantitatively and qualitatively different from that developing in the front legs, which are still connected to the CNS. Hindleg responses lack the coordination and rigour of the front leg responses, and in many cases they must be evoked by stimulation and will not occur spontaneously.

One of the most comprehensive studies of the role of the CNS in DDT symptom development was that of Bromiley and Bard.[29] They observed the actions of DDT in cats and dogs after acute or chronic lesions of various parts of the neuraxis. In intact subjects, the initial symptoms of exposure were rapid, repeated blinking of the eyelids, then fine twitchings of the vibrissae and ears. The disturbance spread caudally to involve the muscles of the neck, shoulder, girdle, forelegs, hips, hindlegs, and trunk. In most cases, a very fine tremor developed and persisted. Later a more gross activity, consisting essentially of repetitive jerking movements, was superimposed on the fine tremor. When intoxication was sufficiently severe, tonic-clonic convulsions occurred.

After decerebration (removal of cortex, diencephalon, and mesencephalon), tremor in response to DDT administration was unaffected. Both the fine and coarse components were readily observable. In some cases convulsive episodes occurred. The authors concluded "that DDT exerts profound effects on parts of the neuraxis situated caudad to the mesencephalon and that those effects result in neuromuscular activity which resembles very closely — in fact is indistinguishable from — that evoked by DDT in the animal without any surgical ablation of central nervous tissue."[29]

In other animals, the cerebellum was removed after DDT symptoms were well developed. Its removal did not modify the symptoms either in intensity or composition. In one case, a dog whose cerebellum had been removed 23 months earlier was given 200 mg/kg of DDT in peanut oil by stomach tube. The subject was found 4 hr later panting heavily and shaking so violently that it had to be removed from its cage. Upon removal, the dog went into a convulsion and died within minutes.

These experiments implicate the spinal cord and brainstem as the primary loci from which the overt symptomology with DDT arises. This is particularly the case for the hyperreflexia and tremors. The presence of a cerebellum, mesencephalon, thalamus, hypothalamus, basal ganglia, limbic system, and cortex are not essential to the appearance of these effects, but they undoubtedly modify the evolution and magnitude of symptom development. Convulsive phenomena rely more heavily upon higher centers, particularly the cortex. Intact subjects exhibit a definitely lower threshold for seizure development than do decerebrate animals.

Bromiley and Bard's[29] findings with spinal cord transections are very similar to those reported in rats by Shankland[47] 15 years later. In cats and dogs, DDT produces tremor in muscles innervated from segments of the spinal cord which have been deprived of all neural connections with the brain. Tremor usually appears first in muscles innervated from regions above the transection. When the level of intoxication is light, tremor may occur in muscles innervated by regions below the transection only reflexly in response to a mechanical disturbance. At more intensive intoxication, tremors occur spontaneously in muscles innervated from levels below the transection, but the tremors may not attain the vigor of the tremor present in muscles innervated from above the transection. In many cases, the quality of the tremor is also different, the fine tremor component being less intense and the coarse component being more prominent.

DDT also alters the reflex activities of the isolated lumbosacral cord. In the cat there is an augmentation of flexor tone, a lowering of the threshold for the flexion reflex, and an irradiation of impulses from proprioceptors of the knee extensors to give a crossed knee-jerk reflex response. In the spinal dog, reflexes involving hindleg extensor muscles become more active and show very long afterdischarges.[29]

3. The "Basic Neural Unit"

These data provide a reasonable picture of the basic neural unit required to evoke DDT symptomology. The critical element is the segment of spinal cord in which the particular peripheral reflex arc is contained. DDT actions at the level of the intrasegmental circuitry suffice to explain the hyperexcitability that can be demonstrated upon stimulation. Effects upon sensory receptors and sensory nerves may contribute also, but they are of less importance in vertebrates than in insects.

The basic neural unit is capable of spontaneously initiating a crude tremor at high DDT exposure levels. However, tremor is normally evoked by afferent activity reaching the spinal segment from higher centers. The quality and quantity of tremor are mainly determined by a restricted section of the neuraxis which contains the extrapyramidal and intraspinal motor systems. Additional modulation of tremoring occurs through various other inputs to the spinal or extrapyramidal systems. These other inputs can arise from a number of different loci including the vestibular system, the cerebellum, basal ganglia, limbic system, or cortex.

Convulsive activity is not initiated by the basic neural unit. Seizures arise from the CNS in response to substances producing excessive neuronal hyperactivity. With DDT, convulsions may develop in subjects lacking the cerebrum and cerebellum. However, the threshold for initiating convulsive activity in these circumstances is very high. In intact animals, these systems play a dominant role in both the initiation and the maintenance of convulsive behavior.

C. Effects on Integrated Central Systems

The activity of higher centers can be modified by DDT and the changes in their activity frequently become manifest indirectly through input to the basic neural unit described above. These effects can be significant and in certain cases have led to the view that DDT has some selectivity for specific areas within the brain.

1. Cerebellum

The cerebellum has been claimed as a specific target for DDT. This claim is based primarily upon the study of Bing et al.[28] They described neurological symptoms in dogs chronically fed 150 to 200 mg/kg/day of DDT which included limb hyperextension, overstepping, and exaggerated placing reactions. These symptoms were reversible and stopped when exposure was stopped. Dogs fed more than 250 mg/kg/day of DDT developed symptoms suggesting cerebellar dysfunction. Animals displayed severe hypermetria, increased muscle tone, and very strong placing reactions. On standing the legs were rigid, hyperextended, and abnormally abducted. Neck and labyrinthine reflexes were exaggerated. The dogs were unable to walk in a straight line and tended to move in a zigzag fashion. Dogs fell frequently, and after falling they had trouble regaining an upright position.

In a related report[34] on the same subjects, pathological findings were described which implicated selective cerebellar damage attributable to chronic DDT exposure. The cerebellum showed a dose-related degree of damage. In those dogs exposed to 150 to 200 mg/kg/day of DDT for 10 to 20 days, a moderate number of Purkinje cells displayed swollen and hyperchromatic cytoplasm and pyknotic nuclei. In dogs receiving 180 to 350 mg/kg/day of DDT the dentate and roof nuclei were also severely affected. The nuclei contained fewer ganglion cells and those present exhibited chromatolysis and vacuolization of the cytoplasm and karyorrhexis or complete loss of nuclei. Haymaker et al.[34] felt that the damage to the cells of the cerebellum was slowly progressive and probably irreparable.

Although many of the symptoms associated with DDT exposure have properties in common with those seen in cerebellar dysfunction, a pathological picture of cerebellar damage has not been supported by other workers. Fitzhugh and Nelson[30] reported no significant pathology of the brain or spinal cord in rats on lifetime exposure to DDT. Lillie et al.[31] fed cats sufficient DDT to cause death in 18 to 65 days. They found some pericellular vacuolation about anterior horn cells and motor nuclei of the medulla, but saw no specific cerebellar pathology. Virgili and Marchiafava[33] described numerous alterations in neuronal and glial morphology that were most pronounced in pontobulbar nuclei with less marked changes observed in the cerebellum of guinea pigs and cats. Pluvinage and Heath[32] observed only a diffuse degeneration of ganglion cells throughout the brain in poisoned cats. Perhaps the most comprehensive pathological study of DDT was carried out by Globus.[35,36] He fed DDT to rats, cats, dogs, and monkeys to produce a variety of exposure levels and durations. Then he did detailed pathology on the brains of all subjects. He found very few changes, none of which could be specifically attributed to DDT.

2. Brain Electrical Activity

DDT will modify the spontaneous and evoked electrical activity from many areas of the CNS, including the cerebellum, cortex, limbic system and various subcortical loci. The most dramatic changes appear at the level of the cerebral cortex and, in freely moving subjects, at the level of the cerebellum.

Crescitelli and Gilman[27] examined the effect of DDT on the electrical activity that could be recorded from the surfaces of the cerebellum and cerebral cortex in cats and monkeys. Recordings made simultaneously from the pyramis or lobulis simplex of the cerebellum and from the motor area of the cortex showed that DDT caused the development of bursts of rhythmic high amplitude waves from both areas. The bursts began and ended simultaneously in cerebellum and cortex. In time, spike-like waves, either singular or grouped, were observed. The spike-like waves and bursts of waves were most prominent in the motor cortex or areas close to it. In other cortical regions these electrical events were only of small amplitude. Definite synchronization of the spike-like activity appeared between the cerebellar vermis and motor cortex. The cerebellar hemispheres gave no indication of such synchronization. When seizures occurred they appeared to arise in the motor cortex and to spread to other areas. Seizures were typical and tonic-clonic in appearance, being similar in structure to seizures evoked by direct electrical stimulation. Although this particular study is frequently used to support the concept of a selective action of DDT upon the cerebellum, Crescitelli and Gilman actually concluded that it was most likely that the motor cortex and the cerebellum were being simultaneously activated by impulses from a third area to which they were both linked. More recent studies[55,56] support their conclusions.

Pollock and Wang[53] carried out a similar study in cats. The subjects were fed 300 to 500 mg/kg DDT in a single meal, then surgically prepared after clearcut symptoms of poisoning were observed. In all subjects, the spontaneous rhythm of the cerebral cortex and the cerebellum was modified. Woolley and Barron[23] extended these earlier studies in two important ways. First, they recorded from subjects (rats) with chronically implanted electrodes in many different brain areas. Secondly, they observed the action of DDT in freely moving subjects over a 24-hr period. DDT affected brain activity from nearly every area examined. Many of these changes mimicked those occurring during stress or arousal and were only indirectly caused by DDT. However, clear and unique changes did develop in cerebral and cerebellar loci. In intact subjects, the amplitude and frequency of cerebellar activity increased. These events occurred earlier than those observed from motor cortex. Spikes appeared in the cerebellar records which were in synchrony with the whole body tremors. Although portions of the changes in cerebellar response can be attributed to the increased proprioceptive input to the cerebellum associated with tremoring and jerking, a direct effect of DDT also exists. DDT markedly potentiated responses evoked in the cerebellum to auditory and visual stimuli, even in paralyzed subjects.[54-56] Woolley[54] has provided further support for a direct action of DDT on the cerebellum by showing that other tremorigenic or convulsive agents do not necessarily potentiate cerebellar evoked activity.

Joy[56,57] studied the actions of DDT on spontaneous and evoked electrical activity in cats intentionally paralyzed so as to eliminate proprioceptive influences. The elimination of proprioceptive influences appreciably reduced the intensity of changes recorded from the cerebellum. Prior to the development of convulsive activity, the most obvious changes were seen in cerebral cortex, particularly in frontal and motor cortical regions. DDT produced a sustained, high amplitude, hypersynchronous electroencephalographic (EEG) pattern in these areas. This hypersynchrony predominated until convulsive episodes occurred. Convulsions were generalized, tonic-clonic seizures in most cases or variants thereof, and did not differ appreciably from other generalized seizures.

DDT also potentiates evoked responses to sensory stimulation.[56,57] Responses evoked in subcortical sensory nuclei and pathways to visual, auditory, or tactile stimuli were not influenced by DDT. However, responses in cortex were greatly increased. The cortical changes were progressive, in the sense that additional enhancement developed at each subsequent synapse in cortex. Thus, the least enhancement was seen at the level of the primary sensory receiving areas, greater enhancement was observed from association and integration areas, and the greatest increases occurred in motor cortex. Such progressive cortical "amplification" provides one basis for the enhanced startle responses and the development of myoclonic jerks in response to sensory stimulation that typically develop after DDT exposure. It undoubtedly contributes as well to the hyperreflexia and tremor through the modulation of the more basic motor unit at the level of the spinal cord.

At lower exposure levels or with chronic exposure, DDT produces changes in electrical activity most consistent with the concept that it generally and nonspecifically increases CNS excitability. Farkas et al.[58,59] observed changes in spontaneous and evoked cortical activity in rats fed DDT at levels of 5 to 40 mg/kg/day for up to 40 days. Subjects exhibited increased amplitude and higher frequency EEGs while exposed to DDT and were capable of following higher frequencies of flashing light. EEGs taken from monkeys fed DDT at 2 mg/kg/day for 18 months also exhibited EEG changes.[60] Unfortunately, in the latter study, recording was done under phencyclidine and pentobarbital anesthesia, a situation producing marked EEG changes in itself, which complicates the interpretation of the observed effects. To date there is still a paucity of data regarding the EEG effects of chronic administration of DDT. Additional evidence of hyperexcitable changes has been provided by an interesting study by Daigneault,[61] in which DDT was found to antagonize the inhibitory actions of acetylcholine or olivocochlear bundle stimulation on the cochlear N1 potential.

3. Behavior

Exposure to high levels of DDT leads to hyperexcitability and convulsions. Concerns must arise as to the possibility that exposure to lower, perhaps asymptomatic, levels may compromise an individual's capacity to meet environmental or other stresses. Chronic exposure to DDT could conceivably result in a reduced threshold in an individual or to an increased probability for the development of undesirable or inappropriate behaviors. A number of attempts have been made to determine interactions developing between DDT exposure and convulsive seizure thresholds, stress responses, and various behaviors.

The effect of DDT exposure on seizure thresholds has not been resolved. Various and usually minimal effects have been reported, often using procedures not well suited to evaluating possible changes in susceptibility. In studies utilizing pentylenetetrazol as the test convulsive agent, Ghazal[62] has reported that acute administration of 200 mg/kg DDT reduced seizure incidence. Enna and Tuttle[63] reported that chronic feeding of 10 ppm DDT to rats slightly increased seizure incidence to a similar pentylenetetrazol challenge. The offspring of mice fed DDT during pregnancy showed similar thresholds and responses to electroshock seizures as did control mice.[64] Other, less relevant information has been reported that examined the effect of DDT upon nonthreshold parameters of seizure genesis. The induction of audiogenic seizures was delayed in offspring of mice fed DDT during pregnancy[65] which suggests maturational effects but not necessarily direct effects upon audiogenic susceptibility. Woolley[66,67] made a comprehensive study of the interaction in rats between DDT and maximal electroshock seizures in which thresholds were not examined but the durations of various components of the seizure were. Rats were given DDT in corn oil and tested at 6-120 hr intervals. The oral doses chosen (150 to 200 mg/kg) were sufficient to produce tremors and hyperexcitability in all subjects. The primary action of DDT was to increase the duration of the tonic flexion phase of the seizure and to decrease the duration of full extension, total extension, total clonus, and the time to recovery of the righting reflex. These changes are not those expected with a convulsive agent. Variable changes in maximal electroshock parameters have also been reported by Sobotka.[68] None of these data provide convincing evidence that DDT exposure does or does not sensitize individuals to convulsive stresses. Additional studies, particularly studies oriented towards assessing sensitivity thresholds in preepileptic or other particularly susceptible subjects, are desirable.

Because of its effects upon motor function, DDT may conceivably alter various measures of behavior through either a direct CNS action modifying behavior or indirectly by deteriorating motor performance. In a systematic study of field behavior in rats, Khairy[69] compared control rats to rats fed 100 to 600 ppm DDT. Testing began about 21 days after exposure and continued for weeks or months. No changes were observed in the speed of locomotion, but the pattern of locomotion was altered by DDT in a manner consistent with an ataxic effect. There was a broadening of the gait and a shortening of the step length. Stress, as measured in a conflict avoidance situation, was not intensified by DDT. On these bases, Khairy concluded that chronic DDT exposure induces an exaggeration of motor phenomena which is sufficient to account for the behavioral findings. No direct effect upon the sensorium or emotionality could be defined. Sobotka[68] found that DDT, when given acutely in doses of 25 mg/kg to mice, increased open field exploratory behavior. Also, the time required for habituation of a behavioral response was increased. These were taken as evidence that DDT facilitates central excitatory processes and thus alters behavior directly. In the same study it was indicated that these doses of DDT do not impair the acquisition or retention of an active avoidance response.

Behavior of progeny from DDT-treated mothers has also been studied.[70,71] No differences were observed in open field behavior in offspring of DDT-treated and control mothers. However, offspring whose mothers received DDT during the second or third trimester exhibited delayed acquisition of a conditioned-avoidance response when tested 30 days after weaning. In a more comprehensive study by Craig and Olgilvie[72] female mice were placed on a diet containing 200 ppm DDT immediately after mating. Litters were manipulated so as to create groups of offspring exposed to DDT during gestation only, during both gestation and lactation, or during lactation only. The young born to and nursed by treated mothers showed signs of DDT toxicity. Mice exposed to DDT during gestation and/or lactation tended to run a maze faster, made more errors during learning, and showed no improvement in performance during retesting.

These various data indicate that DDT can increase CNS excitability sufficiently to produce demonstrable changes in motor-based measures of behavior. The significance of these data relative to the consequences of low-level DDT exposure on personality, emotionality, learning, and memory are not yet clear, and these must await more specific and sensitive analyses.

D. Mechanisms of Action

Analyses of single nerve fibers have demonstrated a specific set of actions of DDT at the level of the nerve membrane. Two types of single nerve preparations have been employed and have yielded similar results: (1) the giant axon of various arthropod species and (2) single nodes of Ranvier from amphibian myelinated nerve. Initial observations in arthropod axons[37,38,73,74] indicated that DDT prolonged the falling phase of the action potential and increased the amplitude and duration of the negative afterpotential (see Figure 1). Associated with the changes in the negative afterpotential, the excitability of the nerve was markedly increased, often sufficient to cause repetitive action potential discharge to single stimuli (see Figures 2 and 3).

The mechanisms of these effects were subsequently clarified by Narahashi and Haas.[75,76] They recorded intracellularly from lobster giant axons perfused in solutions containing DDT. In single axons DDT had no effect upon the resting potential nor on the rising phase or peak amplitude of the action potential. Only the falling phase of the action potential was modified, becoming markedly prolonged in the presence of DDT. A voltage clamp analysis of the fiber revealed that: (1) peak transient current decayed much more slowly after DDT, (2) the steady-state current was reduced in amplitude, and (3) the equilibrium potential of the steady state current was approximately 0 mV rather than near the potassium equilibrium potential (−90 mV) as would be expected in the normal axon. (Figures 4 and 5). The addition of tetrodotoxin, which blocks sodium current, was sufficient to demonstrate that DDT had two major actions. First, sodium conductance, which was activated normally, was not inactivated normally. Inactivation was delayed by many milliseconds. As a result, sodium continued to flow into the neuron during the recovery period of the action potential. This has the effect of maintaining the neuron in a depolarized state during recovery. Second, potassium conductance was depressed by as much as 50% by DDT. This also contributes to a slower repolarization of the membrane. In addition, it makes the membrane more susceptible to spontaneous activation, that is, it facilitates the tendency for repetitive discharge. Similar conductance changes have been observed in cockroach axons by Pichon.[77]

Shanes[78] observed that DDT evoked repetitive discharge of fibers within the sciatic nerve of the frog while only minimally increasing the negative afterpotential. Subsequently, Hille[79] used single nodes of Ranvier from frog sciatic nerve fibers to examine the effects of DDT upon ionic permeabilities. When nodes were perfused with suspensions of DDT, the threshold and initial rising phase of the action potential were unaffected whereas the falling phase was greatly prolonged. Voltage clamp analysis showed that the prolongation was due to a persistence of sodium conductance, analogous to the situation found in arthropod axons. Hille did not find a decrease in potassium conductance in frog nodes of Ranvier, but this effect has been reported by others.[80,81] In Hille's experiments, the kinetics of sodium conductance suggested that a fraction of the total number of sodium pores opening during the action potential were being

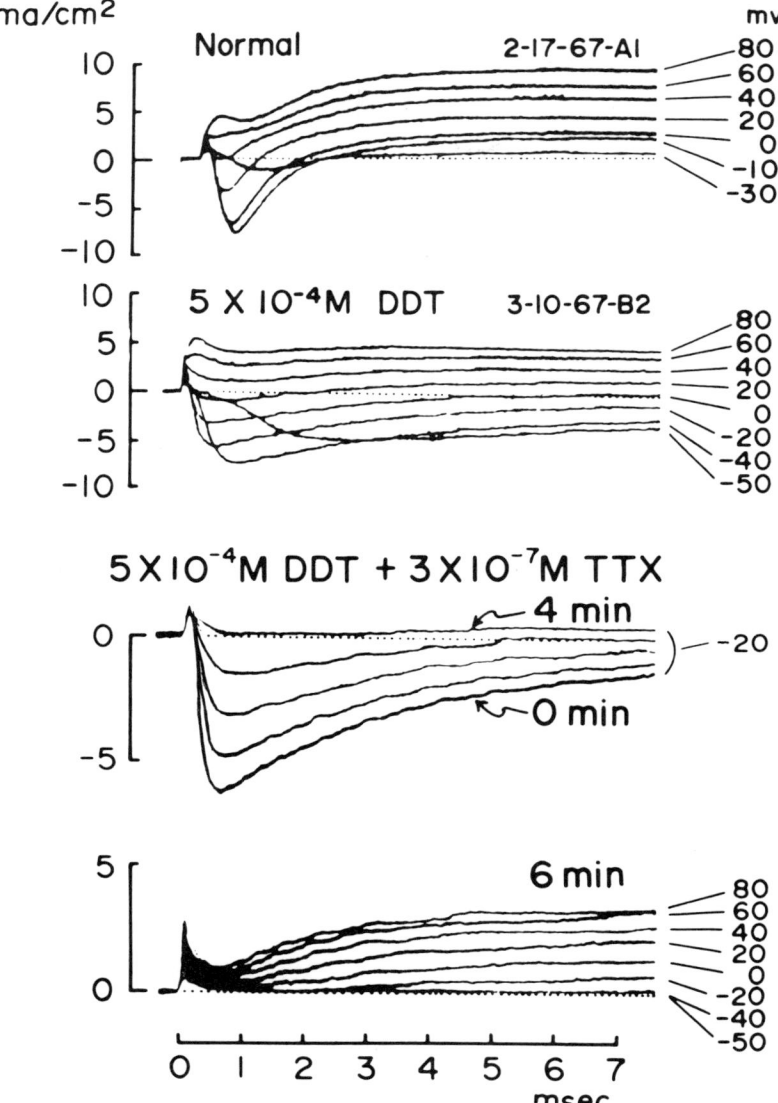

FIGURE 4. Families of membrane currents associated with step depolarizations in a normal lobster giant axon, and another axon treated with DDT 5×10^{-4} M and with DDT and tetrodotoxin (TTX) 3×10^{-7} M. The third set of records shows changes in membrane current during the course of TTX action. The dotted lines in each set refer to the zero base line. (Narahashi, T., and Haas, H. G., *J. Gen. Physiol.*, 51, 177, 1968. With permission.)

held open by DDT. The affected pores shut with a time constant about 45 times longer than normal.[82] Either internal or external perfusion of DDT in crayfish axons is effective in producing these effects. However, the onset is more rapid with internal perfusion which suggests that the sites of action for DDT reside on the inside of the membrane surface.[83]

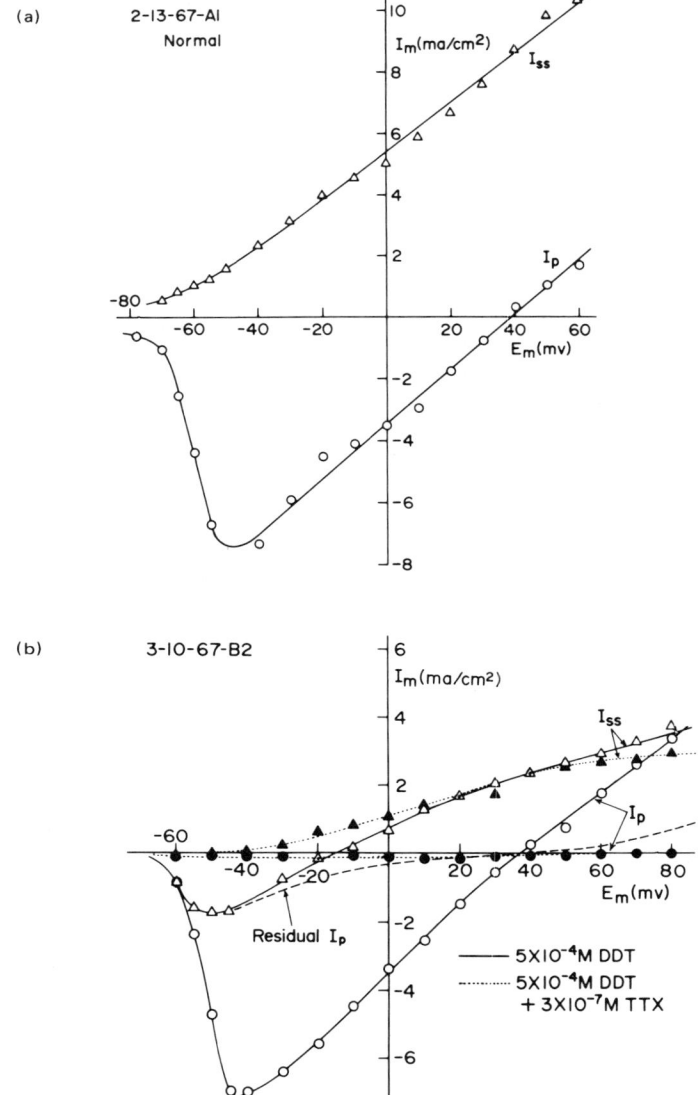

FIGURE 5. Current-voltage relations for peak transient (sodium) current (I_p) and steady-state (potassium) current (I_{ss}) in a normal lobster giant axon, and in another axon treated with DDT 5×10^{-4} M or with DDT and tetrodotoxin (TTX) 3×10^{-7} M. The broken line shows the residual component of the transient current and was obtained by subtracting I_{ss} in DDT plus TTX from I_{ss} in DDT. (Narahashi, T., and Haas, H. G., J. Gen. Physiol., 51, 177, 1968. With permission.)

The changes in ionic conductances which lead to the development of repetitive discharge in excitable membranes are very likely the bases for many symptoms of DDT exposure. In insects, a good correlation exists between exposure levels evoking afterdischarge in the abdominal nerve cord, those producing convulsive behavior, and those which are insecticidal.[42] In vertebrates, they seem particularly relevant to effects observed in peripheral nerve, skeletal muscle, and in the myocardium. It is reasonable to assume that repetitive discharge in axons plays a role in the changes seen in CNS activity, even though direct evidence of this is lacking. It is unclear, however, as to how significant this role actually is. The expression of toxicity in vertebrates occurs at lower levels in intact neuraxes than in isolated portions. The presence of synapses, particularly those associated with a rich neuropil, renders a system particularly susceptible. This may reflect nothing more than the fact that there would be more neuronal elements present. Or it may reflect the fact that presynaptic terminals are of small diameter and may be more susceptible to DDT than larger fibers simply through their geometry.[73] Other, perhaps more subtle, changes may also be occurring which produce effects at the synaptic level at lower exposure than that evoking repetitive discharge in axons. Synaptic effects could involve the same underlying alteration in ionic conductance described for axons, or they may have other causes not demonstrable in axonal preparations.

Some of the effects described for DDT in in vitro systems may be occurring at membrane concentrations unreachable in vivo. Typically DDT and other CHIs are applied in vitro as an alcoholic solution which is rapidly added to aqueous bathing solutions to form an opalescent suspension. The extremely low water solubility of DDT of less than 2 ppb makes true solutions in physiological media of greater than 3×10^{-9} M unattainable. In most reports in the literature, "solutions" of 10^{-3} to 10^{-6} M are described, indicating that they are actually dispersions. Thus, particles of DDT exist which can come in contact with the nerve membrane and produce a localized spot of extremely high concentration. Further, Schneider[84] has shown that the amount of DDT bound to a membrane is limited not by its concentration in the solution but rather by the total amount of DDT present. The nerve membrane behaves more as a separate phase containing a solvent for DDT than as a protein with a limited number of binding sites. Perfusion of DDT "solutions" through chambers containing nerve fibers results in a continuous uptake of DDT by the fiber, limited only by the volume of the bath. As a result, concentrations employed in in vitro studies can not be directly correlated with in vivo concentrations. The true membrane concentrations of CHIs are likely to be greatly underestimated in in vitro studies. The changes observed in such systems may depict effects that only emerge at exposure levels many times greater than can be attained in vivo. It is of interest that the behavior of DDT in vitro has been described as irreversible.[76,79] This can be anticipated in vitro due to the partitioning of DDT into the lipid phase of the membrane and from the extremely low water-solubility. Irreversibility would not be expected in vivo, however, because of the much higher binding capacity of blood elements and plasma proteins for DDT and the presence of vast amounts of other lipid materials available for redistribution.

Structure-activity studies of DDT analogues have demonstrated that various of the membrane effects of DDT can be separated from each other.[43] The p,p'-dinitro derivative of DDT can initiate trains of impulses in the sensory nerve of the cockroach leg and can increase the depolarizing afterpotential in crayfish giant axons, but it lacks insecticidal activity. Dimethyl substituted DDT is insecticidal, produces repetitive firing in sensory nerve but has no effect on the depolarizing afterpotential in crayfish axon. Dimethoxy substituted DDT (Methoxychlor) behaves like DDT whereas the diethoxy substituted compound has no effect on the depolarizing afterpotential. Various of the putative metabolites of DDT share some, but not all, of the effects of DDT. For example, DDT hinders sodium inactivation in amphibian nodes of Ranvier.[80,81] In contrast, the metabolite DDA (dichlorodiphenyl acetic acid) modifies potassium inactivation and the metabolite *bis*-(*p*-chlorophenyl)-acetamide modifies both sodium and potassium conductances. On the same preparation, DDD, DDE, and other metabolites had little or no effect.

The findings of Narahashi and Haas,[75,76] Hille,[79] and others have established the nerve membrane as a primary target for DDT action. In these experiments, effects upon respiration, metabolism, energy supply, or various enzymes within the fiber cytoplasm can be ruled out since DDT works equally well on axons whose cytoplasm has been removed and replaced with artificial perfusion fluids.[83] This membrane action has been studied from a different point of view by using artificial membrane systems or membrane components.[85-87] The binding of active and nonactive DDT analogues to lecithin, cholesterol, or phospholipid monolayers has been found to bear no correlation with their activity on nerve fibers. However, the potassium conductance which is induced by valinomycin in a synthetic membrane is abolished by DDT. In more comprehensive studies of this phenomenon by Hilton et al.,[85-87] a partial correlation was found between the ability of DDT analogues to block valinomycin-induced potassium conductance and to produce physiological effects upon cockroach nerve or crayfish giant axons. Such findings are of interest because of the high affinity of DDT for lipid materials. Other data, however, suggest that membrane proteins may also be significantly affected by DDT.[88]

A hypothesis to explain the molecular basis for DDT action has been advanced by Mullins.[89,90] Mullins suggested that DDT fits into the interspaces formed by juxtaposed membrane macromolecules. The failure of various substituted DDT congeners to have activity would be due to their failure to fit into the same interspace. Holan[91] has examined the model in detail, refined the postulated shape of the binding site, and extended a general theory for the actions of compounds of this type. Metcalf[1] has provided a very readable account of the development and present status of the various theories developed to account for the action of DDT at the membrane level.

Although many of the effects of DDT appear well accounted for on the basis of a membrane effect leading to a change in the kinetics of ion flow through the membrane, other factors may well be involved in the intact neuraxis. As already indicated, the intact neuraxis shows greater sensitivity to DDT than do single axons, and this may be indicative of qualitative rather than quantitative differences in action. Additional effects of DDT upon synaptic activity, transmitter synthesis, kinetics of release or degradation, neurogenic amines or various enzyme systems participating in ion transport, metabolism, oxidative phosphorylation, etc. may exist. DDT has been reported to change the levels or action of many chemical and neurochemical substances, including ammonia,[92] glutamine,[92] GABA, glycine and taurine,[93,94] acetylcholine,[12,24,95-97] norepinephrine,[12,97] dopamine,[12,96] and serotonin.[12,97-100] For more information in this area, the recent review by Hrdina et al.[12] is recommended.

A particularly interesting action of DDT upon membrane coupled ATPases has been reported. Matsumura and Patil[101] reported that DDT inhibited a Na^+,K^+-dependent ATPase preparation prepared from rat brain. Inhibition occurred at DDT concentrations thought to be comparable to those present in cockroach nerves when repetitive activity could be recorded. The correlation found for a number of DDT congeners between insecticidal potency and ATPase inhibition was very high. Subsequently a greater inhibitory action on a Mg^+-dependent ATPase was claimed by Koch et al.[102-104] Matsumura and Narahashi[105] extended these findings by studying ATPase inhibition and electrophysiological changes concurrently. Using lobster nerve, they found that DDT inhibited a portion of the Na^+,K^+-dependent ATPases. Only a portion of the DDT-responsive ATPases could be inhibited by ouabain, suggesting that at least part of these ATPases were not solely involved in active transport systems. Many compounds, including DDT, that are capable of modifying the movement of ions through membranes were found capable of inhibiting this ATPase activity. Subsequently Doherty and Matsumura[106] identified four ATPases with differing sensitivities to DDT and to ionic elements in the axonic fraction of the same nerve material. They concluded that DDT has the potential to inhibit or otherwise interfere with a variety of enzymic reactions that utilize ATP as a substrate. They implicated as a target for DDT a specific

ATPase thought to be involved in the control of the rate of Na⁺ and K⁺ fluxes through the membrane in the presence of ATP. The complete importance of these findings must await clarification of the relationship between these ATPases and conductance regulation in the nerve membrane.

E. Synopsis of Mechanisms, Effects, and Symptom Development

From these observations, a picture emerges as to the production of toxic symptoms following DDT exposure. A major action of DDT occurs at the level of the nerve membrane. There, DDT interferes with the movement of ions through the membrane, prolonging the duration of the recovery phase of the action potential, increasing the excitability of the nerve to stimulation, and eventually inducing repetitive axonal discharge to stimuli which normally would produce only a single response. Important changes in synaptic function occur as well. These may rely upon the same underlying mechanisms or perhaps involve additional, as yet uncovered, actions.

The major symptomology is related to this increased neuronal excitability. The action does not appear to be specific to any particular component of the nervous system, but rather appears to be a generalized effect developing at all levels of the neuraxis. The early emergence of changes in function in one or another component signifies more that this component is particularly sensitive to disruption, not that DDT acts selectively upon it. At some exposure level, effects can be observed at nearly every site within the neuraxis, but the degree of exposure required for the expression of symptoms decreases as more and more of the neuraxis is left intact. Thus, under appropriate circumstances, actions upon muscle, afferent and efferent peripheral axons, spinal, medullary, pontine, cerebellar, and all mesencephalic and prosencephalic portions of the CNS are demonstrable. In insects and other arthropods, peripheral components of the nervous system, particularly receptors and afferent fibers, form an important target. In some species actions at this level may be sufficient to account for most of the symptomology. In vertebrates, however, the central nervous system also plays a critical role.

Tremor and hyperreflexia are produced through a modification by DDT of spinal segmental and supraspinal systems. Segmental control of motor function may be altered in any of three ways: (1) increased afferent input to the segment from its associated proprioceptors by way of actions of DDT upon the receptors or afferent nerve fibers, (2) direct actions upon spinal segmental motor and interneurons, and (3) enhanced input into the segment from suprasegmental systems. The quality and quantity of the tremor can be affected by changes arising in any of these, but in most cases the critical factor is altered suprasegmental input. DDT-induced activity developing at the levels of cerebellar, pontine, basal gangliar, and cortical levels, to name just a few, cause marked changes in the activity coursing to spinal loci via pyramidal and extrapyramidal motor systems.

The convulsive properties are the consequences of a widespread, nonspecific increase of excitability throughout most of the CNS, particularly within higher centers. High cervical section, which removes all peripheral input to the brain via the spinal cord, does not significantly modify the convulsive properties of DDT. The exact genesis of convulsions after DDT exposure remains unknown. However, there are indications which suggest that the convulsive properties of DDT bear resemblance to those produced by other CHIs and probably to those produced by other nonspecific excitatory convulsants, such as pentylenetetrazol and bemegride.[56,57,107] For pentylenetetrazol, a nonspecific excitatory action has been advanced and supported by many studies.[108] Various mechanisms have been proposed including an altered chloride conductance[109] or other altered membrane properties.[110]

The majority of DDT effects involve altered motor function, and the sensorium

Table 2
DOSAGE — RESPONSE TO DDT IN MAN[a]

Dosage (mg/kg/day)	Consequences
0.0004	Dosage of the general population of the U.S. in 1968
0.0025	Dosage of the general population of the U.S. in 1953—1954
0.25	Tolerated by workers for 19 years
0.5	Tolerated by volunteers for 21 months
0.5	Tolerated by workers for 6.5 years
1.5	Tolerated by volunteers for 6 months
6[b]	Moderate poisoning in one man
10[b]	Moderate poisoning in some men
16—286[b]	Prompt vomiting at higher doses, all poisoned, convulsions in some
Unknown[b]	Fatal

[a] Table largely derived from Hayes.[8]
[b] Single dose only.

appears to be relatively unaffected. Although DDT does produce an increase in overt excitability, no great changes in cognitive phenomena, memory, learning, or emotionality have been reported. More significant changes in animal behavior have been reported with other CHIs, however. In man, DDT intoxication results in various sensory alterations including paresthesias and numbing, along with malaise, headache, and delirium. There is remarkably little change in personality or behavior associated with DDT exposure.

F. Neurotoxicity of DDT and Its Analogues to Man

Information about the neurotoxicity of DDT to man has risen from various sources including (1) purposeful exposure by volunteers to known concentrations of DDT; (2) involuntary, accidental exposure to estimable quantities of DDT; (3) involuntary, accidental exposure to unknown or poorly estimable quantities of DDT; or (4) exposure, usually accidental, to chemical mixtures, one component of which is DDT. Only in those data obtained from exposures to DDT, without the complications of solvents or other active chemicals, can the toxic potential of DDT be well defined. Dose-effect data for DDT in man is sparse. However, a reasonable estimate of its toxicity is possible (Table 2). Oral ingestion of DDT is often rapidly followed by vomiting and diarrhea. Thus the actual amount of DDT absorbed may differ markedly from one subject to the next or from one exposure to the next. In the majority of cases in which DDT exposure has been followed, either concise clinical testing and examination are absent or adequate estimation of exposure, blood levels, and/or tissue levels are absent. In many cases, effects are ascribed to DDT when the material ingested actually contained other toxic materials, some more acutely toxic than DDT itself.

1. Acute Exposure

The excellent review of DDT by Hayes[3] considers in great detail the literature on DDT exposure in man up to 1959. That source should be consulted by anyone wishing a comprehensive review of known exposures to DDT up to that time. Only a summary of the material included in that source will be presented here. More comprehensive consideration will be given to cases reported since that time.

A number of studies have reported on the consequences of intentional exposure.

Ingestion of DDT by volunteers indicates that amounts below 10 mg/kg usually produce few or no symptoms, from 10 to 16 mg/kg produce moderate to severe symptoms while exposure to greater than 20 mg/kg can induce convulsions and death. Velbinger's[3,111,112] description of experiments he and two other men participated in are typical. They ingested amounts of DDT from 250 to 1500 mg. Amounts up to 500 mg produced no effects other than variable hyperesthesias of the mouth and lips. At 750 mg (approximately 10 mg/kg), sensory disturbances involving primarily the face developed along with some motor impairment, specifically a reeling motion that occurred when walking was attempted. Following the ingestion of 1500 mg (approximately 20 mg/kg) of DDT, severe paresthesias of the face and mouth developed. Within a few hours equilibrium was impaired, and the subjects experienced dizziness, confusion, and tremoring of the extremities. Associated with these symptoms were a general malaise, headache, and a pronounced feeling of fatigue. By the following day, only a minimal degree of paresthesia and disturbance of equilibrium remained.

At the height of symptoms, a neurological examination revealed dilated pupils which reacted normally to light and displayed normal accomodation. A slight nystagmus was observed. Variable abnormalities in the perception of touch and heat were observed at different levels of the distribution of the trigeminal nerve. The functions of cranial nerves VII, VIII, and XII were normal. Extensive tremoring was present in all extremities. The finger-to-nose test was performed with hesitation and uncertainty, and the degree of tremor was reduced during the movement. Equilibrium impairment was such that the subject could not stand on one leg for any length of time. Symptoms did not persist for more than 2 or 3 days.

The accidental ingestion of DDT is associated with similar symptom development. Garrett's studies[3,113] are typical. In this instance, which occurred during World War II, military prisoners stole a box containing 10% DDT with flour as the diluent. This was used to make biscuits which were subsequently eaten. Twenty-eight men developed signs of poisoning. When first seen, the men were apprehensive and excited and displayed a moderately increased respiration with a noticeable bradycardia. Many had vomited. Numbness and partial paralysis were reported, these being most apparent in the distal portions of the extremities. Some of the men had mild convulsions. Proprioception and vibratory sensation were diminished or lost in the fingers and toes. The knee jerk was hyperactive in 8 of the 25 men. No deaths occurred, and, within 48 hr, only 8 men were still suffering any symptoms. After 3 weeks, three men reported some persisting weakness of both hands and feet, which was still being reported as long as 5 weeks after exposure. Subsequently, the men were transferred and lost from further study. While the persistence of the symptoms may have been purposefully extended by the men to avoid returning to work, the bulk of effects reported are typical of those caused by acute exposure to DDT.

The majority of cases of accidental or intentional exposure to DDT are actually exposures to materials containing many substances. In these cases it is frequently difficult to determine which of the toxicants is actually responsible for the various symptoms. Interactions may occur from combinations of toxic substances which result in unique effects. Solvents employed, such as kerosene, xylene, methylcyclohexanone, and substances mixed with DDT such as thallium, thiocyanates, etc., are also toxic. Such mixtures may be more toxic than the DDT they contain. In addition, solvents and emulsifiers might potentiate the toxicity of the insecticide by enhancing absorption from the intestinal tract or skin. This mixed effect is exemplified in the description by Cunningham and Hill[114] of a 2-year old child who drank an unknown quantity of fly spray which contained 5% DDT. The other ingredients of the spray were unknown. The child became unconscious after 1 hr and had a generalized convulsion. Convulsions continued, and the child was hospitalized and treated with barbiturates and other sedatives which controlled the convulsions. Convulsive episodes occurred on day 4 and day 21 as well. The child was unable to hear on day 12, but hearing slowly returned to normal. By 2 ½ months, no neurological or psychiatric signs remained. In this case, the convulsive episodes are consistent with DDT intoxication, but the intensity and persistence of symptoms is atypical. It is not possible to determine whether DDT or other unknown materials, or perhaps even anoxic damage, was the cause of the other presenting symptoms.

The major route of acute exposure to toxic levels of DDT is oral. Far fewer complications have been reported with either dermal or respiratory exposures, even to massive amounts. Cameron and Burgess[115] noted no reactions in soliders wearing woolen underwear, vests, and shirts impregnated with 1% DDT. The garments were worn day and night for 18 to 26 days, during which time the men on numerous occasions underwent heavy exercise sufficient to produce copious sweating.

Two men were subjected to intermittent exposure of DDT, 35 $\mu g/\ell$ of air, for 1 hr daily for 6 days.[3,116] Both subjects suffered moderate irritation of the nose, throat, and eyes. The concentration of DDT was sufficient to form a white deposit on the nasal hairs of both men. One man was stripped to the waist while the other had bare arms and shoulders. There were no symptoms or aftereffects other than for the irritation described. Laboratory tests and physical examination revealed no changes.

2. Chronic Exposure

Chronic toxicity to DDT may emerge slowly and show certain additional complications not part of the acute poisoning syndrome. It may be characterized by loss of weight, anorexia, mild anemia, muscular weakness, and tremors. Anxiety, nervous tension, hyperexcitability, and fear are commonly expressed by the patient. Myoclonic jerks may occur,[117] but these are more commonly observed with cyclodienes and lindane. Electroencephalographic abnormalities may be encountered even when no neurological or clinical evidence of abnormality is seen.[118] The most common abnormalities include the appearance of sharp waves; excessive theta waves; spike and wave complexes; and low voltage, rhythmic spikes.

3. Persisting Neurological Sequelae

The persistence of DDT toxicity following acute ingestion is generally only 1 to 2 days in duration. Immediate symptoms are most often related to the solvent employed, and frequently 3 to 6 hr elapse between exposure to DDT and the emergence of its symptoms. Peak acute effects are manifest after 6 to 12 hr, and the toxicity disappears within 1 to 3 days. Exceptions to this have been observed when idiosyncratic responses have developed. These are particularly rare, although a few reports suggest that an idiosyncratic response, usually consisting of variable polyneuropathic phenomena, can occur. However, these are more frequently seen on chronic exposure and may possess an allergic component. Wigglesworth[119] reported a case in which a man experimented with DDT and subsequently developed weakness and aching of his limbs along with anxiety. The symptoms intensified over 14 days, and persisted for nearly a year. Certain factors in this instance, however, raise doubt that DDT should have been implicated as the cause of the persisting illness, even though it is frequently cited in the literature. The case history as presented suggests that psychological factors were heavily involved in the development and expression of symptomology in this instance. Case[120] has reported on a more substantive incidence in which DDT may have been responsible for persistent effects. Volunteers were exposed to living quarters in the tropics where the walls and furniture were essentially covered with an oily paint containing 2% DDT. The volunteers were exposed for a total of 96 hr. During this time they complained of tiredness, heaviness, aching of the limbs, and mental sluggishness.

Very minor changes were noted upon neurological examination. A return to preexposure status required 26 to 33 days. Also, in the case reported by Garrett,[113] various sensory and motor abnormalities were being reported by 3 of 28 men up to 5 weeks after exposure to DDT. Such reports are very rare, however, and are much the exception with acute exposures to DDT.

Chronic exposure to DDT has been implicated as a cause for various allergic and hypersensitivity reactions. Although most of these are dermatologic, some polyneuropathies have been described in which exposure to DDT has been present.[3,117,121-128] In none of these reports is there an uncomplicated exposure to DDT alone. The patient has been exposed inevitably to many different chemicals simultaneously, some of which have also been implicated as producing polyneuropathies. For example, the seven cases of peripheral neuritis described by Campbell,[121] which are frequently referenced to implicate DDT as a cause of polyneuropathy, report on patients exposed to a proprietary formulation whose exact makeup was unknown, but which did contain ortho- and para-dichlorobenzene, DDT, and pentachlorophenol. Campbell believed it most likely that pentachlorophenol was the responsible agent. The peripheral neuritis was of a mixed sensory-motor type. The following example is taken from his paper, Case 3:

A housewife aged 44 suffered acute epigastric pain early in November, 1949. It lasted for three days, and on the third day of the illness, she developed weakness of the legs with gross ataxia and pins-and-needles in all four limbs. The symptom appeared to improve about a week after the onset, although by this time a facial paralysis on the right side had developed. I saw her three weeks later and found a right facial paralysis and a glove-and-stocking anaesthesia to all forms of sensation anywhere between the ankle and just above the knee. There was marked ataxia in arms and hands, with diminution in the grip of both hands. She made a slow but steady recovery, and when seen six months later there was no abnormal physical sign. ....inquiry was made about the use of insecticide. The patient had used the (proprietary formulation) three days before the onset of her illness, and she admitted to using about a bottle on old furniture, having mopped it on with a cloth. Her hands were covered with abrasions from gardening, and, during treatment of the furniture, she noticed a stinging pain in these abrasions from contact with the solution.*

In most cases in the literature in which DDT has been associated with peripheral neuropathies, mixed exposure to insecticides occurred, often other CHIs. Onifer and Whisnant[123] described a progressive neuritis that developed in a butcher who worked in a room in which an electric vaporizer consisting of a glass cup containing crystals composed of 33% lindane and 66% DDT were placed. The patient had worked in the room approximately 40 hr a week for 3 ½ months. The patient first experienced a zone of paresthesia on the medial aspect of his right knee. The neuropathy progressed over the next few days until it was manifest by profound weakness, staggering gait, ptosis of the right eyelid, diploplia, and various paresthesias of the fingers and toes. A neurological examination demonstrated cerebellar ataxia coupled with cranial and spinal neuritis. The condition gradually improved, and the patient was essentially normal after 1 month.

Two provocative cases of polyneuropathy, subsequent to insecticide exposure, are described by Jenkins and Toole.[117] In one case, a 13-year-old boy exposed to aldrin and DDD for 2 months developed a progressive neuropathy which persisted for 6 to 9 months. A reexposure to aldrin the subsequent year induced a return of symptoms suggesting a probable relationship between aldrin and the neurotoxicity. In the second instance, a male farmer contaminated himself with a concentrated mixture of endrin and DDT. A blocked and defective spray nozzle came off in his hand, wetting him and his clothing with the mixture. He worked for the rest of the day without washing or changing clothes. The same night he felt ill and experienced a slight aching of his

---

* From Cambell, A. M. G., *Br. Med. J.*, 2, 415, 1945. With permission.

limbs. This increased over the next few days with the concomitant development of muscular weakness, especially in the hands. He was eventually hospitalized. Neurological examination uncovered marked weakness in all limbs. The gait was slapping, but not ataxic. Sensation was only minimally altered. Within 4 weeks, the symptoms had abated, and the man returned to work. Whether the patient was reexposed and what the consequences of reexposure were, was not indicated.

One of the most suggestive reports linking DDT to chronic neurological abnormalities was reported by Stone and Gladstone.[122] They encountered a subject who formulated aerosol bombs containing mostly DDT with small amounts of pyrethrum and piperonal butoxide, and the propellant, Freon®. The patient stated that he had become careless in filling the bombs and had been exposed for 4 years to vapors. He claimed that he had had weakness, poor appetite, and sleeplessness for about 4 years which suddenly had become worse and required attention. When seen, he complained of severe fatigue, difficulty in talking, glare phenomena, photophobia, blurring of vision, and a feeling that he was swimming or walking on air. He had a positive Romberg sign, decreased deep reflexes in all extremities, clumsiness, and tenderness along all nerve trunks. He made an excellent recovery in 14 days, reported back to work and claimed to have felt better than ever during the past 4 years. It was not stated whether he took any precautions to prevent reexposure.

A particularly strong indictment of DDT, and other insecticides, was made by Biskind.[129-133] He claimed that the outbreak of a puzzling syndrome manifesting recurrent and debilitating symptoms was associated with the exposure to DDT. He implicated DDT as a causative agent for "X" disease in cattle (a disease now associated with exposure to chlorinated naphthylenes) and for "virus — X" disease in man. It was claimed that exposure to DDT led to the emergence of a bizarre syndrome which resembled other ailments in individual details but which had never been known to occur in its entirety prior to the introduction of the CHIs. The syndrome was claimed to have occurred repeatedly in hundreds of instances studied by Biskind[132] following known exposure to DDT and to related compounds. It was reported to have been observed over and over again in the same patients, each time following known exposure to these insecticides. In the same publication, it was suggested that the sudden increases in the incidences of certain diseases, including polio, lung cancer, heart disease, and retrolenticular fibroplasia in premature infants, could be related to the increased use of CHIs. No support for any of these claims was forthcoming, either from the U.S. or from abroad,[134] and today DDT is not directly implicated as a causative factor in any of these situations.

Although these various data suggest a possible link between DDT and the emergence of delayed polyneuropathic symptoms, the exact role of DDT remains unsettled. First, such cases are extremely rare. If DDT is a cause of delayed polyneuropathies, then the incidence of such a response is exceedingly low. Second, in those cases reported, chemicals other than DDT have also been present. Third, the studies, in the majority of cases, lack vital data essential to the linking of DDT and the neuropathy. Inadequate determination of exposure is common. In most cases, only testimonial evidence of exposure is provided. No information is provided on other workers or individuals with similar exposure histories. No follow-up material is provided to indicate whether subsequent reexposure occurred, and, if it did, what the consequences were. Subsequent to his review of the literature through 1959, Hayes[3] concluded that

...DDT has not been established beyond the shadow of a doubt as the cause for a single case of aplastic anemia, agranulocytosis, purpura, polyneuropathy or dermatitis. On the contrary, these diseases are known to occur in a few susceptible individuals following sensitization to a wide variety of chemicals including some common drugs. It therefore appears reasonable to assume that DDT is capable of causing these conditions even though not every case attributed to that compound may be valid. Certainly one is forced to conclude (1) that the incidence of these conditions following exposure to DDT is extremely low and (2) the incidence of these conditions irrespective of cause has shown no detectable increase since the introduction of DDT.*

Although there are some data which suggest that polyneuropathies may occur in higher incidence to other chlorinated hydrocarbon insecticides, such as lindane,[127] Hayes' statement remains appropriate for DDT today. In the 2 decades since Hayes' conclusions, no new important evidence has emerged to implicate DDT as a cause of polyneuropathies. Certainly nothing like the propensity of some of the organophosphorous insecticides to induce delayed neurotoxicity exists for the CHIs, as a whole, or for DDT in particular.

VI. NEUROTOXICITY OF THE CYCLODIENE AND HEXACHLOROCYCLOHEXANE DERIVATIVES

The various hexachlorocyclohexane and cyclodiene derivatives (HCH-CYCs) have been observed to possess neurotoxic properties sufficiently different from DDT and its analogues to warrant their separate consideration. These differences are correlated with differences in their proposed modes of action as well. DDT has a clear and distinctive activity on sensory nerve fibers and peripheral receptors, and these effects play an important role in the expression of symptoms in most species. The HCH-CYCs show little or none of these peripheral actions. The CNS is the only important locus of action, and the neurological consequences following exposure results entirely from actions at this level. The hexachlorocyclohexanes are appropriately reviewed along with the cyclodienes in this section. Not only are the attributes of toxicity similar, but also the hexachlorocyclohexanes exhibit cross tolerance with the cyclodienes,[135] implying very similar underlying mechanisms of action. The variation apparent in the literature for these compounds is not much greater than that observed in the literature for DDT alone. Much of it can be attributed to differences in exposure levels, routes of exposure, vehicles employed, species selected, or by the design and expectations of a particular study. These differences are far outweighed by the similarities that have been described for various members of this family.

Species differ in their susceptibility to the HCH-CYCs. Fish and birds are very susceptible. Among the mammals, the order from most to least sensitive is roughly dog, man, monkey, cat, guinea pig, rabbit, rat, hamster, and mouse. The larger domesticated animals are relatively insensitive.[10,136,137] Young animals are usually more susceptible than older animals, lean animals more susceptible than obese animals, and females may be more susceptible than males in some species, notably the rat.[10,137-139]

A. Signs and Symptoms of Poisoning
1. Acute Exposure

The major symptoms evoked by acute exposure to HCH-CYCs are of central nervous origin. This is true of insects, animals, and man.

a. Insects

Insects show five stages of toxicity following exposure.[140] These include periods of (1) hyperexcitability, (2) incoordinated movements, (3) convulsive periods, (4) prostration and paralysis, and (5) death. There are some notable differences in the responses of insects to the HCH-CYCs as contrasted to DDT. For example, in flies, DDT causes

---

* From Hayes, W. J., Jr., *The Insecticide Dichlorodiphenyltrichloroethane and Its Significance*, Vol. 2, Muller, P., Ed., Birkhauser Verlog, Basel, 1959, 11. With permission.

continual, uncoordinated movements without true convulsions. The wings may function normally until death. In contrast, flies poisoned with lindane or a cyclodiene frequently display fanning movements of the wings insufficient to cause flight. At a late stage of poisoning, the legs are all tightly contracted under the body and the wings are deflected downwards, the insect equivalent to convulsion.[135] Bees poisoned with chlordane exhibit similar behavior.[141] With the HCH-CYCs, exposure typically results in a long period of hyperactivity with running and tumbling activity which is followed by progressive exhaustion and death. The hyperactivity is usually greater and lasts longer than that seen with DDT.

b. Vertebrates

Most vertebrates react similarly when acutely poisoned by the HCH-CYCs. These reactions are clearly different from those produced by DDT.[135] For example, DDT and methoxychlor are characterized by producing widespread tremoring and myoclonic jerking. Convulsions are not frequent, are often purely clonic in nature, and may present as an extension of violent tremoring. With lindane or a cyclodiene, tremoring is minimal and often absent. The outstanding feature of toxicity is convulsions, of sudden onset and extreme violence. Death may occur during convulsions, during the postictal period or recovery may occur. The neurological symptoms appear completely reversible in the vast majority of cases.[10] In man, sudden, precipitous convulsions without prior symptoms can be the first observable sign of HCH-CYCs exposure.[142]

The temporal development and the severity of symptoms depend upon the route of administration employed and the dose. Intravenous administration of convulsive doses produces seizures within 2 to 10 min, preceded by signs of hyperexcitability.[57,143] Oral administration produces symptoms after 1 to 2 hr. In most cases, symptoms progress in a stereotyped manner through phases of hyperexcitability to convulsions, followed by recovery or death. For example, mice given LD_{50} doses of dieldrin, orally exhibit hyperexcitability, piloerection, salivation, and other signs of autonomic stimulation during the first hr. Tremoring, when it develops, is intermittent and more akin to myoclonus than to the continual tremor seen after DDT. In time, the myoclonus becomes so severe that the mice adopt very characteristic postures with all four feet widely spread and all limbs extended. Some prop themselves into corners of the cage and maintain a constant push against the cage through their extended limbs. Intermittent, severe clonus occurs abruptly without warning, and the mouse may rear and paw the air near his face, looking much like a shadowboxer. This process is eventually terminated by a seizure which often starts as an especially severe jerk or series of jerks accompanied by vocalization. The mouse begins to run rapidly around the cage only to eventually collapse onto his side. The seizure which follows develops rapidly and usually exhibits four components: (1) a short phase of clonic activity, (2) the development of a tonic flexion followed by a transition to (3) a tonic extension phase. The mouse may die at this point or progress to (4) a final clonic period followed by a postictal prostration. Those that survive the seizure may experience one of three fates. They may remain stuperous and inactive for a period, then recover without further incidence. They may die during the postictal period. They may recover and repeat the process up to and including additional seizures. Similar phenomena have been reported in various species for aldrin and dieldrin,[10,143] endrin,[10,143] Telodrin®,[10,137] toxaphene,[136] endosulfan,[139,144] and lindane.[138,145]

Because of their relative potencies, exposure to the HCH-CYCs by the dermal or respiratory routes is more frequently associated with toxicity than is exposure to DDT by those routes.[10,136,137,146] The symptoms are like those observed with oral administration.

2. Chronic Exposure

With chronic exposure to the cyclodienes or hexachlorocyclohexanes, neurological effects are frequently observed after a variable delay from the onset of exposure. The same components of hyperexcitability may be present, but they may not follow as simple a progression as is seen with acute exposure. Additional effects may also develop which require time for their maturation. As one example, Worden[137] found that feeding telodrin or endrin to rats at levels of 25 to 100 ppm resulted in dose-related effects which included hypersensitivity, audiogenic seizures, swelling of the subcutaneous tissues of the head, staring eyes, bloody incrustations over the eyelids, sporadic mild convulsions lasting about 30 sec, and in fatal cases, violent convulsions leading to death. The symptomology was most marked in the earliest stages of the study and was less apparent with time. Although part of this impression may stem from the fact that death removed the most susceptible animals from the study, it may also imply a true adaptive or metabolic tolerance.

In addition to neurotoxicity, chronic exposure to high levels of the HCH-CYCs can produce other notable changes. These include increased liver weights, weight loss, impaired survival, impaired mating performance, and impaired progeny survival.[10,136,138,147-152] These, of course, are not unique and are observed with a variety of chemicals at near lethal doses. The one major exception to this is the propensity of lindane to induce acute hepatic porphyria in humans and animal species.[149]

B. Sites of Action

The primary locus of action of the HCH-CYCs is the CNS. Peripheral nerves, sensory organs, or effectors are either not affected or are affected at doses many times higher than those eliciting central effects. In those cases where skeletal muscular and autonomic function have been found to be modified by exposure,[46,153-158] the primary cause is intensified nervous activity which originates within the CNS. There is some evidence for a cholinergic action of aldrin,[153] which was once thought to be associated with an inhibition of cholinesterase. However, direct evaluation of cholinesterase activity in exposed subjects does not support this interpretation.[154,159] Although the reasons for the effect remain uncertain, they may be related to an observation of Shankland and Schroeder,[159] in insects, that dieldrin leads to the spontaneous and excessive release of presynaptic stores of acetylcholine from identified cholinergic synapses.

In contrast to DDT, peripheral actions of the HCH-CYCs are relatively unimportant. These compounds are frequently inactive in in vitro systems in which DDT produces typical effects. For example, dieldrin does not induce repetitive activity in the lateral line organ or cutaneous touch receptors of the toad, nor in their associated nerve.[49] Dieldrin does not modify the function of myelinated nerve fibers nor alter ion currents as measured from single nodes of Ranvier.[48] Direct application of dieldrin suspensions to squid giant axons produces no appreciable effects on membrane currents, the resting potential, or upon action potential amplitudes.[160] In rats and mice, chronaxie in phrenic nerve is not altered by dieldrin.[161] Dieldrin does not change the frequency or amplitude of miniature end-plate potentials nor the characteristics of the end-plate potential in frog motor end-plates.[162] In insect nerves, isolated from the CNS so that only sensory impulses would arise spontaneously, the HCH-CYCs may induce spontaneous discharge, but only after a long delay.[44] In contrast to DDT analogues, the discharges tend to be sporadic and of lower voltage and frequency. All these data indicate that receptors and peripheral nerve fibers are not significantly affected by levels of HCH-CYCs that have marked effects in the CNS. The increases that are sometimes observed in intact animals in peripheral nervous activity are more a passive reflection of events generated in the CNS.

Skeletal muscle is affected by high levels of cyclodiene compounds in a manner similar to that produced by DDT. In isolated muscle fibers, dieldrin suspensions of 5×10^{-4} M increase the duration of the action potential and also increase the amplitude of the negative afterpotential. The resting membrane potential is unaffected. Unlike DDT, dieldrin does not bring about repetitive discharge.[46] Muscles of rats exposed to dieldrin or telodrin exhibit higher twitch tensions than do controls. Tetanus also develops at lower stimulus frequencies than in controls.[163] Chlordane, however, does not seem to produce similar effects.[164] Ibrahim[46] has reported that a loss of tension occurs during tetanic contractions in muscles of dieldrin treated rats. These observations may help to explain the findings of Khairy[165] who observed a progressive deterioration in muscular efficiency in rats fed 25 ppm or more of dieldrin. It is likely that factors other than those encountered at the level of the muscle fiber are involved. For example, Fink[158] has reported that single doses of dieldrin (1/8 to 1/2 the LD_{50} dose) caused a prolonged dose-related loss of motor coordination in a rotating rod test. The response characteristics of a nerve-tibialis muscle preparation changed little as compared to controls. This suggests an important central nervous contribution to the performance impairment.

All of the cyclodienes and hexachlorocyclohexanes are capable of increasing spontaneous activity within the insect CNS,[166] and they are more effective there than in the periphery. The frequency of spontaneous discharge in the central nerve cord of the cockroach is increased and the synaptic afterdischarge is greatly prolonged.[43,167,168] Figure 6 shows the results of applying a suspension of dieldrin (HEOD) to the desheathed sixth abdominal ganglion of the cockroach. After a delay of about 1 hr spontaneous activity increases. This increase develops initially as infrequent bursts of spikes of low amplitude. The frequency and amplitude of spikes increases during the ensuing hours. The response to presynaptic stimulation becomes greatly enhanced until, eventually, hundreds of spikes are recruited and the response lasts well over 100 msec.[168]

The failure of the HCH-CYCs to be active in peripheral nerve preparations has most often been interpreted to mean that these compounds produce their effects entirely through CNS mechanisms. Other interpretations have been suggested to account for the lack of activity in in vitro preparations of peripheral nervous tissue. One such was proposed by Wang et al.[167] who reported that a metabolite of dieldrin, aldrin-6,7-trans-diol (ATD), was both more effective and faster-acting than dieldrin in producing spontaneous activity and synaptic afterdischarge in cockroach ganglia. ATD was found to be active on peripheral nerve where dieldrin had no effect.[49,162,169,170] At synapses, ATD first enhanced transmitter release, then blocked synaptic transmission. This was proposed as a reasonable explanation of the sequencing of symptoms in dieldrin poisoning: hyperexcitability and convulsions followed by paralysis and death.[171] On these bases, it was suggested that the metabolism of dieldrin to ATD constituted an activation rather than an inactivation step. Aldrin-6,7-transdiol was proposed as responsible both for the insecticidal and toxic effects of dieldrin and aldrin.[171]

For other reasons, however, it does not seem possible to support this hypothesis. Although ATD does possess clear and unique neurological actions of its own, it is relatively less toxic than dieldrin to intact subjects. In insects,[168] rabbits,[172] and cats,[173] dieldrin is 20 to 100 times more toxic than ATD. In careful studies comparing dieldrin and ATD on cockroach ganglia, Schroeder et al.[168] found that dieldrin was much more potent even though slower acting than ATD in causing synaptic afterdischarge and elevated spontaneous activity. Both compounds were taken up into nerve tissue about equally following injection. ATD was found to penetrate into nervous tissue with the same relative ease as dieldrin. It was not possible in this study to account for the activity of dieldrin on the basis of its conversion to its metabolite, ATD, in nervous tissue. In cats, intravenous administration of dieldrin results in convulsions within 1 to 2 min, while intravenous doses of ATD, even at ten times higher concentration, are

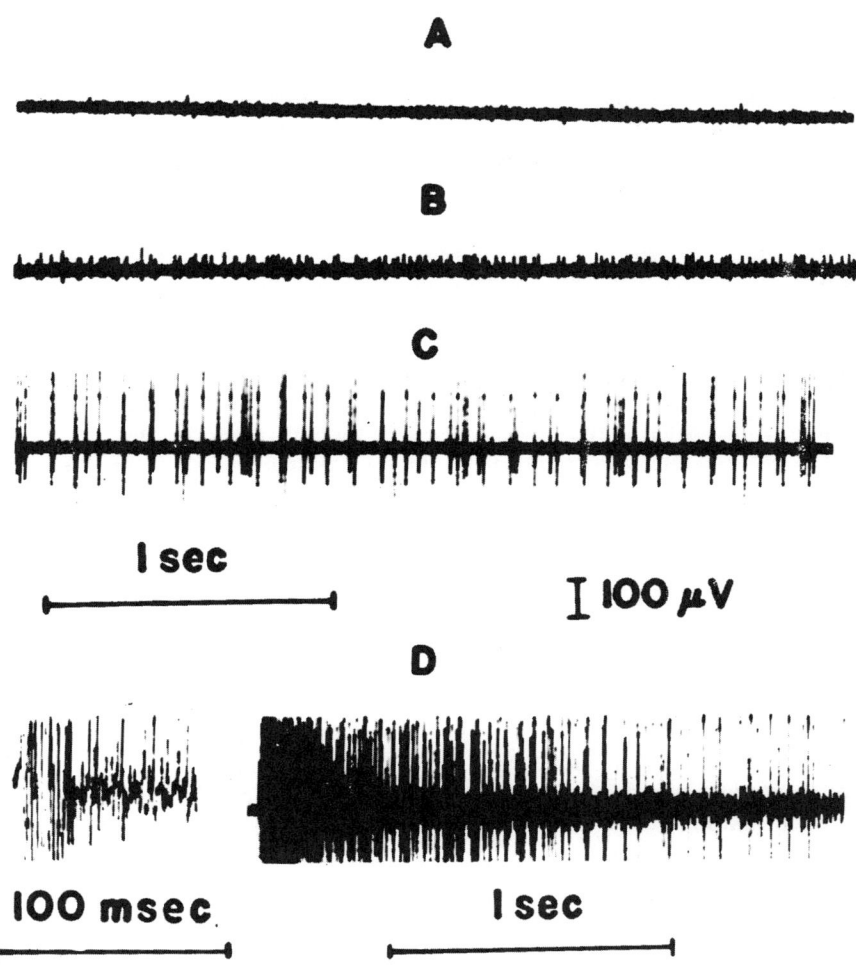

FIGURE 6. Activity in ventral abdominal connectives emanating from the sixth abdominal ganglion of a cockroach. A: Normally quiet appearance of excised desheathed, but otherwise untreated, ganglion. B: Activity involving spikes of about 70 μV and less appearing after 1.5 hr in dieldrin 1×10^{-6} M. C: Spontaneous giant fiber spikes of about 400 μV after 2.5 hr in dieldrin. D: Examples of after-discharge (abnormally prolonged responses to presynaptic stimulation) in giant fibers which appeared before (left) and after (right) giant spikes appeared spontaneously. (Schroeder, M. E., Shankland, D. L., and Hollingsworth, R.M., *Pest. Biochem. Physiol.*, 7, 403, 1977. With permission.)

ineffective. In fact ATD at these doses lacks evidence of hyperexcitability, even with sensitive EEG and evoked potential analyses.[173] These data suggest that dieldrin is directly active and that metabolism of dieldrin to ATD represents primarily a detoxification process. The bulk of evidence for dieldrin and other HCH-CYCs indicates that the parent compound is responsible for the majority of toxic effects. Certain compounds, however, such as aldrin, isodrin, heptachlor, and endosulfan are rapidly oxidized in vivo to dieldrin, endrin, heptachlor epoxide, and endosulfan sulfate, respectively.[11] These metabolites do contribute significantly to the overall toxicity.

C. Effects on Integrated Central Systems
1. Brain Electrical Activity

In mammals, the HCH-CYCs produce a relatively nonspecific increase in neuronal excitability similar to that observed for pentylenetetrazol-type convulsants.[57] Electroencephalographic recordings from freely moving subjects present a sequence of changes which include an initial increase in beta activity, then the development of high voltage 3 to 5/sec spike and wave sequences. At high exposures, tonic-clonic generalized seizure activity accompanies convulsions.[174-176] In anesthetized preparations, similar changes evolve even though they may be modified by the anesthetic. In most chronic studies in which EEGs have been examined, no effects have been reported.[60,150] However, the failure to observe changes in the EEG may be accounted for in part by insufficient exposures or by inadequate quantification methods. Persistent increases in the amount of beta wave activity, lasting up to 1 year, have been reported in monkeys administered a single dose of 4 mg/kg dieldrin.[176]

The EEG effects of some cyclodienes and lindane have been characterized in intact, paralyzed cats under controlled conditions in which proprioceptive feedback was eliminated.[56,57] A stereotypic sequence of stages evolved which included a phase of hypersynchrony during which high voltage, 1 to 3/sec waves develop broadly over the surface of the cortex. This was followed by the development of rhythmic bursts of 2 to 5/sec spikes and waves. Isolated spikes were also common (Figure 7). Seizures were of a generalized, bilateral, and symmetrical tonic-clonic type. Similar EEG modifications were observed with all of the cyclodienes and lindane. The most notable differences between compounds was in their relative potencies and in their time to onset of effects. Lindane and the epoxide compounds, dieldrin and endrin, were the most potent and most rapid acting. The nonepoxides, aldrin, and heptachlor, were less potent and exhibited a delay between administration and symptom onset which could not be abolished by administering greater amounts.

2. Central Sensory-Motor Reflex Status

In intact animals, a prominent feature of poisoning by the HCH-CYCs is the development of an inappropriate motor response to sensory stimuli. Sudden sensory input results in progressively exaggerated reflexive movements which become frank myoclonic jerks. These may precipitate a convulsive seizure. Worden[137] reported that in rats fed telodrin during a 37-day study, the noises produced during the cleaning of the room in which subjects were housed frequently elicited convulsive behavior. Stroboscopic light stimulation is also effective in precipitating seizures during exposure.[145] Even in paralyzed subjects, where the proprioceptive consequences of motor activity are eliminated, tactile, auditory, and stroboscopic light stimulation are effective precipitants of convulsive activity.[56,57]

The neuronal substrate underlying this hyperresponsiveness has been studied by Joy.[107,177] The sensory, central integrative, and motor components of a complex reflex arc, utilizing either somatosensory or visual stimuli, were analyzed in cats exposed to dieldrin (Figure 8). No significant changes were seen in response to stimulation at any level along the afferent limb from the periphery up to the level of the cerebral cortex. At that level, however, response modification occurred. This was found to involve both a greater number of cortical cells responding to the stimulus and an increase in the number of action potentials generated per cell. The increased responsiveness was progressive at successive synapses in a cortical polysynaptic pathway. As a consequence, the number of motor cortex pyramidal cells activated by sensory stimulation and the intensity of their response was increased by several fold. Confirmation of this effect was obtained by recording from units within the pyramidal tract (motor outflow pathway from cortex). There, dieldrin increased the probability that a given unit would respond to a set intensity of sensory input and also increased the number and frequency of action potentials generated by that unit. This effect was dose dependent, and correlated well with the observed changes in motor function displayed by intact, poisoned animals.

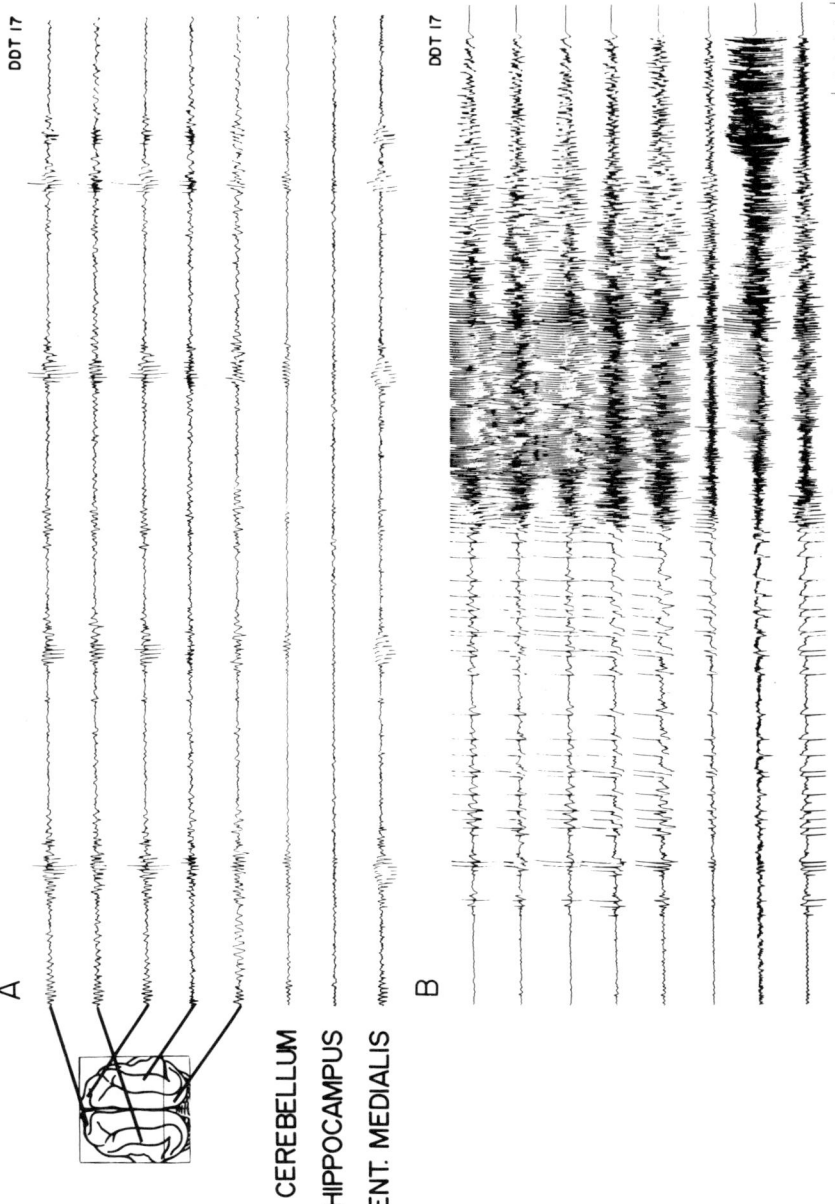

FIGURE 7. Preconvulsive and convulsive changes in brain electrical activity after dieldrin administration to a cat. Figurine insert at top indicates the location from which cortical recordings were taken. Bottom three lines depict electrical activity from cerebellum, hippocampus and thalamic nucleus, centralis medialis, respectively. A: Electrical activity 20 min after intravenous administration of dieldrin 2 mg/kg. B: Electrical activity 10 min after the administration of an additional 2 mg/kg. Hypersynchronous bursts of 3 to 5 cycle/sec activity and spikes are most evident in cortex and the thalamic lead, less so in hippocampus and cerebellum. Seizure discharge (middle of record B) is distributed bilaterally and symmetrically across the cortex. Seizure activity in the hippocampus is delayed in onset and persists after the seizure has terminated elsewhere. Calibrations for both A and B are shown to the right of B: for leads 1 to 6, 500 µV, lead 7, 200 µV and lead 8, 100 µV. Time marks indicate 1 sec intervals. (From Joy, R. M., *Neuropharmacol.*, 12, 63, 1973. With permission.)

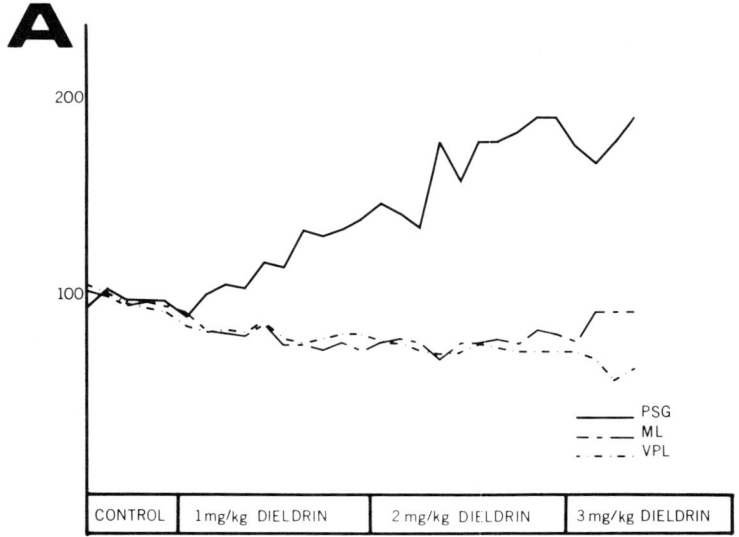
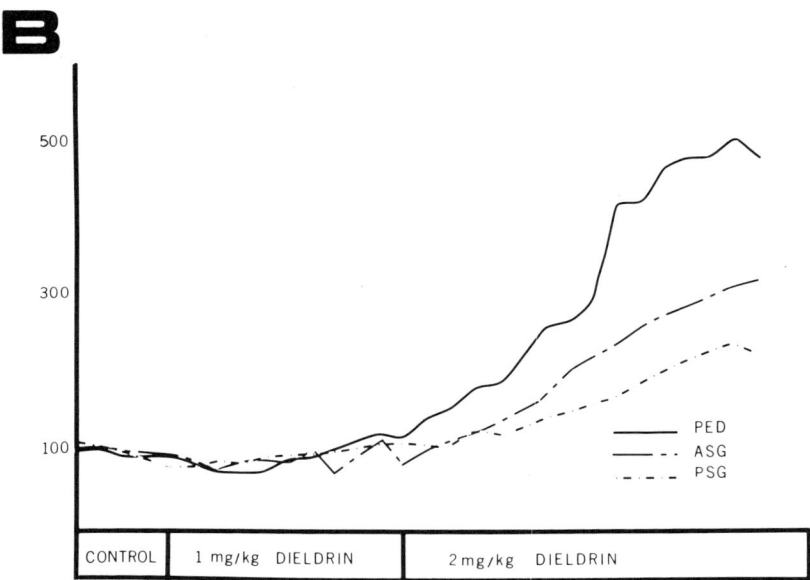

FIGURE 8. Effect of dieldrin on primary evoked-response amplitudes in sensory and motor pathways in a cat. Graphs indicate the changes in amplitude, as percentages of control amplitudes, during dieldrin administration. Average evoked-response amplitudes were determined during each 2 min time period and are plotted continuously for the time period starting 10 min before dieldrin injection until the first appearance of convulsive seizures. A — Evoked-response amplitudes recorded along the somatosensory pathway to sciatic nerve stimulation. B: Evoked-response amplitudes recorded in sensorimotor sites to stimulation of somatosensory relay nucleus, ventralis posterolateralis, in the thalamus. Abbreviations used: ML = medial lemniscus, VPL = thalamic nucleus ventralis posterolateralis, PSG = posterior sigmoid gyrus (primary somatosensory cortex), ASG = anterior sigmoid gyrus (primary motor cortex), and PED = axons of corticofugal fibers in the cerebral peduncle. (From Joy, R. M., *Toxicol. Appl. Pharmacol.*, 42, 137, 1977. With permission.)

3. Behavior

Overt signs of CNS hyperexcitability are readily apparent after exposure to sufficient concentrations of the cyclodienes or hexachlorocyclohexanes. It can be anticipated that sensitive testing methods would unmask changes at lower exposure levels. It can also be anticipated that altered susceptibility to epileptiform activity and modifications in behavior might be consequences of such exposure. Dieldrin has been observed to precipitate seizures in sheep exposed to stroboscopic light stimulation at frequencies of 11 to 14 Hz.[174] These same subjects did not convulse prior to ingesting dieldrin. The seizures were elicited at exposure levels which produced no overt evidence of clinical toxicity. Natoff and Reiff[161] have reported that acute exposure to dieldrin reduced the threshold for pentylenetetrazol-induced tonic seizures in mice. However, when mice were fed dieldrin on a chronic basis at levels designed to produce equivalent hepatic dieldrin concentrations, no changes in the pentylenetetrazol threshold were found. Woolley[66] observed changes in maximal electroshock seizure components in rats administered 45 to 55 mg/kg dieldrin orally 6 hr before testing. The duration of the initial phase of tonic flexion and the total seizure duration were increased by dieldrin. Postictal recovery time was shortened. It is unfortunate that no measure of seizure threshold had been employed in this or similar studies during acute dieldrin intoxication, as threshold would seem to be a more appropriate measure with which to assess changes in seizure susceptibility. Electroshock thresholds were employed by Al-Hachim[178] on offspring of mice treated with aldrin, 2 or 4 mg/kg/day for 7 days during the third trimester of pregnancy. The progeny, when tested at 38 days of age, exhibited significantly higher thresholds than did controls. The various changes described do not yet provide a clear picture of the consequences of low-level cyclodiene exposure on seizure susceptibility, and clarification would be desirable.

Modification of behavior has been observed in many exposure situations. At high levels of exposure some very detrimental effects have been reported upon mating and progeny-caring behavior. Wynn[152] has reported that dieldrin produces nest desertion and other dysfunctional behavior in mallard ducks. In mice, dieldrin at 10 ppm decreased the tendency of mothers to nurse. At exposure levels above 15 ppm, pups were killed or simply neglected.[152] At lower exposure levels, various behavioral deficits have been described. Desi[179] dosed rats with lindane daily at levels of 2.5 to 50 mg/kg orally. Rats on the lowest exposure levels ran faster in a maze test and committed more errors than did controls. Higher exposure resulted in an increased running time and a high percentage of errors. Operant procedures demonstrated that even the lowest exposure level of 2.5 mg/kg/day produced performance deficits suggestive of increased irritability.

A number of behavioral studies have been reported by van Gelder and colleagues regarding dieldrin exposure to sheep and monkeys. In initial studies,[180] sheep were exposed to dieldrin at levels of 10 to 25 mg/kg/day. No effects were noted during the acquisition or extinction phase of a conditioned avoidance task nor on a detour problem. However, dieldrin did impair the learning and relearning of a visual discrimination task. More detailed testing showed that dieldrin at 2.5 mg/kg/day had no effect.[181] Similar alterations have been reported with auditory discrimination tasks.[182] In monkeys trained on a visual nonspatial successive discrimination task, exposure to dieldrin at 0.1 mg/kg/day resulted in a failure to learn the task. Control subjects and monkeys exposed to 0.01 mg/kg/day were able to learn the task without difficulty.[183]

In a test for potential effects of aldrin exposure in utero, progeny of rats fed 2 to 4 mg/kg/day of aldrin during the third trimester of pregnancy were tested on a conditioned avoidance task between days 30 to 37 of age. Exposed offspring acquired and performed the task as well as nonexposed offspring.[178] These various data clearly indicate that exposure to sufficiently high levels of cyclodiene or hexachlorocyclohexanes can result in behavioral changes and associated deficits in learning and performance.

It is also apparent that simple learning paradigms require higher exposure for disruption than do more complex learning tasks. More data is desirable in this area to define accurately the types of deficits involved and to better estimate exposure levels below which various behaviors are unaffected.

D. Mechanisms of Action

The axon is relatively resistant to the effects of HCH-CYCs exposure, and it does not appear to be an important target. In both insects[159,167] and higher animals[56,57,107,177,184] the synapse is a primary target. Those synapses characterized by having high numbers of converging presynaptic elements whose net activity serves to modulate the excitability of the postsynaptic cell are most sensitive. At these synapses, the response of the postsynaptic cell to afferent input increases, leading to a lowered threshold of excitation and to an increase in the number and frequency of action potentials generated as a response. This process "avalanches" along polysynaptic pathways, ultimately producing responses 10 to 100 times more intense than normal. This exaggerated neuronal activity eventually is transmitted by way of motor outflow to the various organs, glands, and muscles. Stimulation results in inappropriately exaggerated responses, first transient, but eventually sustained in a convulsive seizure.

Mechanisms responsible for the hyperresponsiveness of a postsynaptic cell to its synaptic input, orient around two general views: (1) inhibitory activity is depressed or blocked, or (2) excitatory input is increased. Convulsants such as strychnine, picrotoxin, and bicuculline appear to block inhibitory activity, whereas convulsants such as pentylenetetrazol and bemegride appear to increase excitatory activity.[108] The latter mechanism is not universally accepted, however, and a blockade of GABA-mediated inhibition as a mechanism for pentylenetetrazol was proposed[185-187] and later refuted.[188,189] An interesting mechanism has been proposed by Pellmar and Wilson[109] to explain the actions of pentylenetetrazol. They found that, in *Aplysia*, pentylenetetrazol greatly reduces chloride-dependent synaptic responses while not affecting sodium or potassium-dependent ones. This would have a selective effect upon inhibitory postsynaptic potentials, because chloride channels are involved in inhibitory but not in excitatory postsynaptic events.

It is not possible to decide at present whether the convulsive properties of the cyclodienes and hexachlorocyclohexanes result from a blockade of inhibition or from an increase in excitatory activity. Dieldrin, which is the only compound to have been studied to any degree, shares sufficient properties with pentylenetetrazol[56,57,107,184] to suggest that both of these compounds act by very similar, and perhaps equivalent, mechanisms. This hypothesis is strengthened by the fact that dieldrin and pentylenetetrazol are antagonized by the same anticonvulsant compounds in exactly the same ratios of concentrations.[56] In one case it has been claimed that dieldrin blocks postsynaptic inhibition in the isolated toad spinal cord.[169] There, dieldrin increased polysynaptic activity and eliminated the inhibition of synchronized reflex discharge normally produced by paired stimulus presentation. In cats, Joy[177] has described similar changes to paired stimulus presentation in sensory, association, and motor cortex. These changes appeared more consistent with, and were interpreted as being due to, an increase in excitatory drive into the cortical area. In motor cortex, recurrent collateral inhibition, produced by antidromic stimulation of axons in the pyramidal tract, was unaffected by dieldrin.[184] There is no direct evidence to date which would support a mechanism involving the blockade of inhibitory function in the CNS as a basis for the observed effects.

In their analysis of the actions of dieldrin on cockroach abdominal ganglia, Shankland and Schroeder[159] found evidence for a presynaptic site of action. They investigated the cercal nerve—giant fiber synapse which is a cholinergic synapse. The increased spontaneous activity and the prolonged afterdischarge to cercal nerve stimulation evoked by dieldrin was blocked in the presence of atropine, hemicholinium, and magnesium ion. The blockade by atropine can be explained by its competitive blockade of postsynaptic muscarinic receptors. The effectiveness of magnesium ion and hemicholinium suggest a presynaptic locus of action for dieldrin. From such evidence, Shankland and Schroeder proposed that dieldrin acts by potentiating both evoked and spontaneous release of acetylcholine from the presynaptic terminal. The mechanisms suggested to account for the hyperresponsiveness, convulsive episodes, and ultimate paralysis and death were (1) periods of enhanced transmitter release, (2) spontaneous transmitter release, and (3) eventual depletion of transmitter from the terminal. Although of questionable relevance, similar presynaptic actions have been described for aldrin-6,7-transdiol, a dieldrin metabolite, at the frog motor end-plate.[162]

Most subcellular theories regarding the action of the HCH-CYCs center around their potential interaction with the neuronal membrane. Because the synapse, and perhaps more specifically the presynaptic nerve terminal, are much more sensitive than is the axon, general membrane interactions, such as those proposed for DDT, appear less relevant. Nevertheless, a mechanism of action for lindane has been proposed by Mullins[89] and Soloway[190] on these bases. Soloway has further extended these concepts to the cyclodiene compounds in general and should be consulted by those interested in molecular modeling. Good summary reviews have been provided by O'Brien,[9] Matsumura,[11] and Burt.[191]

In addition to the action of the cyclodiene group upon synaptic function, additional effects exist which may be relevant to their production of neurotoxicity. Effects at the level of neurogenic amines and various enzyme systems participating in ion transport, metabolism, oxidative-phosphorylation, etc. have been described. Changes in the concentration, synthesis, or kinetics of many substances have been reported, including ammonia,[92] glutamine,[92,192] aspartic and glutamic acid,[192] phospholipids[192,193] P-450,[194] GABA, glycine and taurine,[192,195] acetylcholine,[12,24,98,196-198] norepinephrine,[12,195-200] serotonin,[12,95,98,196-199,200-202] and dopamine.[195,199,200] For more complete details, the reader is referred to the review of Hrdina et al.[12] It should be kept in mind that many of the changes reported may not be specific to the exposure per se. Similar alterations in a variety of chemical and neurochemical substances have been observed to accompany hyperexcitability and convulsions from a number of different causes.[203,204]

Koch and colleagues[103] have provided evidence for an action of the cyclodienes upon various ATPases. Initial studies indicated that both Na^+,K^+-ATPase, associated with cation transport through membranes, and Mg^{++}-ATPase, associated with oxidative-phosphorylation, were inhibited by the cyclodienes and lindane. In a comparative study of ATPases from different sources, they found chlordane and lindane to have their greatest inhibitory effects upon Na^+, K^+-ATPase from nerve tissue.[102] In contrast to their neurotoxicity, however, epoxides were found to be less potent than their nonepoxide congeners.[104] This may relate to their relative lipid/water partition coefficients, which is decreased by epoxidation.[205] If such is the case, then a nonspecific mechanism for ATPase inhibition is suggested. Additional studies[206,207] have indicated variable results with ATPases from different sources. Although the picture remains unclear as to the importance of these data, they are intriguing. However, the concentrations employed in these in vitro studies are almost always beyond the solubility range of the insecticide, and the same concerns must be raised here as in the case of DDT (see previous section on the mechanism of action of DDT). It is possible that some of these effects may be occurring by way of mechanisms operating only at concentrations many times greater than those attained in intact subjects.

E. Neurotoxicity of the Cyclodiene and Hexachlorocyclohexane Derivatives in Man

Because the HCH-CYCs are toxic at lower exposure concentrations than are the DDT analogues, toxic manifestations in man have been more frequently encountered. Various sources have provided information concerning: (1) voluntary, purposeful exposure, usually with suicidal intent, to high concentrations ingested orally; (2) involuntary, accidental exposure to estimable quantities by the oral, inhalation, or dermal route; (3) involuntary, accidental exposures to unknown quantities; and (4) exposures to mixtures of chemicals, one or more of which are cyclodienes or hexachlorocyclohexanes. In many cases poisoning has occurred without the complication of solvents or other materials. Here, the symptomology is typical and can be clearly related to the identified chemical. The symptoms of toxicity in man are similar for all of these substances, and a separate discussion by compound is unnecessary.

The preponderance of the symptoms developing after exposure are the direct result of the CNS actions. Nonneurological sequelae also occur. Large doses induce nausea and vomiting. Repeated, chronic exposure may induce histopathological changes in liver and kidney. The compounds and their metabolic products are stored in fat. The mobilization of these stores during intensive activity and starvation may cause toxic symptoms to recur.[16] In subjects dying during acute exposure, dilation of blood vessels, and petechial hemorrhages may be demonstrable. These may result from convulsions induced by the toxicant.

Intentional and unintentional exposure to any of these compounds produces symptomology that depends upon the amount and duration of exposure. Although dose-response data for man is limited, it is possible to provide partial data for certain of them, such as dieldrin (Table 3). Occupational exposures to dieldrin at levels 1000 times higher than the exposure of the general population have been tolerated without obvious effect for up to 15 years. Chronic exposure to levels above 0.03 mg/kg/day may cause signs and symptoms of convulsive or nonconvulsive intoxication in some people. Acute exposure to 20 to 40 mg/kg has been associated with convulsions, followed by recovery. Higher exposures may be fatal.

Early symptoms of acute poisoning include evidence of motor hyperexcitability which may include myoclonic jerking and convulsions. Headache, nausea, vomiting, general malaise, and dizziness may or may not be experienced. As a result of extensive studies of workers occupationally exposed to aldrin, dieldrin, endrin, and telodrin, Jager[10] suggested that three syndromes of poisoning could be differentiated, based upon the agent involved and the exposure history. The first of these is characterized as a convulsive intoxication with few or no prodromi. These result from one or a few gross overexposures. This syndrome occurs most frequently with the highly toxic and not very persistent compounds. The majority of endrin poisonings described in the literature fit into this category. With compounds having a lower acute toxicity, but a higher persistence, an accumulative intoxication can occur. Clinically, this results in a syndrome of headache, dizziness, drowsiness, hyperirritability, general malaise, nausea, anorexia, and occasional vomiting. At times, muscle twitchings, myoclonic jerks, and convulsions occur. This type of intoxication is typical of many poisonings reported with aldrin, dieldrin, and telodrin. In workers using aldrin, dieldrin, and telodrin an acute over-exposure, in itself not significant, may cause an abrupt onset of convulsions superimposed upon a subclinical accumulative intoxication of the type just described. In these circumstances the minor increases in insecticide level in the blood, caused by the acute exposure, induce the convulsive intoxication.

1. Acute Exposure

The most consistent response to the ingestion of toxic amounts of all the HCH-CYCs is the development of convulsive seizures. These are often sudden and abrupt in onset, and they may occur without warning or any prodromal symptoms. In some

Table 3
DOSAGE—RESPONSE TO DIELDRIN IN MAN[a]

Dosage (mg/kg/day)	Consequences
0.00004—0.00024	Dosage of general population of U.S. during 1961—1967
0.00008—0.00030	Dosage of general population of U.K. during 1961—1968
0.0031	Tolerated by volunteers for 2 years without indication of enzyme induction
0.0084—0.0170	Tolerated by workers for 3—12 years
0.0175	Long-term occupational exposure with no detectable increase in hepatic microsomal enzyme activity
0.0332	Tolerated by workers for up to 15 years
0.0332 and up	Daily intakes of this order for long periods of time may cause signs and symptoms of convulsive or nonconvulsive intoxication in some men: intakes of up to twice this dose are known not to cause signs and symptoms of intoxication in some men
26—44[b,c]	Convulsions and recovery
70[b]	Estimated fatal single dose

[a] Table largely derived from Hayes.[8]
[b] Single dose only.
[c] Compound was aldrin, which is rapidly metabolized to dieldrin.

subjects, convulsions may be preceded by a variable period of time in which hyperexcitability, myoclonic jerks, and general malaise may be experienced. The following cases are typical of those in the literature.

Spiotta[208] reported the outcome of a case in which a 23-year-old farmer drank a Coca-Cola® bottle full of an aldrin mixture containing 1.8 g of aldrin (25.6 mg/kg) in an attempted suicide. Immediately thereafter, he panicked and was rushed to a local hospital. The patient had a convulsion about 20 min after ingesting the aldrin. Before medication became effective, he had a total of three convulsive episodes, one sufficiently severe to dislocate his shoulder. Pentobarbital was given for the next 4 days and no further convulsions occurred. He was markedly restless and semistuperous, hyperthermic, tachycardic, and hypertensive 5 days after he had ingested the aldrin. An EEG taken 2½ weeks after ingestion revealed bursts of high amplitude, hypersynchronous 2 to 3 per sec waves symmetrical over both hemispheres. The EEG changes improved with time, but some abnormality remained up to 6 months later. Recovery occurred slowly, and the patient was discharged 6 weeks after ingestion. No evidence of persisting effects were noted.

Starr and Clifford[209] described a case in which a 2½-year-old-girl ingested about 1.5 g of lindane. She began convulsing ½ hr later and had intermittent grand mal seizures which were controlled with phenobarbital. The girl improved rapidly and was released from the hospital the following day.

Garrettson and Curley[210] described a similar toxic reaction in a 4-year-old boy who, along with his 2-year-old sister, drank from a bottle containing 5% dieldrin. When the children were first observed, the 2-year-old was dead and the older child was in status epilepticus. After hospitalization, anticonvulsant drug administration broke the status. The last convulsion occurred 7½ hr after ingestion. The boy's condition immediately improved. He regained his equilibrium within 3 days. No symptoms which could be related to the exposure were observed in the ensuing year.

A number of acute poisonings by endrin, occurring as a contaminant of flour, have been reported.[142,211,212] The cases described by Coble et al.[142] are remarkable in that some of the victims were not even aware that they had had a convulsive episode. In one case, a woman was brought to the hospital because of a tender lump on her forehead. She had no explanation for it and could not remember how it got there. At the time of hospitalization she was oriented but could not remember much of the morning's events. Shortly thereafter, she suddenly became speechless and had a generalized tonic-clonic convulsion lasting about 2 min. With anticonvulsant therapy, no more convulsions were experienced and all symptoms disappeared within a week. Two neighbors also experienced sudden, convulsive episodes without prodromal effects. The cause of all three poisonings was eventually traced to bread made with endrin-contaminated flour.

Essentially similar descriptions have appeared in the literature relating to acute poisoning by aldrin,[213] endrin,[214-216] and lindane.[128,217] Probably the best single source reviewing the clinical literature on aldrin, dieldrin, endrin, and telodrin is Jager.[10]

Poisoning on an epidemic scale has occurred with endrin. Weeks[212] described four explosive outbreaks of acute poisoning occurring in Qatar and Saudi Arabia. Altogether 874 persons were hospitalized and 26 died. In all four situations the cause of poisoning was determined to be flour which had become contaminated with endrin.

2. Chronic Exposure

Two types of chronic exposure syndromes have been described by Jager.[10] In one of these, exposure occurs constantly and results in a slow accumulation of the insecticide coupled with a progressing symptomology. In the second type of chronic exposure, the intake of insecticide remains below that required for overt symptoms. However, the subject is rendered extremely sensitive to additional acute exposure. Both types of chronic poisoning have been reported for aldrin,[10,218] thiodan,[219] dieldrin,[10,14,16] and endrin.[10,216]

Hayes'[14] description of toxicity occurring in spraymen applying dieldrin exemplifies the consequences of chronic exposure to relatively small daily amounts of dieldrin. In certain programs carried out in South America, 10 to 20% of spraymen had been poisoned. The earliest observed poisoning occurred after about 3 months of exposure, but in the majority of cases, exposures of 8 months or longer were required. The mildest clinical illnesses were characterized by headache, which was frequently persistent and unresponsive to drugs, blurred vision, dizziness, slight involuntary movements, sweating, insomnia, nausea, and general malaise. More severe illness was characterized by these same symptoms plus myoclonic jerking involving one or more limbs, sometimes resulting in the patient falling to the ground. In some cases the myoclonic movements were accompanied by momentary loss of consciousness. In the most severe cases, one or more epileptic convulsions occurred. Convulsions were reported by over half of those patients who sought medical aid. One sprayman had more than 30 convulsions.

Winthrop and Felice[14] examined about half of the spraymen in Venezuela. Even in those without obvious illness, individuals exposed to dieldrin for 30 or more weeks complained of certain symptoms more often than did nonexposed subjects. These complaints included headache, blurred vision, diploplia, tinnitus, dizziness, involuntary muscular movements, sweating, difficulty in sleeping, bad dreams, alteration of reflexes, incoordination, nystagmus, muscular fibrillation, and changes in personality.

Kozantzis et al.[218] described 4 cases of toxic reactions in industrial workers exposed to aldrin who developed symptoms associated with a change in job status. The change increased their exposure to aldrin. One subject had been employed for 21 years at the plant and had been reassigned to handling aldrin concentrate. Subsequently he began to notice an involuntary jerking of his hands and forearms. This consisted of a rapid flexion movement which had caused him to drop his cup of tea on more than one occasion. He had also vomited on a number of occasions and complained of being chronically irritable and insomnic. An EEG taken at that time showed alpha irregularity with occasional localized and more frequent generalized discharges of irregular slow and sharp waves. He was removed from exposure, and his symptoms showed prompt improvement. The EEG also became more normal. Similar experiences have been reported for workers exposed to thiodan,[219] dieldrin,[10] and aldrin.[10]

Gupta[220] observed the consequences of chronic exposure to lindane and aldrin by members of a small Indian village community. Twelve people and five domestic animals developed toxic reactions from eating grain mixed with aldrin dust and lindane powder for 6 to 12 months. Two people and two of the animals died from convulsions. The primary symptomology was restricted to frequent myoclonic jerking and generalized grand mal seizures. EEGs were done on six individuals, and in every case they contained abnormal amounts of slow and sharp waves. Bursts of symmetrically distributed high amplitude waves were also present. These changes tended to revert after exposure was discontinued. However, symptoms, including myoclonic jerking, were still reported by all people involved for up to a year following the exposure. Some individuals complained of persisting memory loss and irritability. One child was diagnosed as hyperkinetic and two were diagnosed as mentally retarded after recovery from exposure. Their status before exposure was not indicated.

3. Persisting Neurological Sequelae

In a number of reports involving either short- or long-term exposure, persisting neurological sequelae have been described. The period over which these occur appears highly variable but does seem to bear a relationship to the severity of symptoms occurring in acute intoxication or to the duration of exposure in more chronic ones.

Changes in the EEG occur during intoxication, and some degree of abnormality may persist for weeks or months. Workers exposed to HCH-CYCs exhibit a greater degree of abnormal EEG activity than do nonexposed controls.[118,221] During intoxication, bilateral, synchronous theta wave activity and occasional bilateral synchronous spike and wave complexes are typical. The great majority of abnormalities described are bilateral and symmetrical and are reminiscent of changes described for myoclonic, photomyoclonic, and certain petit mal variant epilepsies.[10,14,208,209,213,218,220,221] The severity of the changes regresses with time, but EEG abnormalities persist long after after objective signs of intoxication are gone. Although the use of routine EEG evaluations as a preventative measure for determining exposure in industrial workers has been largely replaced by blood level determinations, the EEG is still relied upon in cases of suspected intoxication.[10]

In those cases where adequate preexposure EEGs are available, EEG patterns have been seen to return eventually to normal. It has been estimated that normalization may require up to 2 to 3 months for endrin, a year for dieldrin, and more than a year for telodrin.[10] In one retrospective study of children poisoned 18 months to 14 years prior, no significant EEG changes were observed.[222]

Prolonged recovery, as measured by the time to complete disappearance of clinical symptoms, has been observed. While the involvement of subjective factors must be considered before attributing all symptom complaints to the insecticide exposure, many do appear to be so related. In acute exposures, such as those reported by Spiotta,[208] Garrettson and Curley,[210] Bell,[213] Starr and Clifford,[209] and Coble et al.[142] symptoms usually disappeared after a few days and never persisted beyond 2 weeks. The only exceptions to this were EEG abnormalities and a permanent amnesia for various time periods preceding and during convulsive episodes.

With more chronic exposures, symptoms may be more persistent. Hayes[14] has reported that spraymen developing symptoms of intoxication to dieldrin showed initial improvement when removed from further exposure. However, some men required up to 105 days to recover. One sprayman suffered a recurrence of convulsions 84 days after his last exposure. The latter may or may not have any relevance to the previous exposure. Jager[10] has evaluated all cases of insecticide-related toxicity occurring at the Shell Nederland Chemie in Rotterdam, the Netherlands, between 1954 and 1970. A total of 54 insecticide intoxications occurred among workers exposed to aldrin, dieldrin, endrin, and telodrin. In every case, complete recovery was reported. However, symptoms persisted in accordance with the time required for blood levels to fall. Thus, workers intoxicated by telodrin complained of persisting symptoms for periods up to 6 months whereas workers poisoned with endrin or dieldrin were symptom free after a few weeks. EEG changes outlasted reports of objective symptoms on the part of the patient, and outlasted any clinical or laboratory findings. Essentially similar reports have appeared for other exposed populations.[118,218,219]

Other data suggest that chronic exposure to these substance may produce persisting alterations in neurological and psychological functions. Muminov[223] observed selective hearing losses in cases of acute and chronic poisoning involving either organochlorine or organophosphate agents. Those individuals with predominately organochlorine exposure had reduced perception of low frequency tones. Schuettmann[127] has implicated lindane as a responsible factor in the development of hematopoietic, hepatic, and nervous system pathologies. He observed polyneuritis in workers exposed to lindane. Psychological disorders including anxiety, irritability, insomnia, and motor pathology were also detected. The nervous system effects are reminiscent of those found by Hayes[14] in spraymen using dieldrin.

Gupta[220] reported a number of persisting abnormalities in family members chronically poisoned with aldrin and lindane. Two family members died as a result of convulsions and two children were subsequently classified as retarded and hyperkinetic after recovery. Family members still complained of myoclonic jerking, irritability, and memory loss 1 year after exposure was stopped.

The CHIs and/or the convulsive consequences of exposure to them may be particularly dangerous to children. A number of reports suggest that the exposure or its consequences may lead to irreversible damage. Gupta[220] reported the presence of mental retardation and hyperkinesia in children exposed to aldrin and lindane. Jacobziner and Raybin[224] reported two cases of endrin poisoning in infants, one of whom died in a state of decerebrate rigidity and the other of whom exhibited permanent brain damage. This was thought to be the consequence of the anoxia and trauma associated with prolonged seizure activity. Angle et al.[222] made a retrospective study of children poisoned by various substances and found that those who had had convulsive episodes showed a significant deficiency in Binet scores with associated learning and behavioral difficulties. All of the children poisoned by CHIs sufficient to produce convulsions displayed major difficulties.

Isolated cases of latent polyneuropathies have been related to CHI exposures. Onifer and Whisnant[123] described a case involving exposure to DDT and lindane from a commercial vaporizor which was implicated as the cause of a polyneuropathy developing in a 33-year-old butcher. The subject had been working for 2 weeks in a closed room containing a DDT-lindane vaporizor. He experienced a progression of neurological symptoms including paresthesias, incoordination, muscular weakness, staggering gait, and fatigue. The increasing severity of these symptoms finally led to his hospitalization. The symptomology persisted for about a week, then began to recede. He was nearly normal 1 month later. No confirmation that DDT or lindane were causative agents was available. It was not indicated whether reexposure occurred when the patient returned to work.

A case in which partial confirmation about cause exists was reported by Jenkins and Toole.[117] The subject was a 13-year-old boy who had worked in his father's fields which had been sprayed with aldrin. He worked barefooted and had also carelessly applied DDD powder on numerous occasions. He was symptom free for 3 months, until he began to be awakened from his sleep by uncontrollable jerkings of his body. Subsequently he developed paresthesias, became rapidly fatigued, and complained of difficulty in talking and swallowing. Muscle weakness progressed to complete facial diplegia and loss of movement in the lower extremities. Recovery was slow. He was able to walk only after about 3 months and total recovery required 9 to 10 months. The following summer he reentered the fields after they had again been sprayed with aldrin. Within days he again developed weakness, myoclonic jerks, and sensory disturbances similar to those experienced the year before. Removal from exposure was sufficient to eliminate the symptoms. The same authors reported a second case in which DDT and endrin were implicated, without confirmation, as the cause of a peripheral neuropathy developing in a 31-year-old farmer. The neuritis receded in 6 to 8 weeks, and the farmer returned to work. His subsequent exposure history was not reported.

VII. NEUROTOXICITY OF CHLORDECONE (KEPONE®)

Chlordecone (1,1a,3,3a,4,4a,5b,6-decachlorooctahydro 1,3,4-methano-2H-cyclobuta (c,d) pentalen-2-one) is a complex structured chlorinated hydrocarbon insecticide developed by the Allied Chemical Corporation in the early 1950s and registered as a pesticide in 1955. Small amounts were made up to until the mid-1960s and, between 1966 and 1973, Allied became the sole producer of this chemical. It manufactured 50,000 to 200,000 kg annually until 1973. In 1973, the Allied Chemical Corporation contracted out the manufacture of this agent to a new firm, Life Science Products Company of Hopewell, Va., which became the sole producer of chlordecone in the world, the annual production reaching a level of 400,000 kg of greater than 90% purity by 1975.[225,226]

CHLORDECONE (KEPONE®)

Little of this chemical was used in the U.S., the major home use being in a formulation at a concentration of 0.125% in ant and roach traps. Some 99% of the world production was exported to West Germany where it was used in the formulation of a pesticide mixture, Kelevan®, which was exported to Central and South America for use in controlling the banana borer weevil. During the manufacture by Life Science Products Company, chlordecone synthesis was not required by law to be registered with the U.S. Environmental Protection Agency (EPA) on the basis that EPA considered the pure compound (undiluted) as a chemical under shipment to another manufacturer and not as a pesticide.

Chlordecone is synthesized by the reaction of hexachlorocyclopentadiene with sulfur trioxide in the presence of an antimony pentachloride catalyst with a subsequent hydrolytic step to produce a ketone. It is a white-to-tan colored powder which is hygroscopic, forming a hydrate at room temperature and normal humidity. It is nonvolatile, will sublime and decomposes at 350°C. It has low water solubility (0.4 g/100 ml at 100°C) but is soluble in acidic or alkaline solutions and in the more polar organic solvents (acetone, diethylether, lower aliphatic alcohols) but not in nonpolar solvents such as benzene or n-hexane.[227,228]

A. Toxicity to Animals

Chlordecone is readily absorbed from the gastrointestinal tract of laboratory animals, and absorption through the skin has been reported.[229] A constant tremor syndrome appeared in mice within 4 weeks of feeding a diet containing 30 ppm or higher and these tremors disappeared within 4 weeks following withdrawal of the treated diet. At a dietary level of 40 ppm, the liver doubled in size in 60 to 90 days and histological examination revealed focal necrosis, cellular hypertrophy, hyperplasia, congestion, and liposphere formation. The enlarged livers decreased in size when the treatment was withdrawn. Following feeding studies in rats for 2 weeks, the distribution of chlordecone in tissues was measured. The liver, adrenals, and lungs contained the highest concentrations (10- to 60-fold higher than blood), while the brain and adipose tissue contained levels approximately 8- and 10-fold higher than the blood.[230] While neurological problems were observed in animals treated with chlordecone, the main interest for some time centered around the effects on the liver. The observed hepatomegaly was accompanied by the marked induction of hepatic microsomal monooxygenase enzymes.[231] Chlordecone was a potent nonspecific inducing agent; the no-effect level for induction based on O-demethylase activity in mice being less than 10 ppm fed over a 14 day interval. Fed to Japanese quail at a level of 200 ppm in the diet, chlordecone caused marked cellular changes in the livers (lipid-like inclusions) and adrenal glands (hypertrophy of both cortical and medullary cells).[232] These tissue observations correlate well with the high tissue levels found in the rodent studies, suggesting that the liver and adrenals are prime target organs.[230] Hepatomas have been reported in chlordecone-treated rats and mice.[233] Malignant tumors were also found in other organs than the liver, and female animals appeared to develop malignant tumors more readily than males.

In studies designed to characterize the onset, peak, and duration of effects following single oral doses of chlordecone in corn oil to rats, tremors and abnormal gait were the major early features of toxicity.[234] Both developed within 4 hr of treatment, reached a peak 2 days after treatment, had greatly diminished after 2 weeks, and were absent after 49 days. Startle responses were exaggerated and followed a similar time course. Muscle weakness developed during the second week and was still increasing in intensity at 49 days although recovery was obviously occurring by the end of 6 months. The conclusion of these investigators was that, while neurological effects are the most severe and rapidly developing manifestation of poisoning, muscle weakness was the most persistent outward sign of toxicity in the rats. The oral LD_{50} values for rats (132 mg/kg for males, 126 mg/kg for females), male rabbits (71 mg/kg), and dogs (approximately 250 mg/kg), have been established.[235] In all cases, death was preceded by the development of severe tremors. In longer term feeding studies carried out in rats, the principal histopathological findings in rats were degenerative changes in liver cells (enlargement, fatty infiltration, hyperplasia), glomerulosclerosis in the kidneys, and testicular atrophy. This investigation indicated also that the nervous system, the reproductive system, and the liver were the major targets of chlordecone toxicity.

The major physiologic effects resulting from the ingestion of sublethal levels of chlordecone, exclusive of the liver abnormality and the tremor syndrome, involved the reproductive processes. This was shown in mice by Huber[229] in 1965 and by Good et al.[236] in 1965. Data collected by vaginal smears, hormone bioassays, histologic examinations, and matings revealed that the female hormonal system was disturbed. Primarily, the results showed constant estrus, the development of large follicles, the absence of corpora lutea in the ovaries, and failure to reproduce, all indications of prolonged stimulation of follicle stimulating hormone and estrogen and insufficient production of luteinizing hormone.[229] Reproductive problems have also been encountered in laying hens and in Japanese quail treated with chlordecone.[232,237] Chlordecone increased cellular proliferation, cytodifferentiation, and tubular gland formation in the oviducts of immature quail, but no cellular changes were noted in the oviducts from laying birds. Ovarian tissue from treated birds contained more primary oocytes and smaller follicles than did tissue from control birds. In male Japanese quail, chlordecone had a marked influence on the structure and function of the testes.[232,238] Chlordecone produced both enlarged (fluid-filled) and atrophic testes. In enlarged testes, the seminiferous tubules were dilated, the germinal epithelium was severely disrupted, and sperm production was affected. In atrophic testes, the seminiferous tubules were reduced, spermatogenesis appeared suppressed, and the germinal epithelium contained abnormal sperm. The excurrent ducts contained either few sperm or were packed with degenerating cellular debris. Structural reparations appeared to occur when the chlordecone-containing diet was withdrawn, and increased spermatogenesis recurred in some birds. In others, degeneration and necrosis was observed in the tissues, and the reproductive capabilities of the birds were severely limited. It will be seen, in a later section, that significant reproductive problems occurred in man as well.[239]

As was indicated earlier, the adrenal glands appear to be a prime target organ for chlordecone. Hypertrophy of adrenal cortical and medullary cells was noted in chlordane-treated quail.[232] In rats treated with chlordecone (200 ppm) for 8 days, the adrenals were removed for catecholamine analysis and ultrastructural examination.[240] The total catecholamine content was reduced by some 54% and there was a selective decrease in epinephrine of 63% and an increase in norepinephrine of 28%. The biochemical data were supported by electron microscopy in that the catecholamine-containing granules of the epinephrine cells, but not of the norepinephrine cells, were depleted. The observed differences were speculated to result from effects on mitochondrial and/or granular Mg^{2+}-ATPase and on adrenocorticosteroid levels. Studies have demonstrated a marked inhibition of Na^+, K^+-ATPase, and Mg^{2+} ATPase (the oligomycin-sensitive mitochondrial enzyme) in channel catfish brain by chlordecone.[241] The mitochondrial Mg^{2+}-ATPase of rat liver was also inhibited by this agent.[242] The ATPases and K^+-phenylphosphatase of mouse brain synaptosomes were also inhibited.[243] The overall importance of this inhibitory effect of chlordecone awaits elucidation since it may only be one of many toxic effects attributed to the agent. However, the essentiality of the ATPases in the maintenance of neural function and conduction is well known. In a recent paper, End et al.[244] demonstrated that the administration of a tremorigenic dose of chlordecone (40 mg/kg, po) to rats inhibited both succinate- and ATP-supported calcium uptake in brain mitochondria in vitro. They also demonstrated that the levels of chlordecone in brain mitochondria of treated animals were comparable to those attained by exposing mitochondria in vitro to a 2×10^{-6} M inhibitory concentration of chlordecone. The results suggested that mitochondrial inhibition might be a mechanism for chlordecone toxicity and involve energy metabolism and cellular calcium distribution.[244,245] While the mechanism of action of chlordecone is not all that clear, an involvement of catecholamines is suspected since propranolol, a β-adrenergic blocking agent, was found to be effective therapy for the tremor seen in chlordecone-exposed workers.[226]

B. Toxicity to Humans

What has become known as the "Hopewell epidemic" occurred during the summer of 1975. This is the only known human intoxication with this compound and resulted as a consequence of the industrial exposure of a number of workers, some 133 individuals, employed during a 17-month period of chlordecone synthesis under very unsatisfactory hygienic and health conditions. The epidemic has been succinctly described by Cannon et al.[225] and Taylor et al.[226,239] Since it was determined that considerable chlordecone had been discharged both into the air and into the local sewage system, the Virginia State Health Department, the EPA, and the Center for Disease Control (CDC) all became involved in an extensive epidemiological study of the magnitude of the problem. The factory was closed, but it was found that chlordecone had also been discharged into the James River for some time. This resulted in the contamination of this river system and the tidal areas of the bottom end of Chesapeake Bay. In consequence, chlordecone was bioaccumulated in the aquatic life and seriously affected the local fishing industry. The clinical aspects of the chlordecone intoxication have been described in detail in several papers but a brief summary is in order.[225,226,246]

Patients with prolonged exposure to chlordecone developed a nervousness, tremors, chest pains, weight loss, arthralgia, skin rash, mental changes, opsoclonus, muscle weakness, ataxic gait, incoordination, and slurred speech.[246] In severe cases, the tremor was present at rest and in all cases was increased by use of the affected limb. The hands were chiefly involved but fine tremor of the head and trembling of the entire body (called "Kepone shakes" by the workers) have been observed. A greatly exaggerated startle response was also noted in severely affected people. Another prominent complaint was visual difficulty, characterized by an inability to fixate and focus. Erythematous macular skin eruptions were also seen. Personality changes were observed, the most common being irritability, difficulty with recent memory, and mild depression which, in some cases, approached frank disorientation. In addition, semen analysis in four patients revealed severe impairment of spermatogenesis. Oligospermia was seen in 13 men, the severity ranging from no motile sperm in 2 workers to an intermediate sperm count in the rest which was lower than normal values.[226] In all, it is remarkable how closely the syndrome observed in the workers compares with the signs and symptoms observed in animal studies.

The average latency interval between the start of employment at the factory and the onset of symptoms was 6 weeks. The severity of the neurological symptoms was proportional to the "dose" and the duration of exposure. Higher attack rates were seen in production personnel than in those not directly involved in production. Analysis of blood samples from workers for chlordecone revealed a range from 0.009 to 11.8 ppm with a mean level of 2.53 ppm.[225] One worker was hospitalized in a confused and disoriented state and was found to have the highest blood chlordecone level (33.0 ppm). He exhibited auditory and visual hallucinations and presented a startle response which consisted of a series of brief but incapacitating, whole-body myoclonic jerks.[226] Former workers, who had left employment primarily because of the working conditions, had a mean blood level of 1.22 ppm of chlordecone. Two wives of workers were found to have had objective tremor, possibly associated with their washing of agent-contaminated clothing.[225] The large interindividual variation in blood levels and symptomatology prevented the determination of a more precise dose-response relationship. Approximately 50% of the total work force investigated was moderately ill and a few required extended hospitalization for several months. The biological half-life of chlordecone in untreated workers was approximately 165 days.[226] In the subsequent 18 months after detection, most symptoms and signs gradually disappeared but, even after 4 years, several workers continued to manifest a mildly incapacitating tremor.[226] There are no data presently available which suggest whether additional delayed effects of chlordecone, including carcinogenic and reproductive sequelae, might occur. However, long-term, clinical follow-up of the exposed workers will be continued.

C. Neurotoxicity

While it is obvious from the symptomatology outlined above that the toxicity manifested itself primarily as a neurologic syndrome, routine neurologic studies including electroencephalograms, electromyograms, skull X-rays, lumbar punctures, and radionuclide scans have all been normal or have demonstrated nonspecific abnormalities.[225] Muscle biopsies from six patients showed an increased lipofusion and lipid-like droplets as well as a predominance of type I fibers.[247] Examination of biopsies of the sural nerve from five patients was conducted.[246] Under light miroscopy, all specimens contained increases in endo- and perineural connective tissue. A relative decrease in the population of small myelinated and unmyelinated axons was noted. There was no evidence of myelin disintegration or of myelin phagocytosis. With electron microscopy, a number of abnormalities were visible and the significant findings in peripheral nerves included:

1. Accumulation of elongated, crystalloid, and electron-dense rods and short, laminated, and parallel membranous inclusions within Schwann cell cytoplasm
2. Redundant Schwann cell cytoplasmic folds
3. Prominent endoneural collagen pockets
4. Vacuolization of unmyelinated fibers associated with an interdigitation of stacks of Schwann cell cytoplasmic processes
5. Focal degeneration of axons characterized by condensation of neurofilaments, neurotubules, dense mitochondria, and clusters of dense bodies
6. Focal interlamellar splitting of myelin sheath and formation of "myelin bodies"
7. Complex infolding of inner mesaxonal membranes into axoplasm
8. Decreased number of unmyelinated fibers[246]

The involvement of unmyelinated fibers and smaller myelinated fibers may partially explain the clinical symptoms and the electrophysiological and electromyographic features. Martinez et al.[246] have suggested that the changes in the nerves may be caused by a disturbance in the metabolism of Schwann cells and that the observed toxicity may be due to a metabolite rather than a direct action of chlordecone on nerve fibers or cellular elements of nerves. Despite the impressive list of abnormal observations cited in the previous paragraph, these degenerative changes are all nonspecific and common to other toxic polyneuropathies.

D. Treatment

Most of the chlordecone which is eliminated is excreted in the feces, and high concentrations have been found in the bile.[230,248] An obvious enterohepatic circulation was detected since only about 5 to 10% of the chlordecone secreted in the bile was eliminated in the feces.[249] If the secreted chlordecone could be trapped in the intestinal lumen by a nonabsorbable agent which could bind it, then the elimination of this agent from the body could be accelerated. With this hypothesis, Boylan et al.[230] tried cholestyramine, an anionic exchange resin which they found would bind chlordecone in vitro, and successfully demonstrated in studies with rats that this agent enhanced chlordecone excretion in the feces and dramatically reduced tissue levels of the agent. When cholestyramine was administered to chlordecone-exposed workers, the enhanced fecal excretion of bound agent reduced the blood half-life of the chemical to approximately 80 days from 165 days, thereby demonstrating that cholestyramine offered a practical solution to the detoxification of these victims.[248]

REFERENCES

1. Metcalf, R. L., A century of DDT, *J. Agric. Food Chem.*, 21, 511, 1973.
2. Knipling, E. F., The greater hazard — insects or insecticides, *J. Econ. Entomol.*, 46, 1, 1953.
3. Hayes, W. J., Jr., Pharmacology and toxicology of DDT, in *DDT — The Insecticide Dichlorodiphenyltrichloroethane and Its Significance,* Vol. 2, Muller, P., Ed., Birkhauser Verlag, Basel, 1959, 11.
4. Carson, R., *Silent Spring,* Houghton, Mifflin, Boston, 1962, 368.
5. Rudd, R. L., *Pesticides and the Living Landscape,* University of Wisconsin Press, Madison, 1964, 320.
6. President's Science Advisory Committee, *Use of Pesticides,* U. S. Government Printing Office, Washington, D.C., May 15, 1963, 25.
7. Wurster, C. F., Aldrin and dieldrin, *Environment,* 13, 33, 1971.
8. Hayes, W. J., Jr., *Toxicology of Pesticides,* Williams & Wilkins, Baltimore, 1975, 580.
9. O'Brien, R. D., *Insecticides Action and Metabolism,* Academic Press, New York, 1967, 332.
10. Jager, K. W., An epidemiological and Toxicological study of long-term occupational exposure, in *Aldrin, Dieldrin, Endrin and Telodrin,* Elsevier, New York, 1970, 234.
11. Matsumura, F., *Toxicology of Insecticides,* Plenum Press, New York, 1975, 503.
12. Hrdina, P. D., Singhal, R. L., and Ling, G. M., DDT and related chlorinated hydrocarbon insecticides: pharmacological basis of their toxicity in mammals, *Adv. Pharmacol. Chemother.,* 12, 31, 1975.
13. DeBruin, A., *Biochemical Toxicology of Environmental Agents,* Elsevier, Amsterdam, 1976, 1544.
14. Hayes, W. J., Jr., Dieldrin poisoning in man, *Public Health Rep.,* 72, 1087, 1957.
15. Fairchild, E. G., Ed., *Agricultural Chemicals and Pesticides.* A subfile of the NIOSH Registry of toxic effects of chemical substances, National Institute of Occupational Safety and Health, U.S. Department of Health, Education and Welfare Public Health Service, Cincinnati, July 1977.
16. Hayes, W. J., Jr., Clinical Handbook on Economic Poisons, A Public Health Services Publ., No. 476, U.S. Government Printing Office, Washington, D.C., 1963, 144.
17. **American Medical Association Committee on Pesticides,** Pharmacologic and toxicologic aspects of DDT (Chlorophenothane USP), *J.A.M.A.,* 145, 728, 1951.
18. Tobias, J. M. and Kollros, J. J., Loci of action of DDT in the cockroach (Periplaneta americana), *Biol. Bull.,* 91, 247, 1946.
19. Henderson, G. L. and Woolley, D. E., Ontogenesis of drug-induced tremor in the rat, *J. Pharmacol. Exp. Ther.,* 175, 113, 1970.
20. Smith, M. I. and Stohlman, E. F., The pharmacologic action of 2,2 bis (p-chlorophenyl) 1,1,1 trichloroethane and its estimation in the tissues and body fluids, *Public Health Rep.,* 59, 984, 1944.
21. Philips, F. S. and Gilman, A., Studies on the pharmacology of DDT (2,2 bis-(para-chlorophenyl)-1,1,1 trichloroethane). I. The acute toxicity of DDT following intravenous injection in mammals with observations on the treatment of acute DDT poisoning, *J. Pharmacol. Exp. Ther.,* 86, 213, 1946.
22. Stohlman, E. F. and Lillie, R. D., The effect of DDT on the blood sugar and of glucose administration on the acute and chronic poisoning of DDT in rabbits, *J. Pharmacol. Exp. Ther.,* 93, 351, 1948.
23. Woolley, D. E. and Barron, B. A., Effects of DDT on brain electrical activity in awake, unrestrained rats, *Toxicol. Appl. Pharmacol.,* 12, 440, 1968.
24. St. Omer, V. V. and Ecobichon, D. J., The acute effect of some chlorinated hydrocarbon insecticides on the acetylcholine content of rat brain, *Can. J. Physiol. Pharmacol.,* 49, 79, 1971.
25. Black, W. D. and Ecobichon, D. J., A pharmacodynamic study of DDT in rabbits following acute intravenous administration, *Can. J. Physiol. Pharmacol.,* 49, 45, 1971.
26. Stein, A. A., Comparative toxicology of methoxychlor, in *Pesticides Symposia,* Diechmann, W. B., Ed., Halos, Miami, 1970, 225.
27. Crescitelli, F. and Gilman, A., Electrical manifestations of the cerebellum and cerebral cortex following DDT administration in cats and monkeys, *Am. J. Physiol.,* 147, 127, 1946.
28. Bing, R. J., McNamara, B., and Hopkins, F. H., Studies on the pharmacology of DDT (2,2 bis-parachlorophenyl-1,1,1-trichloroethane). The chronic toxicity of DDT in the dog, *Bull. Johns Hopkins Hosp.,* 78, 308, 1946.
29. Bromiley, R. B. and Bard, P., Tremor and changes in reflex status produced by DDT in decerebrate, decerebrate-decerebellate and spinal animals, *Bull. Johns Hopkins Hosp.,* 84, 414, 1949.
30. Fitzhugh, O. G. and Nelson, A. A., The chronic oral toxicity of DDT (2,2-*bis*(*p*-chlorophenyl) 1,1,1-trichloroethane), *J. Pharmacol. Exp. Ther.,* 89, 18, 1947.
31. Lillie, R. D., Smith, M. I., and Stohlman, E. F., Pathologic action of DDT and certain of its analogs and derivatives, *Arch. Pathol.,* 43, 127, 1947.
32. Pluvinage, R. J. and Heath, J. W., Neural effects of DDT poisoning in cats, *Proc. Soc. Exp. Biol. N.Y.,* 63, 212, 1946.

33. Virgili, R. and Marchiafava, G., Quadri anatomo-patologici nell'intossicazione acuta sperimentale da DDT con particolare referimento al sistema nervosa, *Riv. Malariol.*, 28, 107, 1949.
34. Haymaker, W., Ginzler, A. M., and Ferguson, R. L., The toxic effects of prolonged ingestion of DDT on dogs with special reference to lesions in the brain, *Am. J. Med. Sci.*, 212, 423, 1946.
35. Globus, J. H., DDT (2,2-bis(p-chlorophenyl) 1,1,1-trichloroethane) poisoning, *Trans. Am. Neurol. Assoc.*, 73, 202, 1948.
36. Globus, J. H., DDT (2,2-bis(p-chlorophenyl) 1,1,1-trichloroethane) poisoning, *J. Neuropath.*, 7, 418, 1946.
37. Narahashi, T. and Yamasaki, T., Behaviors of membrane potential in the cockroach giant axons poisoned by DDT, *J. Cell. Comp. Physiol.*, 55, 131, 1960.
38. Narahashi, T. and Yamasaki, T., Mechanism of increase in negative afterpotential by dicophanum (DDT) in the giant axons of the cockroach, *J. Physiol. (London)*, 152, 122, 1960.
39. Roeder, K. D. and Weiant, E. A., The site of action of DDT in the cockroach, *Science*, 103, 304, 1945.
40. Roeder, K. D. and Weiant, E. A., The effect of DDT on sensory and motor structures in the cockroach leg, *J. Cell. Comp. Physiol.*, 32, 175, 1948.
41. Yeager, J. F. and Munson, S. C., Physiological evidence of a site of action of DDT in an insect, *Science*, 102, 305, 1945.
42. Uchida, M., Naka, H., Irie, Y., Fujita, T., and Nakajima, M., Insecticidal and neuroexciting actions of DDT analogs, *Pestic. Biochem. Physiol.*, 4, 451, 1974.
43. Narahashi, T., Effects of insecticides on excitable tissues, *Adv. Insect. Physiol.*, 8, 1, 1971.
44. Lalonde, D. I. V. and Brown, A. W. A., The effect of insecticides on the action potentials of insect nerve, *Can. J. Zool.*, 32, 74, 1954.
45. Eyzaguirre, C. and Lilienthal, J. L., Jr., Veratrinic effects of pentamethylenetetrazol (Metrazol) and 2,2-bis(p-chlorophenyl) 1,1,1-trichloroethane (DDT) on mammalian neuromuscular function, *Proc. Soc. Exp. Biol. Med.*, 70, 272, 1949.
46. van den Bercken, J., Akkermans, L. M. A., and van Langen, R. G., The effect of DDT and dieldrin on skeletal muscle fibers, *Eur. J. Pharmacol.*, 21, 89, 1973.
47. Shankland, D. L., Involvement of spinal cord and peripheral nerves in DDT-poisoning syndrome in albino rats, *Toxicol. Appl. Pharmacol.*, 6, 197, 1964.
48. van den Bercken, J., The effect of DDT and dieldrin on myelinated nerve fibers, *Eur. J. Pharmacol.*, 20, 205, 1972.
49. Akkermans, L. M. A., van den Bercken, J., and Versluijs-Helder, M., Comparative effects of DDT, allethrin, dieldrin and aldrin-transdiol on sense organs of *Xenopus laevis*, *Pestic. Biochem. Physiol.*, 5, 451, 1975.
50. van den Bercken, J. and Akkermans, L. M. A., Negative temperature coefficient of the action of DDT in a sense organ, *Eur. J. Pharmacol.*, 16, 241, 1971.
51. Isaacson, P. A., Action of DDT on the peripheral nervous system of the grass frog *(Rana pipiens)*, *Curr. Sci.*, 18, 530, 1968.
52. Tripod, J., Le point d'attaque du DDT chez la grenouille, *Arch. Int. Pharmacodyn.*, 74, 343, 1947.
53. Pollock, G. H. and Wang, R. I. H., Synergistic actions of carbon dioxide with DDT in the central nervous system, *Science*, 117, 596, 1953.
54. Woolley, D. E., Toxicological and pharmacological studies of visual and auditory potentials evoked in the cerebellum of the rat, *Proc. West. Pharmacol. Soc.*, 11, 69, 1968.
55. Woolley, D. E., Some aspects of the neurophysiological basis of insecticide action, *Fed. Proc.*, 35, 2610, 1976.
56. Joy, R. M., Electrical correlates of preconvulsive and convulsive doses of chlorinated hydrocarbon insecticides in the CNS, *Neuropharmacology*, 12, 63, 1973.
57. Joy, R. M., Convulsive properties of chlorinated hydrocarbon insecticides in the cat central nervous system, *Toxicol. Appl. Pharmacol.*, 35, 95, 1976.
58. Farkas, I., Desi, I., and Kemeny, T., The effect of DDT in the diet on the resting and loading electrocortigram record, *Toxicol. Appl. Pharmacol.*, 12, 518, 1968.
59. Desi, I., Farkas, I., and Kemeny, T., The early effects of low DDT doses on the nervous system in animal experiments, *Experientia*, 24, 51, 1968.
60. Santolucito, J. A. and Morrison, G., EEG of Rhesus monkeys following prolonged low-level feeding of pesticides, *Toxicol. Appl. Pharmacol.*, 19, 147, 1971.
61. Daigneault, E. A., Correlation of prolonged insecticide exposure and the activity of the inhibition of cochlea N1 potential, *Toxicol. Appl. Pharmacol.*, 21, 495, 1972.
62. Ghazal, A., Chlorphenothane (DDT) and glutamine in pentylenetetrazole (Cardiazole) convulsions in the rat, *Toxicol. Appl. Pharmacol.*, 6, 627, 1964.
63. Enna, S. J. and Tuttle, W. W., Effect of chronic low doses of DDT on metrazol induced convulsions, *Fed. Proc. Fed. Am. Soc. Exp. Biol.*, 26, 428, 1967.

64. Al-Hachim, G. M. and Fink, G. B., Effect of DDT or parathion on the minimal electroshock seizure threshold of offspring from DDT- or parathion-treated mothers, *Psychopharmacologia,* 13, 408, 1968.
65. Al-Hachim, G. M. and Fink, G. B., Effect of DDT or parathion on audiogenic seizures of offspring from DDT- or parathion-treated mothers, *Psychol. Rep.,* 20, 1183, 1967.
66. Woolley, D. E., Effects of DDT and of drug-DDT interactions on electroshock seizures in the rat, *Toxicol. Appl. Pharmacol.,* 16, 521, 1970.
67. Woolley, D. E., Effects of acute and chronic exposure to DDT and of DDT-drug interactions on experimental seizure responses, in *Pesticides Symposia,* Diechmann, W. B., Ed., Halos, Miami, 1970, 21.
68. Sobotka, T. J., Behavioral effects of low doses of DDT, *Proc. Soc. Exp. Biol. Med.,* 137, 952, 1971.
69. Khairy, M., Changes in behaviour associated with a nervous system poison (DDT), *Q., J. Exp. Psychol.,* 11, 84, 1959.
70. Al-Hachim, G. M. and Fink, G. B., Effect of DDT or parathion on condition avoidance response of offspring from DDT or parathion-treated mothers, *Psychopharmacologia,* 12, 424, 1968.
71. Al-Hachim, G. M. and Fink, G. B., Effect of DDT or parathion on open-field behavior of offspring from DDT- or parathion-treated mothers, *Psychol. Rep.,* 22, 1193, 1968.
72. Craig, G. R. and Ogilvie, D. M., Alteration of T-maze performance in mice exposed to DDT during pregnancy and lactation, *Environ. Physiol. Biochem.,* 4, 189, 1974.
73. Welsh, J. H. and Gordon, H. T., The mode of action of certain insecticides on the arthropod nerve axon, *J. Cell. Comp. Physiol.,* 30, 147, 1947.
74. Shanes, A. M., Electrical phenomena in nerve. II. Crab nerve, *J. Gen. Physiol.,* 33, 75, 1949.
75. Narahashi, T. and Haas, G. H., DDT: interaction with nerve membrane conductance changes, *Science,* 157, 1438, 1967.
76. Narahashi, T. and Haas, H. G., Interaction of DDT with the components of lobster nerve membrane conductance, *J. Gen. Physiol.,* 51, 177, 1968.
77. Pichon, Y., Effets du DDT sur la fibre nerveuse isoléc d'insecte. Étude en courant et en voltage imposés, *J. Physiol. (Paris),* (Suppl. 1), 162, 1969.
78. Shanes, A. M., Electrical phenomena in nerve III. Frog sciatic nerve, *J. Cell. Comp. Physiol.,* 38, 17, 1951.
79. Hille, B., Pharmacological modifications of the sodium channels of frog nerve, *J. Gen. Physiol.,* 51, 199, 1968.
80. Arhem, P. and Frankenhaeuser, B., DDT and related substances: effects on permeability properties of myelinated Xenopus nerve-fibre. Potential clamp analysis, *Acta. Physiol. Scand.,* 91, 502, 1974.
81. Arhem, P., Frankenhaeuser, B., Gothe, R., and O'Bryan, P., DDT and related substances on myelinated nerve: effects on permeability properties, *Acta. Physiol. Scand.,* 91, 130, 1974.
82. Dubois, J. M. and Bergman, C., Asymmetrical currents and sodium current in Ranvier nodes exposed to DDT, *Nature (London),* 266, 741, 1977.
83. Shrager, P. G., Macey, R. I., and Strickholm, A., Internal perfusion of crayfish giant axons: action of tannic acid, DDT and TEA, *J. Cell. Physiol.,* 74, 77, 1969.
84. Schneider, R. P., Mechanism of inhibition of rat brain (Na + K)-adenosine triphosphatase by 2,2-bis(p-chlorophenyl)-1,1,1-trichloroethane (DDT), *Biochem. Pharmacol.,* 24, 939, 1975.
85. Hilton, B. D. and O'Brien, R. D., Antagonism by DDT of the effect of valinomycin on a synthetic membrane, *Science,* 168, 841, 1970.
86. Hilton, B. D. and O'Brien, R. D., The effects of DDT and its analogs upon lecithin and other monolayers, *Pestic. Biochem. Physiol.,* 3, 206, 1973.
87. Hilton, B. D., Bratkowski, T. A., Yamada, M., Narahashi T., and O'Brien, R. D., The effects of DDT analogs upon potassium conductance in synthetic membranes, *Pestic. Biochem. Physiol.,* 3, 14, 1973.
88. Barnola, F. V., Camejo, G., and Villegas, R., Ionic channels and nerve membrane lipoproteins: DDT-nerve membrane interaction, *Int. J. Neurosci.,* 1, 309, 1971.
89. Mullins, L. J., Structure-toxicity in hexachlorocyclohexane isomers, *Science,* 122, 118, 1955.
90. Mullins, L. J., The use of models of the cell membrane in determining the mechanism of drug action, in *A Guide to Molecular Pharmacology-Toxicology Part I,* Featherstone, R. M., Ed., Marcel Dekker, New York, 1973, 1.
91. Holan, G., New halocyclopropane insecticides and the mode of action of DDT, *Nature (London),* 221, 1025, 1969.
92. St. Omer, V., Investigations into mechanisms responsible for seizures induced by chlorinated hydrocarbon insecticides: the role of brain ammonia and glutamine in convulsions in the rat and cockerel, *J. Neurochem.,* 18, 365, 1971.
93. Kar, P. P. and Matin, M. A., Possible role of cerebral amino acids in acute neurotoxic effects of DDT in mice, *Eur. J. Pharmacol.,* 25, 36, 1974.

94. Matin, M. A., Kar, P. P., and Anand, M., Modification of p,p'-DDT induced convulsions by changes in the level of cerebral γ-aminobutyric acid in mice, *J. Neurochem.*, 27, 979, 1976.
95. Hrdina, P. D., Singhal, R. L., Peters, D. A. V., and Ling, G. M., Role of brain acetylcholine and dopamine in acute neurotoxic effects of DDT, *Eur. J. Pharmacol.*, 15, 379, 1971.
96. Hrdina, P. D., Maneekjee, A., Kacew, S., Peters, D. A. V., and Singhal, R. L., Acetylcholine and dopamine in rat striatum during acute DDT poisoning, *Proc. Can. Fed. Biol. Soc.*, 14, 67, 1971.
97. Hrdina, P. D., Singhal, R. L., Peters, D. A. V., and Ling, G. M., Some neurochemical alterations during acute DDT poisoning, *Toxicol. Appl. Pharmacol.*, 25, 276, 1973.
98. Hrdina, P. D, Singhal, R. L., Peters, D. A. V., and Ling, G. M., Comparison of the chronic effects of p,p'-DDT and alpha-chlordane on brain amines, *Eur. J. Pharmacol.*, 20, 114, 1972.
99. Peters, D. A. V., Hrdina, P. D., Singhal, R. L., and Ling, G. M., The role of brain serotonin in DDT-induced hyperpyrexia, *J. Neurochem.*, 19, 1131, 1972.
100. Hwang, E. C. and Van Woert, M. H., DDT-induced neurotoxic syndrome: animal model of myoclonus, *Pharmacologist*, 19, 199, 1977.
101. Matsumura, F. and Patil, K. C., Adenosine triphosphatase sensitive to DDT in synapses of rat brain, *Science*, 166, 121, 1969.
102. Koch, R. B., Cutkomp, L. K., and Do, F. M., Chlorinated hydrocarbon insecticide inhibition of cockroach and honey bee ATPases, *Life Sci.*, 8, 289, 1969.
103. Koch, R. B., Chlorinated hydrocarbon insecticides: inhibition of rabbit brain ATPase activities, *J. Neurochem.*, 16, 269, 1969.
104. Cutkomp, L. K., Yap, H. H., Cheng, E. Y., and Koch, R. B., ATPase activity in fish tissue homogenates and inhibitory effects of DDT and related compounds, *Chem.-Biol. Interact.*, 3, 439, 1971.
105. Matsumura, F. and Narahashi, T., ATPase inhibition and electrophysiological change caused by DDT and related neuroactive agents in lobster nerve, *Biochem. Pharmacol.*, 20, 825, 1971.
106. Doherty, J. D. and Matsumura, F., DDT effects on certain ATP related systems in the peripheral nervous system of the lobster *Homarus americanus*, *Pestic. Biochem. Physiol.*, 5, 242, 1975.
107. Joy, R. M., Alteration of sensory and motor evoked responses by dieldrin, *Neuropharmacology*, 13, 93, 1974.
108. Esplin, D. W. and Zablocka-Esplin, B., Mechanisms of action of convulsants, in *Basic Mechanisms of the Epilepsies*, Jasper, H. H., Ward, A. A., and Pope, A., Eds., Little, Brown, Boston, 1969, 167.
109. Pellmar, T. C. and Wilson, W. A., Synaptic mechanism of pentylenetetrazole: selectivity for chloride conductance, *Science*, 197, 912, 1977.
110. Joy, R. M. and Wong, D. L., Evoked and spontaneous corticofugal multiple unit activity during different phases of metrazol infusion in cats, *Neuropharmacology*, 17, 281, 1978.
111. Velbinger, H. H., Zur Frage der 'DDT'-toxizität fur Menschen, *Dtsch. Gesundheitswes.*, 2(11), 335, 1947.
112. Velbinger, H. H., Beitrag zur tozikologie des 'DDT'-wirkstoffes Dichlordiphenyltrichlormethylmethan, *Pharmazie*, 2, 268, 1947.
113. Garrett, R. M., Toxicity of DDT for man, *J. Med. Assoc. State Ala.*, 17, 74, 1947.
114. Cunningham, R. E. and Hill, F. S., Convulsions and deafness following ingestion of DDT, *Pediatrics*, 9, 745, 1952.
115. Cameron, G. R. and Burgess, F., The toxicity of 2,2-*bis* (*p*-chlorophenyl) 1,1,1-trichloroethane (DDT), *Br. Med. J.*, 1, 865, 1945.
116. Neal, P. A., von Oettingen, W. F., Smith, W. W., Malmo, R. B., Dunn, R. C., Moran, H. E., Sweeney, T. R., Armstrong, D. W., and White, W. C., Toxicity and potential dangers of aerosols and dusting powders containing DDT, *U.S. Suppl. Publ. Health Rep. S.*, 177, 1, 1944.
117. Jenkins, R. B. and Toole, J. F., Polyneuropathy following exposure to insecticides, *Arch. Int. Med.* 113, 691, 1964.
118. Mayersdorf, A. and Israeli, R., Toxic effects of chlorinated hydrocarbon insecticides on the human electroencephalogram, *Arch. Environ. Health*, 28, 159, 1974.
119. Wigglesworth, V. B., A case of DDT poisoning in man, *Br. Med. J.*, 1, 517, 1945.
120. Case, R. A. M., Toxic effects of 2,2-*bis*-(p-chlorophenyl)-1,1,1-trichloroethane (DDT) in man, *Br. Med. J.*, 2, 842, 1945.
121. Campbell, A. M. G., Neurological complications associated with insecticides and fungicides, *Br. Med. J.*, 2, 415, 1952.
122. Stone, T. T. and Gladstone, L., DDT (dichlorodiphenyltrichloroethane), *J.A.M.A.*, 145, 1342, 1951.
123. Onifer, T. M. and Whisnant, J. P., Cerebellar ataxia and neuronitis after exposure to DDT and Lindane, *Proc. Mayo Clinic*, 32, 67, 1957.
124. McLeod, J. G., Peripheral neuropathy caused by drugs and toxic substances, *Aust. N. Z. J. Med.*, 3, 268, 1971.

125. Timofeyeva, N. T., (A case of poisoning from combined action of toxic compounds), *Vrach. Delo*, 4, 140, 1971 (in Russian).
126. Garcin, R., Ginsbourg, M., Godlewski, St., and Emile, J., Intoxication par le D.D.T.: syndrome méningo-encéphalique aigu regressif avec decharges cloniques diffuses et mouvements désordonnés "en salves" des globes oculaires, *Rev. Neurol.*, 113, 559, 1965.
127. Schuettmann, W., Klinishe Beobachtungen zur chronischen Toxizitaet der Chlorkohlenwasserstoffpestizide (Clinical observations on the chronic toxicity of organochlorine pesticides), *Z. Gesamte Hyg.*, 17, 12, 1972.
128. Kolyada, I. S. and Mikhal'chenkova, O. F., (Acute organochlorine pesticide poisoning), *Gig. Tr. Prof. Zabol.*, 17(5), 43, 1973.
129. Biskind, M. S., DDT poisoning a serious public health hazard, *Am. J. Dig. Dis.*, 16, 73, 1949.
130. Biskind, M. S., DDT poisoning and the elusive "virus X": a new cause for gastro-enteritis, *Am. J. Dig. Dis.*, 16, 79, 1949.
131. Biskind, M. S., DDT poisoning and X disease in cattle, *J. Am. Vet. Med. Assoc.*, 114, 20, 1949.
132. Biskind, M. S., Public health aspects of the new insecticides, *Am. J. Dig. Dis.*, 20, 331, 1953.
133. Biskind, M. S. and Bieber, I., DDT poisoning — a new syndrome with neuropsychiatric manifestations, *Am. J. Psychother.*, 3, 261, 1949.
134. Plichet, A., Le D.D.T. serait-il responsable de certaines gastro-enterites?, *Presse Med.*, 57, 1121, 1949.
135. Busvine, J. R., Houseflies resistant to a group of chlorinated hydrocarbon insecticides, *Nature (London)*, 174, 783, 1954.
136. A. M. A. Committee on Pesticides, Pharmacologic properties of toxaphene, a chlorinated hydrocarbon insecticide, *J.A.M.A.*, 149, 1135, 1952.
137. Worden, A. N., Toxicity of telodrin, *Toxicol. Appl. Pharmacol.*, 14, 556, 1969.
138. Solomon, L. M., Gamma benzene hexachloride toxicity. A review, *Arch. Dermatol.* 113, 353, 1977.
139. Gupta, P. K., Endosulfan-induced neurotoxicity in rats and mice, *Bull. Environ. Contam. Toxicol.*, 15, 708, 1976.
140. Roan, C. C. and Hopkins, T. L., Mode of action of insecticides, *Ann. Rev. of Entomol.*, 6, 333, 1961.
141. Eckert, J. E., Toxicity of some of the newer chemicals to the honeybee, *J. Econ. Entomol.*, 41, 487, 1948.
142. Coble, Y., Hildebrandt, P., Davis, J., Raasch, F., and Curley, A., Acute endrin poisoning, *J.A.M.A.*, 202, 489, 1967.
143. Walsh, G. M. and Fink, G. B., Comparative toxicity and distribution of endrin and dieldrin after intravenous administration in mice, *Toxicol. Appl. Pharmacol.*, 23, 408, 1972.
144. Gupta, P. K. and Chandra, S. V., The toxicity of endosulfan in rabbits, *Bull. Env. Contam. Toxicol.*, 14, 513, 1975.
145. Hulth, L., Larsson, M., Carlsson, R., and Kihlstrom, J. E., Convulsive action of small single oral doses of the insecticide lindane, *Bull. Env. Contam. Toxicol.*, 16, 133, 1976.
146. Hanig, J. P., Yoder, P. D., and Krop, A., Convulsions in weanling rabbits after a single topical application of 1% lindane, *Toxicol. Appl. Pharmacol.*, 38, 463, 1976.
147. Keplinger, M. L., Diechmann, W. B., and Sala, F., Effects of combinations of pesticides on reproduction in mice, in *Pesticide Symposia*, Diechmann, W. B., Ed., Halos, Miami, 1970, 125.
148. Fitzhugh, O. G., Nelson, A. A., and Quaife, M. L., Chronic oral toxicity of aldrin and dieldrin in rats and dogs, *Fed. Cosmet. Toxicol.*, 2, 551, 1964.
149. Gralla, E. J., Fleischman, R. W., Luthra, Y. K., Hagopian, M., Baker, J. R., Esber, H., and Marcus, W., Toxic effects of hexachlorobenzene after daily administration to beagle dogs for one year, *Toxicol. Appl. Pharmacol.*, 40, 227, 1977.
150. Walker, A. I. T., Stevenson, D. E., Robinson, J., Thorpe, E., and Roberts, M., The toxicology and pharmacodynamics of dieldrin (HEOD): two-year oral exposures of rats and dogs, *Toxicol. Appl. Pharmacol.*, 15, 345, 1969.
151. Treon, J. F. and Cleveland, F. P., Toxicity of certain chlorinated hydrocarbon insecticides for laboratory animals, with special reference to aldrin and dieldrin, *J. Agric. Food Chem.*, 3, 402, 1955.
152. Virgo, B. B. and Bellward, G. D., Effects of dietary dieldrin on offspring viability, maternal behavior, and milk production in the mouse, *Res. Comm. Chem. Pathol. Pharmacol.*, 17, 399, 1977.
153. Gowdey, C. W., Graham, A. R., Seguin, J. J., Stavraky, G. W., and Waud, R. A., A study of the pharmacological properties of the insecticide aldrin, *Can. J. Med. Sci.*, 30, 520, 1952.
154. Gowdey, C. W., Graham, A. R., Seguin, J. J., and Stavraky, G. W., The pharmacological properties of the insecticide dieldrin, *Can. J. Biochem. Physiol.*, 32, 498, 1954.
155. Gowdey, C. W. and Stavraky, G. W., A study of the autonomic manifestations seen in acute aldrin and dieldrin poisoning, *Can. J. Biochem. Physiol.*, 33, 272, 1955.
156. Reins, D. A., Holmes, D. D., and Hinshaw, L. B., Acute and chronic effects of the insecticide endrin on renal function and renal hemodynamics, *Can. J. Physiol. Pharmacol.*, 42, 599, 1964.

157. Emerson, T. E., Jr., Brake, C. M., and Hinshaw, L. B., Cardiovascular effects of the insecticide endrin, *Can. J. Physiol. Pharmacol.*, 42, 41, 1964.
158. Fink, G. B., Interaction of lead, DDT and dieldrin on motor coordination and neuromuscular function, *Pharmacologist,* 16, 207, 1974.
159. Shankland, D. L. and Schroeder, M. E., Pharmacological evidence for a discrete neurotoxic action of dieldrin (HEOD) in the American cockroach, *Periplaneta americana* (L), *Pestic. Biochem. Physiol.*, 3, 77, 1973.
160. van den Bercken J. and Narahashi, T., Effects of aldrin-transdiol — a metabolite of the insecticide dieldrin — on nerve membrane, *Eur. J. Pharmacol.*, 27, 255, 1974.
161. Natoff, I. L. and Reiff, B., The effect of dieldrin (HEOD) on chronaxie and convulsion thresholds in rats and mice, *Br. J. Pharmacol. Chemother.*, 31, 197, 1967.
162. Akkermans, L. M. A., van den Bercken, J., van der Zalm, J. M., and van Straaten, H. W. M., Effects of dieldrin (HEOD) and some of its metabolites on synaptic transmission in the frog motor end-plate, *Pestic. Biochem. Physiol.*, 4, 313, 1974.
163. van Genderen, H. and Ibrahim, T. M., The action of dieldrin on striated muscle of the rat, *Acta Physiol. Pharmacol. Neerl.*, 13, 193, 1965.
164. Santolucito, J. A. and Whitcomb, E., Mechanical response of skeletal muscle following oral administration of pesticides, *Toxicol. Appl. Pharmacol.*, 20, 66, 1971.
165. Khairy, M., Effects of chronic dieldrin ingestion on the muscular efficiency of rats, *Br. J. Ind. Med.*, 17, 146, 1960.
166. Wang, C. M. and Matsumura, F., Relationship between the neurotoxicity and *in vivo* toxicity of certain cyclodiene insecticides in the German cockroach, *J. Econ. Entomol.*, 63, 1731, 1970.
167. Wang, C. M., Narahashi, T., and Yamada, M., The neurotoxic action of dieldrin and its derivatives in the cockroach, *Pestic. Biochem. Physiol.*, 1, 84, 1971.
168. Schroeder, M. E., Shankland, D. L., and Hollingworth, R. M., The effects of dieldrin and isomeric aldrin diols on synaptic transmission in the American cockroach and their relevance to the dieldrin poisoning syndrome, *Pestic. Biochem. Physiol.*, 7, 403, 1977.
169. Akkermans, L. M. A., van den Bercken, J., and Versluijs-Helder, M., Excitatory and depressant effects of dieldrin and aldrin-transdiol in the spinal cord of the toad *(Xenopus laevis) Eur. J. Pharmacol.*, 34, 133, 1975.
170. Akkermans, L. M. A., van den Bercken, J., and van der Zalm, J. M., Effects of aldrin-transdiol on neuromuscular facilitation and depression, *Eur. J. Pharmacol.*, 31, 166, 1975.
171. Akkermans, L. M. A., van der Zalm, J. M., and van den Bercken, J., Is aldrin-transdiol the active form of the insecticide dieldrin?, *Arch. Int. Pharmacodyn.*, 206, 363, 1973.
172. Korte, F. and Arent, H., Metabolism of insecticides: IX (1). Isolation and identification of dieldrin metabolites from urine of rabbits after oral administration of dieldrin-^{14}C, *Life Sci.*, 4, 2017, 1965.
173. Joy, R. M., Contrasting actions of dieldrin and aldrin-transdiol, its metabolite, on cat CNS function, *Toxicol. Appl. Pharmacol.*, 42, 137, 1977.
174. Van Gelder, G. A., Sandler, B. E., Buck, W. B., and Karas, G. G., Convulsive seizures in dieldrin exposed sheep during photic stimulation, *Psychol. Rep.*, 24, 502, 1969.
175. Hyde, K. M. and Falkenberg, R. L., Neuroelectrical disturbance as indicator of chronic chlordane toxicity, *Toxicol. Appl. Pharmacol.*, 37, 499, 1976.
176. Burchfiel, J. L., Duffy, F. H., and Sim, V. M., Persistent effects of sarin and dieldrin upon the primate electroencephalogram, *Toxicol. Appl. Pharmacol.*, 35, 365, 1976.
177. Joy, R. M., The alteration by dieldrin of cortical excitability conditioned by sensory stimuli, *Toxicol. Appl. Pharmacol.*, 38, 357, 1976.
178. Al-Hachim, G. M., Effect of aldrin on the condition avoidance response and electroshock seizure threshold of offspring from aldrin-treated mothers, *Psychopharmacologia*, 21, 370, 1971.
179. Desi, I., Neurotoxicological effect of small quantities of lindane, *Int. Arch. Arbeitsmed.*, 33, 153, 1974.
180. van Gelder, G. A., Sandler, B. E., Buck, B., Maland, J. B., and Karas, G. G., Behavioral and electrophysiological effects of dieldrin in sheep, *Ind. Med.*, 38, 64, 1969.
181. Schnorr, J. K., Effects of dieldrin on learning and retention of a visual discrimination task in sheep, *Diss. Abstr. Int. B*, 33 (8), 3395, 1973.
182. Elsberry, D. D., Effect of dieldrin on auditory detection behavior and central auditory mechanisms in sheep, *Diss. Abstr. Int. B*, 33 (8), 4028, 1973.
183. Smith, R. M., Cunningham, W. L., and Van Gelder, G. A., Dieldrin toxicity and successive discrimination reversal in squirrel monkeys *(Saimiri sciureus)*, *J. Toxicol. Environ. Health*, 1, 737, 1976.
184. Joy, R. M., Comparative effects of convulsants on the antidromic cortical response to pyramidal tract stimulation, *Neuropharmacology*, 14, 869, 1975.
185. Loscher, W. and Frey, H.-H., Effect of convulsant and anticonvulsant agents on level and metabolism of γ-aminobutyric acid in mouse brain, Naunyn-Schmiedeberg's, *Arch. Pharmacol.*, 296, 263, 1977.

186. Hayes, A. G., Gartside, I. B., and Straughan, D. W., Effects of four convulsants on the time course of presynaptic inhibition and its relation to seizure activity, *Neuropharmacology,* 16, 725, 1977.
187. MacDonald, R. L. and Barker, J. L., Specific antagonism of GABA-mediated postsynaptic inhibition in cultured mammalian spinal cord neurons: a common mode of convulsant action, *Neurology,* 28, 325, 1978.
188. Hill, R. G., Simmonds, M. A., and Straughan, D. W., A comparative study of some convulsant substances as γ-aminobutyric acid antagonists in the feline cerebral cortex, *Br. J. Pharmacol.,* 49, 37, 1973.
189. Straughan, D. W., Convulsant drugs: amino acid antagonism and central inhibition, *Neuropharmacology,* 13, 495, 1974.
190. Soloway, S. B., Correlation between biological activity and molecular structure of the cyclodiene insecticides, in *Advances in Pest Control Research,* Vol. 6, Metcalf, R. L., Ed., John Wiley & Sons, New York, 1965, 85.
191. Burt, P. E., Biophysical aspects of nervous activity in relation to studies on the mode of action of insecticides, *Pestic. Sci.,* 1, 88, 1970.
192. Waseda, Y., Studies on the mechanism of toxic actions of organochloric pesticides, *Nippon Hoigaku Zasshi,* 25, 64, 1971.
193. Hokin, M. R. and Brown, D. F., Inhibition by gamma-hexachlorocyclohexane of acetylcholine-stimulated phosphatidylinositol synthesis in cerebral cortex slices and of phosphatidic acid-inositol transferase in cerebral cortex particulate fractions, *J. Neurochem.,* 16, 475, 1969.
194. Clausen, J. and Konat, G., Enzymic and behavioral changes induced in mice fed polychlorinated biocides followed by starvation, *Experientia,* 28, 902, 1972.
195. Sharma, R. P., Brain biogenic amines: depletion by chronic dieldrin exposure, *Life Sci.,* 13, 1245, 1973.
196. Hrdina, P. D., Peters, D. A. V., and Singhal, R. L., Role of noradrenaline, 5-hydroxytryptamine and acetylcholine in the hypothermic and convulsive effects of alpha-chlordane in rats, *Eur. J. Pharmacol.,* 26, 306, 1974.
197. Hrdina, P. D., Singhal, R. L., and Peters, D. A. V., Changes in brain biogenic amines and body temperature after cyclodiene insecticides, *Toxicol. Appl. Pharmacol.,* 29, 119, 1974.
198. Peters, D. A. V., Hrdina, P. D., Singhal, R. L., and Mazurkiewicz-Kwilecki, I., Comparison of the acute and chronic effects of chlordane and DDT on brain biogenic amines, *Fed. Proc. Fed. Am. Soc. Exp. Biol.,* 32, 320, 1973.
199. Sharma, R. P., Influence of dieldrin on serotonin turnover and 5-hydroxyindole acetic acid efflux in mouse brain, *Life Sci.,* 19, 537, 1976.
200. Wagner, S. R. and Greene, F. E., Effect of acute and chronic dieldrin exposure on brain biogenic amines of male and female rats, *Toxicol. Appl. Pharmacol.,* 29, 119, 1974.
201. Kohli, K. K., Chandrasekaran, V. P., and Venkitasubramanian, T. A., Stimulation of serotonin metabolism by dieldrin, *J. Neurochem.,* 28, 1397, 1977.
202. Willhite, C. C. and Sharma, R. P., Effect of acute dieldrin exposure on brain serotonin, 5-hydroxyindole acetic acid and monoamine oxidase in three animal species, *Fed. Proc.,* 36 (3), 356, 1977.
203. Stone, W. E., Action of convulsants: neurochemical aspects, in *Basic Mechanisms of the Epilepsies,* Jasper, H. H., Ward, A. A., and Pope, A., Eds., Little, Brown, Boston, 1969, 184.
204. Tower, D. B., *Neurochemistry of Epilepsy,* Charles C Thomas, Springfield, Ill., 1060, 335.
205. Park, K. S. and Bruce, W. N., The determination of the water solubility of aldrin, dieldrin, heptachlor and heptachlor epoxide, *J. Econ. Entomol.,* 61, 770, 1968.
206. Yap, H. H., Desaiah, D., Cutcomp, L. K., and Koch, R. B., *In vitro* inhibition of fish brain ATPase activity by cyclodiene insecticides and related compounds, *Bull. Env. Contam. Toxicol.,* 14, 163, 1975.
207. Desaiah, D. and Koch, R. B., Inhibition of fish brain ATPases by aldrin-transdiol, aldrin, dieldrin and photodieldrin, *Biochem. Biophys. Res. Commun.,* 64, 13, 1975.
208. Spiotta, E. J., Aldrin poisoning in man, *Arch. Ind. Hyg. Occup. Med.,* 4, 560, 1951.
209. Starr, H. G. and Clifford, N. J., Acute lindane intoxication — a case study, *Arch. Environ. Health,* 25, 374, 1972.
210. Garrettson, L. K. and Curley, A., Dieldrin — studies in a poisoned child, *Arch. Environ. Health,* 19, 814, 1969.
211. Davies, G. M. and Lewis, I., Outbreak of food poisoning from bread made of chemically contaminated flour, *Br. Med. J.,* 2, 393, 1956.
212. Weeks, D. E., Endrin food poisoning. A report on four outbreaks caused by two separate shipments of endrin-contaminated flour, *Bull. W.H.O.,* 37, 499, 1967.
213. Bell, A., Aldrin poisoning: a case report, *Med. J. Aust.,* 47, 698, 1960.
214. Kuebrich, W. and Urban, I., (Endrin poisoning — neurophysiology of epileptic seizures), *Z. Aerztl. Fortbild.,* 67, 1076, 1973 (in German).

215. Tosa, Y., Yasugi, N., Okada, N., Nagami, H., and Seki, R., (A case of survival in endrin poisoning), *Nippon Noson Igakkai Zasshi,* 19(4), 370, 1971 (in Japanese).
216. Reddy, D. B., Edward, V. D., Abraham, G. J. S., and Rao, K. V., Fatal endrin poisoning, *J. Indian Med. Assoc.,* 46, 121, 1966.
217. Eskenasy, J. J., Status epilepticus by dichlorodiphenyltrichloroethane and hexachlorocyclohexane poisoning, *Rev. Roum. Neurol.,* 9, 435, 1972.
218. Kazantzis, G., McLaughlin, A. I. G., and Prior, P. F., Poisoning in industrial workers by the insecticide aldrin, *Br. J. Ind. Med.,* 21, 46, 1964.
219. Ely, T. S., MacFarlane, J. W., Galen, W. P., and Hine, C. H., Convulsions in thiodan workers, *J. Occup. Med.,* 9, 35, 1967.
220. Gupta, P. C., Neurotoxicity of chronic chlorinated hydrocarbon insecticide poisoning — a clinical and electroencephalographic study in man, *Indian J. Med. Res.,* 63, 601, 1975.
221. Hoogendam, I., Versteeg, J. P. J., and Vlieger, M., Electroencephalograms in insecticide toxicity, *Arch. Environ. Health,* 4, 92, 1962.
222. Angle, C. R., McIntire, M. S., and Meile, R. L., Neurologic sequelae of poisoning in children, *J. Pediatr.,* 73, 531, 1968.
223. Muminov, A. I., (Functional condition of the hearing organ in persons with pesticide intoxication), *Vestn. Otorinolaringol.,* 34, 33, 1972.
224. Jacobziner, H. and Raybin, H. W., Poisoning by insecticide (endrin), *N. Y. State J. Med.,* 59, 2017, 1959.
225. Cannon, S. B., Veasey, J. M., Jackson, R. S., Burse, V. W., Hayes, C., Straub, W. E., Landrigan, P. J., and Liddle, J. A., Epidemic Kepone poisoning in chemical workers, *Am. J. Epidemiol.,* 17, 529, 1978.
226. Taylor, J. R., Selhorst, J. B., and Calabrese, V. P., Chlordecone, in *Experimental and Clinical Neurotoxicology,* Spencer, P. S. and Schaumberg, H. H., Eds., Williams & Wilkins, Baltimore, 1980, chap. 28.
227. Martin, H., Ed., *Pesticide Manual,* Worcester British Crop Protection Council, 1974, 101.
228. Stecher, P. G., Ed., *The Merck Index. An Encyclopedia of Chemicals and Drugs,* 8th ed., Merck, Rahway, N. J., 1968.
229. Huber, J., Some physiologic effects of the insecticide Kepone in the laboratory mouse, *Toxicol. Appl. Pharmacol.,* 7, 516, 1965.
230. Boylan, J. J., Egle, J. L. and Guzelian, P. S., Cholestyramine: use as a new therapeutic approach for chlordecone (Kepone) poisoning, *Science,* 199, 893, 1978.
231. Fabacher, D. L. and Hodgson, E., Induction of hepatic mixed function oxidase enzymes in adult and neonatal mice by Kepone and mirex, *Toxicol. Appl. Pharmacol.,* 38, 71, 1976.
232. Eroschenko, V. P. and Wilson, W. O., Cellular changes in the gonads, livers and adrenal glands of Japanese quail as affected by the insecticide Kepone, *Toxicol. Appl. Pharmacol.,* 31, 491, 1975.
233. Reuber, M. D., Carcinogenicity of Kepone, *J. Toxicol. Environ. Health,* 4, 895, 1978.
234. Egle, J. L., Jr., Guzelain, P. S., and Borzelleca, J. F., Time course of the acute toxic effects of sublethal doses of chlordane (Kepone), *Toxicol. Appl. Pharmacol.,* 48, 533, 1979.
235. Larson, P. S., Egle, J. L., Jr., Hennigar, G. R., Lane, R. W., and Borzelleca, J. F., Acute, subacute and chronic toxicity of chlordecone, *Toxicol. Appl. Pharmacol.,* 48, 29, 1979.
236. Good, E. E., Ware, G. W., and Miller, D. F., Effects of insecticides on reproduction in the laboratory mouse. I. Kepone, *J. Econ. Entomol.,* 53, 751, 1965.
237. Naber, E. and Ware, G., Effect of Kepone and mirex on reproductive performance in the laying hen, *Poultry Sci.,* 44, 875, 1965.
238. Eroschenko, V. P., Alterations in the testes of Japanese quail during and after the ingestion of the insecticide Kepone, *Toxicol. Appl. Pharmacol.,* 43, 535, 1978.
239. Taylor, J. R., Selhorst, J. B., Houff, S., and Martine A. J., Chlordecone intoxication in man, *Neurology,* 28, 626, 1978.
240. Baggett, J. McC., Thureson-Klein, A., and Klein, R. L., Effects of chlordecone on the adrenal medulla of the rat, *Toxicol. Appl. Pharmacol.,* 52, 313, 1980.
241. Desaiah, D. and Koch, R. B., Inhibition of ATPase activity in channel catfish brain by Kepone and its reduction product, *Bull. Environ. Contam. Toxicol.,* 13, 153, 1975.
242. Desaiah, D., Ho, I. K., and Mehendale, H., Effect of Kepone and mirex on mitochondrial Mg-ATPase activity in rat liver, *Toxicol. Appl. Pharmacol.,* 39, 219, 1977.
243. Desaiah, D., Mehendale, H., and Ho, I. K., Kepone inhibition of mouse brain synaptosomal ATPase activities, *Toxicol. Appl. Pharmacol.,* 45, 268, 1978.
244. End, D. W., Carchman, R. A., Ameen, R., and Dewey, W. L., Inhibition of rat brain mitochondrial transport by chlordecone, *Toxicol. Appl. Pharmacol.,* 51, 189, 1979.
245. Carmines, E. L., Carchman, R. A., and Borzelleca, J. F., Kepone: cellular sites of action, *Toxicol. Appl. Pharmacol.,* 49, 543, 1979.

246. **Martinez, A. J., Taylor, J. R., Houff, S. A., and Isaacs, E. R.**, Kepone poisoning: cliniconeuropathological study, in *Neurotoxicology,* Vol. 1, Roizin, L., Shiraki, H., and Grčević, N., Eds., Raven Press, New York, 1977, 443.
247. **Martinez, A. J., Taylor, J. R., Dyck, P. J., Houff, S. A., and Isaacs, E.**, Chlordecone intoxication in man. II. Ultrasound of peripheral nerves and skeletal muscle, *Neurology,* 28, 631, 1978.
248. **Cohn, W. J., Boylan, J. J., Blanke, R. V., Fariss, M. W., Howell, J. R., and Guzelain, P. S.**, Treatment of chlordecone (Kepone) toxicity with cholestyramine, *N. Engl. J. Med.,* 298, 243, 1978.
249. **Cohn, W. J., Blanke, R. V., Griffith, F. D., and Guzelain, P. S.**, Distribution and excretion of Kepone (KP) in humans, *Gastroenterology,* 71, 901, 1976.

Chapter 5

ORGANOPHOSPHORUS ESTER INSECTICIDES

Donald J. Ecobichon

I. INTRODUCTION

The study of the ester bond involving phosphoric acid led to the development of an entirely new group of insecticides of almost unlimited structural arrangement, potency, and physicochemical properties. The organophosphorus insecticides account for some 40% of the registered pesticides in the U.S. and the number increases yearly. In 1973, the annual sales of organophosphorus insecticides exceeded 180 million lbs whereas that of the aldrin-toxaphene group was 170 million lbs and declining due to legal actions and restricted usage.[1] The nomenclature of this group of insecticides is confused by the use of a variety of chemical, generic, and trade names. Prior to discussing individual chemicals and their biological effects, the nomenclature and chemical properties should be clarified.

II. CHEMISTRY

A. Nomenclature

The term "organophosphates" has been loosely applied to all compounds containing carbon which are derivatives of an acid containing phosphorus. To satisfy the chemical purists and to prevent the perpetuation of confusion, more precise terminology will be used since, as will be seen, different acids of phosphorus are used in the wide range of insecticides now synthesized. In this chapter, the nomenclature proposed by the International Union of Pure and Applied Science (IUPAC) will be used since the system has an intrinsic logic for users of the English language. Unfortunately, phosphorus nomenclature becomes very difficult among other language groups. For more detailed discussions and comparisons of nomenclature, the reader is referred to the books of O'Brien,[2] Melnikov,[3] Fest and Schmidt,[4] and Eto.[5]

Table 1 presents the nomenclature and structures of the commonly found derivatives of acids of phosphorus. It should be noted that, since all of these esters are derivatives of pentavalent phosphorus acids, the valency suffix of the acid will be "ic" and that of esters will be "ate". The substituents "X", "Y", and "Z" are included in the name of the chemical with an indication of the atom to which they are attached, i.e., O-methyl, O-ethyl, O-isopropyl, etc. as a prefix. Atoms or functional groups other than alkoxy are included by insertion into the key-word, i.e., phosphoro — ate. Two or more identical groups are designated by the prefixes "di", "tri", etc. The names for the free acids are based on the word phosphoro-ic with insertion of the properties as is shown in Table 1. As an example, the insecticide commonly known as *Fenitrothion* and having the structure

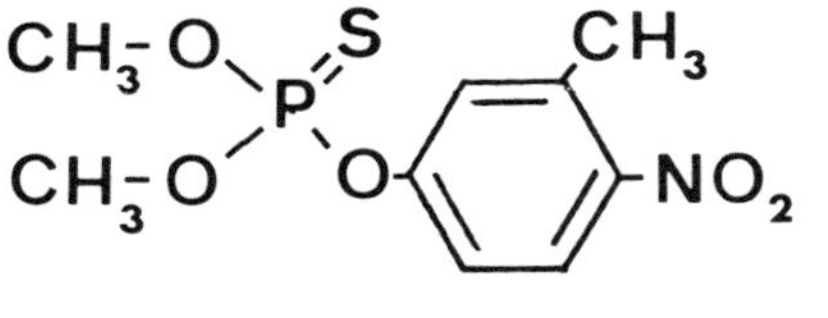

would be called O,O-dimethyl-O-(4-nitro-tolyl) phosphorothioate.

Table 1
STRUCTURE AND NOMENCLATURE OF ORGANOPHOSPHORUS ESTER INSECTICIDES

Phosphate	Phosphonate	Phosphorothioate*
XO–P(=O)(OY)(OZ)	XO–P(=O)(Y)(OZ)	XO–P(=S)(OY)(OZ)

Phosphorodithioate	Phosphorothiolate	Phosphorochloridate / Phosphorofluoridate
XO–P(=S)(OY)(SZ)	XO–P(=O)(OY)(SZ)	XO–P(=O)(OY)(Cl or F)

Phosphoroamidate	Phosphorodiamidate
X$_2$N–P(=O)(OY)(OZ)	(X)N–P(=O)(N(X))(OZ)

a. The term "thioate" distinguishes that the bond between the sulfur and the phosphorus is P=S rather than P—S which would be called "thiolate".

For compounds containing two phosphorus atoms (dimeric phosphoric acid derivatives) there has been no international rule established but a common terminology has been adopted since most compounds of this class are anhydrides having

-P(=O)-O-P(=O)- (phosphoric anhydride)

-P(=S)-O-P(=S)- (thionophosphoric anhydride)

-P(=O)-S-P(=O)- (thiolophosphoric anhydride) bonds.

One often sees "slips of the tongue" back to earlier nomenclature for such compounds as O,O,O,O-tetraethyl phosphoric anhydride which is also known as tetraethylpyrophosphate or TEPP. Similarly, the chemical N,N,N,N-octamethyl phosphorodiamidic anhydride is also known as octamethyl pyrophosphorotetramide octamethyl pyrophosphorodiamidate, bis-N-dimethyl phosphordiamidic anhydride or more simply by the acronym OMPA.

Table 2
EXAMPLES OF THE DIVERSITY OF STRUCTURES OF ORGANOPHOSPHORUS INSECTICIDES

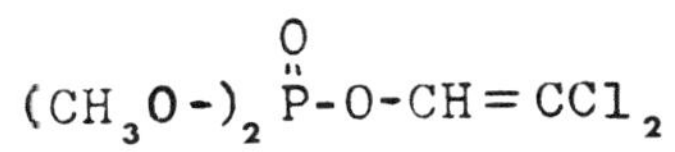

$(CH_3O-)_2 \overset{O}{\overset{\|}{P}}-O-CH=CCl_2$

Dichlorvos

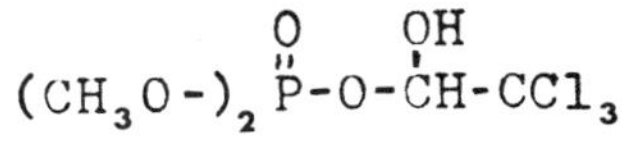

$(CH_3O-)_2 \overset{O}{\overset{\|}{P}}-O-\overset{OH}{\underset{|}{C}H}-CCl_3$

Trichlorfon

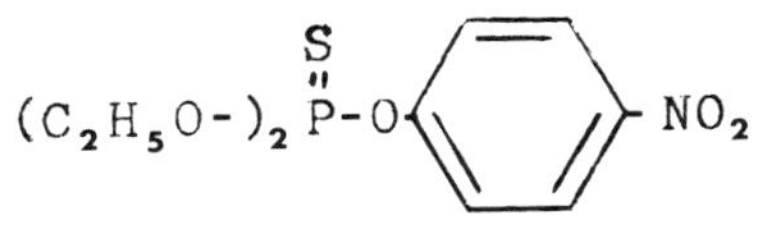

$(C_2H_5O-)_2 \overset{S}{\overset{\|}{P}}-O-\!\!\left\langle\!\bigcirc\!\right\rangle\!\!-NO_2$

Parathion

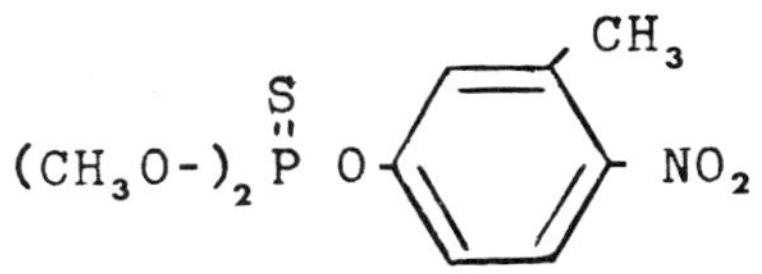

$(CH_3O-)_2 \overset{S}{\overset{\|}{P}}-O-\!\!\left\langle\!\bigcirc\!\right\rangle\!\!-NO_2$ (with CH_3 on ring)

Fenitrothion

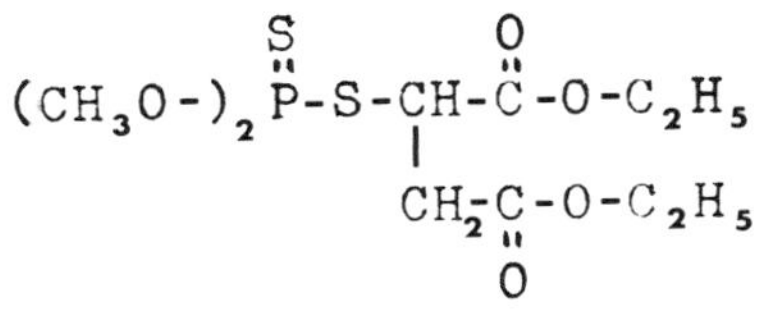

$(CH_3O-)_2 \overset{S}{\overset{\|}{P}}-S-\underset{\underset{O}{\overset{|}{C}H_2-\overset{\|}{C}-O-C_2H_5}}{CH}-\overset{O}{\overset{\|}{C}}-O-C_2H_5$

Malathion

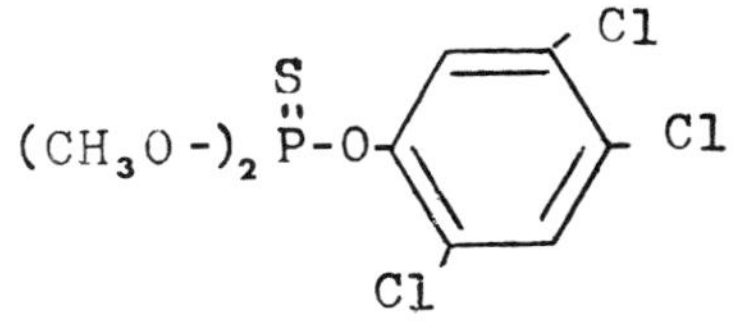

$(CH_3O-)_2 \overset{S}{\overset{\|}{P}}-O-\!\!\left\langle\!\bigcirc\!\right\rangle\!\!-Cl$ (2,4,5-trichlorophenyl)

Ronnel

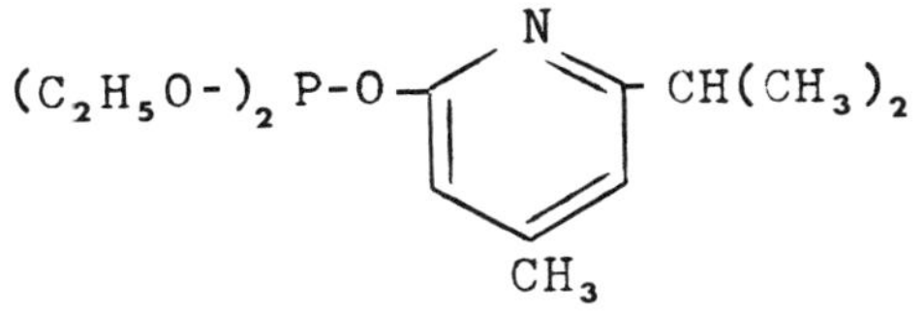

$(C_2H_5O-)_2 P-O-\!\!\left\langle\!\bigcirc\!\right\rangle\!\!-CH(CH_3)_2$ (pyridine with CH_3)

Diazinon

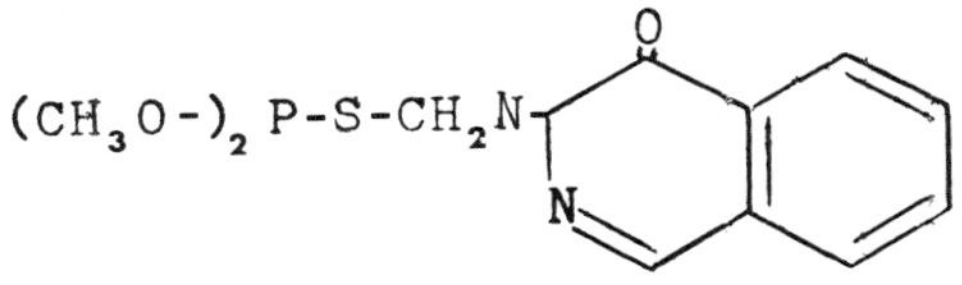

$(CH_3O-)_2 P-S-CH_2-N\!\!\left\langle\!\bigcirc\!\right\rangle$ (benzazine with C=O)

Guthion

$(CH_3O-)_2 \overset{O}{\overset{\|}{P}}-O-\overset{CH_3}{\underset{Cl}{C}}=\overset{}{C}-\overset{O}{\overset{\|}{C}}-N(C_2H_5)_2$

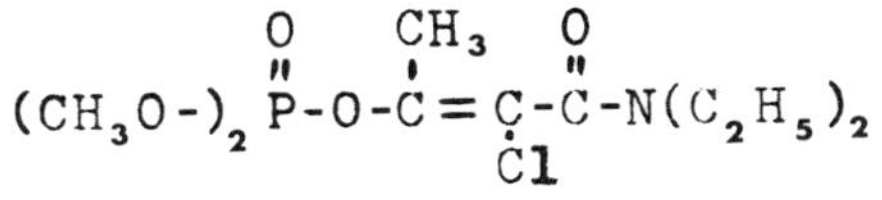

Phosphamidon

Most of the commonly used organophosphorus ester pesticides can be conveniently grouped in broad classes on the basis of the structures shown in Table 1. Examples are shown in Table 2. The substituents X and Y are capable of almost infinite variation and may represent alkyl or aryl groups attached directly to the phosphorus or through an oxygen, nitrogen, or sulfur. The Z substituent may be a halogen (fluorine or chlorine), an aryl group, a phosphate (as in a "pyrophosphate") cyanide, thiocyanate, carboxylate, or almost any phenoxy or thiophenoxy group. The substituents introduce major alterations in the physical and chemical properties of the chemical which are intimately linked to rates of penetration, distribution, metabolic activation, and/or degradation as well as target site interactions and ultimately with the potency and selectivity of the pesticide.

B. Physiochemical Properties

The biological potency of a pesticide is dependent upon a number of factors including (1) penetration and distribution to various tissues of the body, (2) biotransformation, both activation as well as degradation in vivo, and (3) interaction with a structural or functional target site in a tissue. New biologically active agents are frequently discovered by the rational approach of studying a series of analogs and subsequently correlating chemical structure and related physicochemical properties with biological activity. In most cases, pesticidal activity cannot be correlated with physicochemical properties alone since one important factor, metabolic transformation, does not always lend itself to analysis in this manner. However, target site interaction studies, particularly those in vitro involving purified acetylcholinesterases and homologous series of organophosphorus or carbamate esters, have demonstrated the importance of physicochemical parameters.

Phosphorus can exist in either a trivalent or a pentavalent state. The trivalent phosphoric acids and partial acids are very water soluble and readily undergo isomerization into

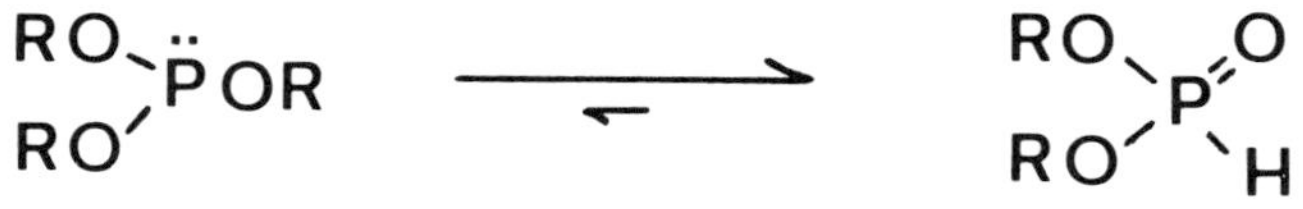

$$(RO)_2\ddot{P}\text{-}OR \rightleftharpoons (RO)_2P(=O)H$$

tautomeric pentavalent acids resulting in a stable phosphoryl form. Di- and trialkyl derivatives are highly reactive nucleophilic agents and are important as intermediates in the synthesis of more complicated organophosphorus compounds but few are biologically active per se as pesticides. In contrast, the esters of pentavalent phosphorus acids are derivatives of either phosphinic (I), phosphonic (II), or phosphoric (III) acids, anhydrides, or of sulfur-containing analogs. These neutral esters or amides are biologically active, lipid soluble, and,

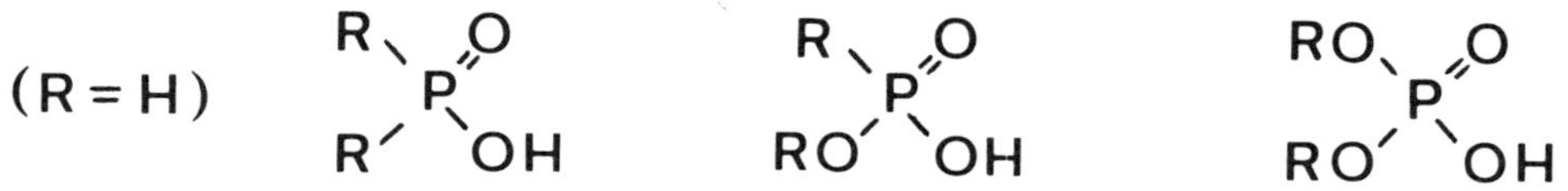

(R = H)

I: $R_2P(=O)OH$ II: $R(RO)P(=O)OH$ III: $(RO)_2P(=O)OH$

because the phosphorus atom is electron deficient, are highly reactive though unstable electrophilic agents. The hydrolysis of these esters, catalyzed by the aqueous medium or by specific enzymes, yields diester products which are biologically inactive. For a detailed discussion of the chemistry of the acids and esters of phosphorus, the reader is referred to the appropriate chapters in the books by Melnikov[3] or Eto.[5]

It is readily apparent that, as one changes substituent groups at X, Y, and Z, one changes the physicochemical properties of the molecule. The influence of substituent groups can be exerted in three ways: electronic, steric, and hydrophobic. The anticholinesterase activity of a compound depends largely on the phosphorylating ability of the esters and esters containing an easily displaced Z substituent were found to be good inhibitors. Aldridge and Davison[6] demonstrated a direct relationship between the first order inhibition constant k_1 for erythrocyte AChE and the hydrolysis rate of diethyl substituted phenyl phosphate esters in buffer. These observations, with those of Schrader, formed the basis for Schrader's "acyl rule" in which the "acyl" group was an anhydride linkage between a substituent and the phosphorus, the leaving group (Z) being fluoride, cyanate, thiocyanate, mercaptide phosphoryloxy, aryloxy, or heterocyclic groups.[7]

$$\begin{array}{c}\text{XO}\\\text{YO}\end{array}\!\!\diagdown\!\!\overset{\text{O}}{\underset{}{P}}\!\!\diagdown\!\!\begin{array}{c}\\\text{Z}\end{array} + \text{H}-\text{A} \longrightarrow \begin{array}{c}\text{XO}\\\text{YO}\end{array}\!\!\diagdown\!\!\overset{\text{O}}{\underset{}{P}}\!\!\diagdown\!\!\begin{array}{c}\\\text{A}\end{array} + \text{H}-\text{Z}$$

Similarly, phosphorylating agents were described by Clark et al.[8] as P-XYZ systems where X, Y, and Z are atoms different from those used in the above nomenclature. In this scheme, the electrons of the P—X bond are accepted by Z which becomes electronegative thereby weakening the P—X bond and resulting in its hydrolysis with the formation of a phosphorus acid.

$$>\!\!\overset{\text{O}}{\underset{}{P}}-\bar{X}-Y=Z \longrightarrow >\!\!\overset{\text{O}}{\underset{}{P}}-X=Y-\bar{Z}^{\ominus}$$

The important aspect, again, is the electronic rearrangement which weakens the phosphorus-substituent group bond resulting in ease of hydrolysis and optimizing biological activity.

Other investigators have utilized electronic factors, one being Hammett. Originally applied to the linear relationship between ionization constants of *meta-* and *para-*substituted benzoic acids and the reactivity of corresponding esters, Hammett's constants, (σ) determine the electron-withdrawing or donating properties of a substituent or an aromatic ring at a reaction center.[9,10] A good correlation has been observed between σ and the logarithm of the I_{50} (molar concentration of inhibitor necessary to produce 50% inhibition) of flybrain AChE for a series of *meta-* and *para-*substituted diethylphenyl phosphates (Figure 1).[11] Such studies have demonstrated the influence exerted by substituent groups on the lability of the P-O-phenyl bond, shifts in the P-O-phenyl stretching frequencies, and the hydrolysis rates.[12]

In studying a series of ethyl *p*-nitrophenyl alkyl-phosphates, Fukuto and Metcalf[13] noted that, in contrast to other results, more than half of the structures did not form a linear relationship when the logarithm of the inhibitor constant (K_e) was plotted against the logarithm of the hydrolysis or solvolysis constant (K_{hyd}). In reexamining the data of Fukuto and Metcalf, Hansch and Deutsch[14] established a good correlation between K_e and Taft's steric constant E_s. This constant is a measure of the ability of a certain structural configuration to form an enzyme-inhibitor complex, the relationship between K_e and E_s suggesting that the larger the substituent group the bulkier the aryloxy group becomes and the greater the steric interference with the overall phosphorylation process. Such studies, conducted with series of analogs, emphasize the importance of steric factors. As has been stated by Fukuto,[12] steric effects are probably more important in interfering with the formation of the enzyme-inhibitor complex than with the actual phosphorylation step.

On the basis that all biocidal phosphorus compounds were neutral esters of pentavalent phosphorus, the importance of the lipid solubility of these chemicals was appreciated in relation to their biological activity. A parameter which has proven as useful with pesticides as with drugs is based on the partition of the chemical between a "lipid" phase (*n*-hexane, octanol, chloroform, benzene, etc.) and an aqueous phase (water or a suitable buffer). This parameter, called the oil-water (O/W) partition coefficient was initially introduced by Meyer[15] and by Overton[16] in an attempt to explain similar biological properties of chemicals with unrelated structures which produced narcosis or sleep in animals.[17] The correlation between the O/W coefficient and biological activity proved best when congeners were compared. Hansch and colleagues[18-21] have developed the π constant which, when used in conjunction with other physical parameters, has led to an improvement in the correlation of biological activity with physicochemical properties. The constant π is analogous to the Hammett σ constant and is defined as:

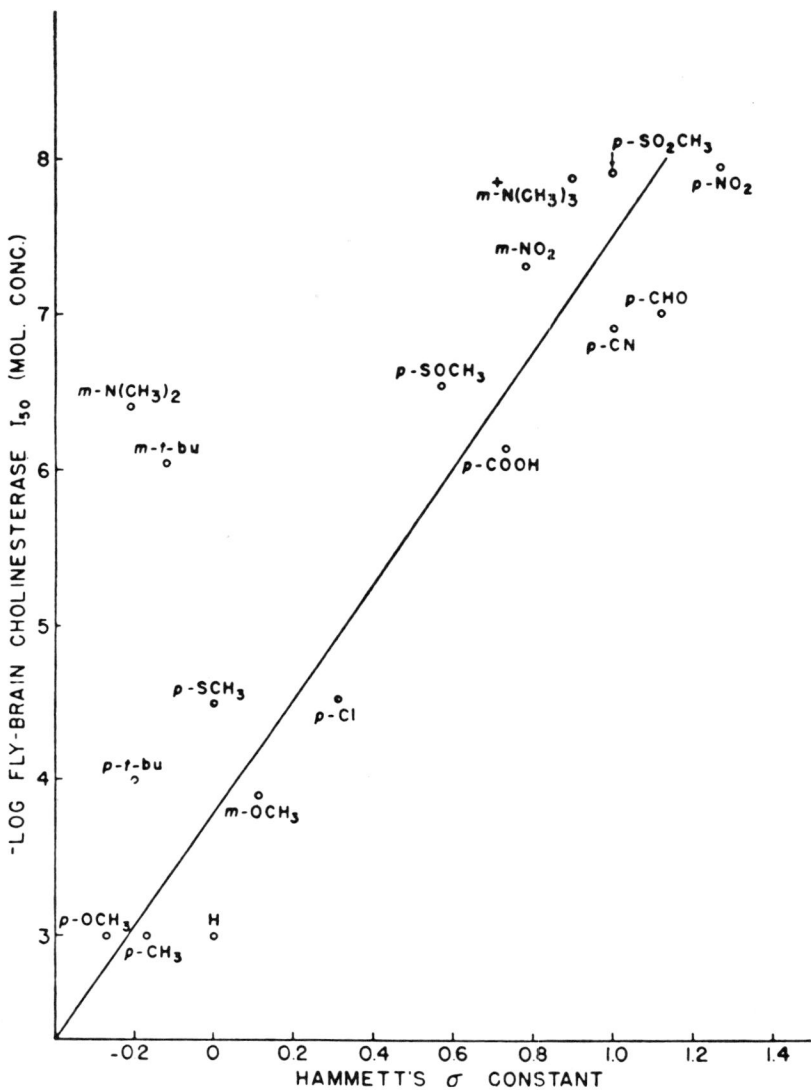

FIGURE 1. The relationship between -log fly brain acetylcholinesterase I_{50} and Hammett's constant σ for *meta*- and *para*-substituted phenyl diethyl phosphates showing that the degree of inhibition is a direct function of the electron-withdrawing capacities of the substituents on the benzene nucleus. (From Fukuto, T. R. and Metcalf, R. L., *J. Agric. Food Chem.*, 4, 930, 1956. With permission.)

$$\pi = \log P_X/P_H$$

where P_H is the octanol-water partition coefficient of the parent or unmodified chemical and P_X is the value for the modified or chemically altered structure. π is a free-energy constant and provides a measure of the relative free-energy change as a result of moving the substituted molecule from one phase to another. Since π is always related to the parent compound for a homologous series, it becomes a constant representing the constituent group and is dependent only on the nature of the substituent. The Hansch model can be applied to enzymes in solution, membranes or cells in suspension, or whole animals and can be used to estimate the influence of change in the molecule on the distribution, penetration, interaction, or binding of a chemical at membranes or active sites. Since π estimates the change in lipophilic character of the molecule brought about by alteration of substituent group(s), it is useful in assessing hydrophobic bonding.

In such a brief description of the interdependence of physicochemical properties and biological activity, one would be remiss if one did not mention the attempts made to integrate the electronic, steric, and hydrophobic properties of the agents and correlate these with biological activity:

biological response (BR) = f (hydrophobicity) + f (electronic) + f (steric)

Hansch[18] has examined this aspect qualitatively as follows:

$$\log BR = k_1 \pi^2 + k_2 \pi + k_3 \sigma + k_4 E_s + k_5$$

where k_1, k_2, k_3, and k_4 are weighting constants for each contributing parameter (π, σ, E_s) discussed above. The π^2 term indicates that the biological response curve is parabolic in nature and implies that highly lipophilic and hydrophilic agents would find it exceedingly difficult to penetrate tissues and interact with an active site. Multiple regression analysis of the components of the equation would be necessary to ascertain the significance of the contribution of each parameter. Excessively high doses would be required to elicit a characteristic biological response since, at one extreme, hydrophilic molecules would be readily eliminated from the body and would have a short biological half life. At the other extreme, highly lipophilic molecules would be compartmentalized in tissues having high lipid content and would be stored.

C. Biotransformation

The difficulty in establishing good correlations between the physicochemical properties and the biological potency of an organophosphorus ester revolves around the contribution made by the ability of the organism to biotransform or metabolize the agents into biologically less active compounds which often possess physicochemical properties quite different from those of the parent chemical. The almost infinite variety of substituent groups which one encounters in the organophosphorus esters has made each one an individual case. An in-depth description of the mechanisms of biotransformation of organophosphorus esters is beyond the scope of this book, but the reader should be aware of the major generalizations and is referred to appropriate recent reviews, such as Eto,[5] Matsumura,[22] Nakatsugawa and Morelli,[23] and Ecobichon,[24] or more specific references mentioned below.

Endogenous and exogenous (xenobiotic) compounds usually undergo metabolic transformation in vivo to less toxic and more polar metabolites which can be eliminated from the organism more readily. Two major, enzyme-catalyzed steps are involved in the detoxification of such chemicals: the primary metabolic processes or

	I		II	
Pesticide	———	Metabolite	———	Excretory product
	Oxidation		Glucuronidation	
	Reduction		Sulfation	
	Hydrolysis		Acetylaton	
	Transfer			

Phase I detoxifications (nonsynthetic processes) involve oxidative, reductive, hydrolytic, and/or other enzymatic reactions and form reactive metabolites (Figure 2) whereas the secondary processes or Phase II detoxifications (synthetic processes) conjugate po-

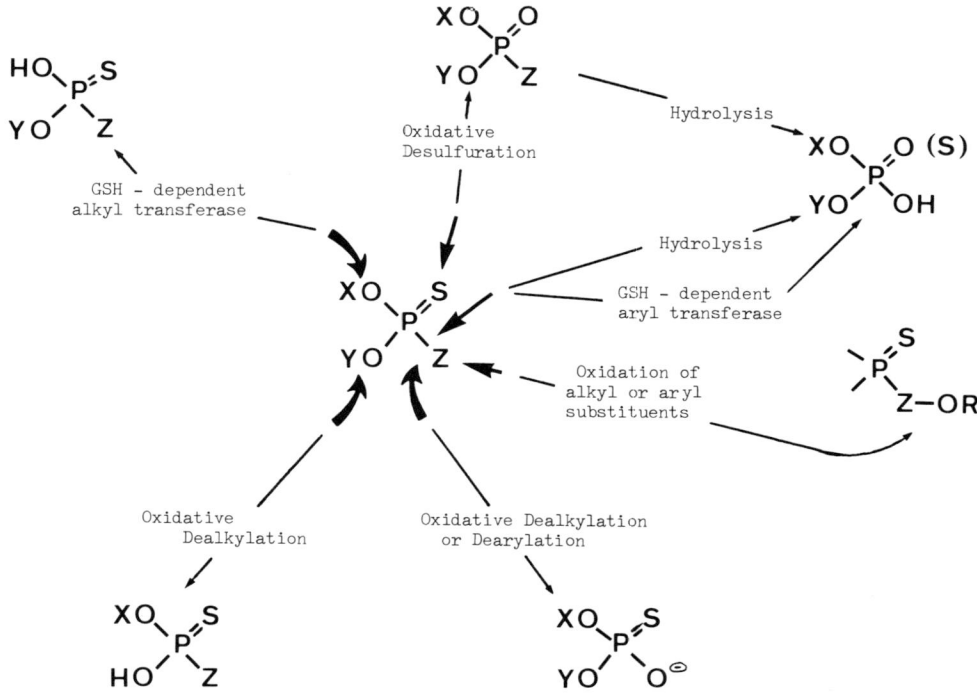

FIGURE 2. A schematic diagram of various Phase I detoxification reactions showing the point of attack on the ester and the product(s) formed as a consequence of oxidative, hydrolytic, and glutathione-dependent alkyl transfer reactions.

lar Phase I metabolites with a natural body substituent to form a product with enhanced water solubility and excretability. These enzyme systems are ubiquitously localized in most tissues of the body and, no doubt, contribute to the overall degradation of the pesticides but the major organ involved in detoxification in mammals is the liver which has the highest specific activity per gram of tissue of any organ of the body.

1. Phase I Detoxifications
a. Oxidative Biotransformation

Most organic pesticides are subject to oxidation by enzymes localized in the smooth and rough endoplasmic reticulum of many body cells. These enzymes, known as microsomal mixed function oxidases or monooxygenases, can catalyze a number of different reactions and utilize the coenzyme reduced nicotinamide adenine dinucleotide phosphate (NADPH), the ubiquitously distributed cytochrome P-450 system, and an NADPH-regenerating system to provide the necessary oxygen and electrons to convert the pesticide into polar compounds.[25-27] These monooxygenases can catalyze an extraordinary variety of different reactions as can be seen in Figure 2, removing alkyl and aryl groups as well as replacing the sulfur in phosphorothionate esters.

1. Oxidative Desulfuration

Diggle and Gage[28] reported that parathion (O,O-diethyl O-4-nitrophenyl phosphorothioate) was a poor inhibitor of cholinesterases in vitro but became quite a potent inhibitor once it had been administered to an animal. This observation led to the hypothesis that a reaction occurred in vivo which activated the compound and eventually led to proof that the incubation of parathion with liver slices resulted in the formation of a potent anticholinesterase agent. Homogenates of liver as well as isolated hepatic microsomes, when fortified with Mg^{2+} ions, NADPH, and nicotinamide, would also form this product when incubated with parathion.[29,30] The product was identified as paraoxon (O,O-diethyl-O-4-nitrophenyl phosphate)

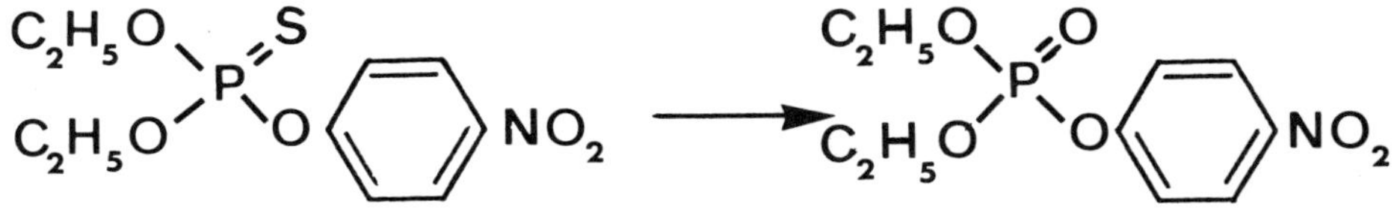

$$\underset{\text{PARATHION}}{(C_2H_5O)_2P(=S)-O-C_6H_4-NO_2} \longrightarrow \underset{\text{PARAOXON}}{(C_2H_5O)_2P(=O)-O-C_6H_4-NO_2}$$

This reaction is an essential feature of the selective toxicity of all phosphorothioate insecticides, a rather weak parent compound being converted to a several hundred-fold more potent anticholinesterase agent by tissue enzymes. In contrast to many of the other reactions to be considered below, this oxidative desulfuration almost always increases the toxicity of the phosphorothioate and hence can be considered an "activation" reaction rather than a detoxifying one. The susceptibility of animals to poisoning by phosphorothioate insecticides is dependent in part upon the rates at which the oxygen analogs are made available to inhibit cholinesterase at critical sites in nervous tissue and upon a dynamic relationship between activation and inactivation of these chemicals by tissue enzymes. As will be seen later, this oxidative reaction is essential for the next stage of biotransformation which is truly a degradative step.[31]

2. Oxidative Dealkylation

Cleavage of the esterified short-chain alkyl substituents of organophosphorus insecticides also plays an important role in detoxification. More than one mechanism exists for such a reaction but oxidative dealkylation involves microsomal monooxygenases which are NADPH- and cyctochrome P-450-dependent and are quite specific in that the removal of methyl groups occurs readily but longer-chain alkyl groups become progressively more unreactive. The products formed include an aldehyde and a monoalkylated derivative.[32,33]

$$(CH_3O)_2P(=S)-Z \longrightarrow (HO)(CH_3O)P(=S)-Z$$

The removal of one alkyl group from a phosphorothioate triester such as methyl parathion (above) results in a dramatic decrease in anticholinesterase activity and subsequent toxicity.

3. Thioether Oxidation

A number of thioether-containing organophosphorus insecticides can be activated to substantially more potent anticholinesterase agents by monooxygenase functions in plant, animal, and insect tissues. As is shown below, such reactions occur in two steps with the formation of a sulfoxide followed by the formation of a sulfone.[34-36]

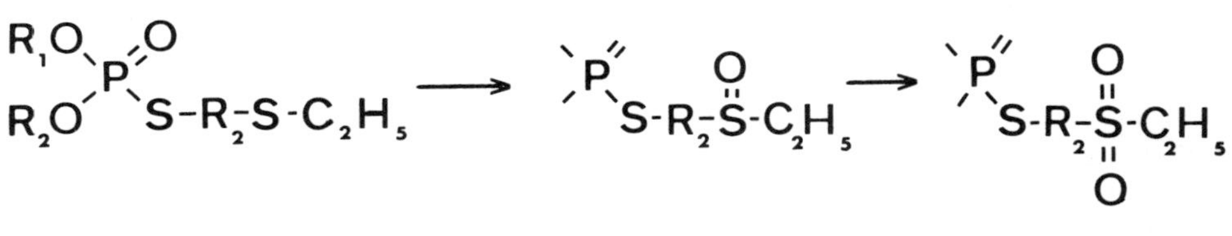

$$(R_1O)(R_2O)P(=O)-S-R_2-S-C_2H_5 \longrightarrow \text{...}-S-R_2-S(=O)-C_2H_5 \longrightarrow \text{...}-S-R_2-S(=O)_2-C_2H_5$$

The initial oxidation, the formation of the sulfoxide, is usually very rapid and the product accumulates in tissues in large concentrations and is probably the primary toxicant. The sulfone forms more slowly but is more potent than the sulfoxide. The I_{50} values (concentration producing 50% inhibition) for disulfoton and its oxidized metabolites were $1 \times 10^{-4} M$ for disulfoton, 7×10^{-5} M for disulfoton sulfoxide and $3.5 \times 10^{-6} M$ for disulfoton sulfone, respectively.[37] Similar changes in potency have been reported for the compound phorate (O,O-diethyl S-[(ethylthio)methyl]phosphorodithioate).[38]

4. Oxidative Dearylation

The cleavage of the P-O-aryl bond by an oxidative mechanism has been reported for such compounds as parathion, methylparathion, fenitrothion, and diazinon, the reactions being dependent upon the presence of NADPH as a cofactor.[39-41] In addition, the compound fonophos (Dyfonate), a phosphonothiolothioate, was also cleaved at the P-S-aryl bond by a monooxygenase. It has also been reported that P-S-alkyl bonds found in malathion and azinophosmethyl can be oxidatively cleaved, although criticism of the results has been raised on the basis that esterases or phosphatases may be involved in this reaction in some species.[42,43]

5. Minor Oxidative Reactions

In addition to the oxidative reactions which markedly affect the toxicological properties of the organophosphorus esters, the microsomal monooxygenases catalyze a number of reactions involving substituents or side groups resulting in: (1) aromatic ring hydroxylation, (2) deamination, (3) alkyl and N-hydroxylation, (4) N-oxide formation, (5) N-dealkylation. The reader is referred to more detailed discussions of these reactions to be found in Eto[5] and Matsumura.[22]

b. Glutathione Transferases

An enzyme which could catalyze the dealkylation of organophosphorus triesters was found localized in the postmicrosomal fraction of liver. This soluble, cytoplasmic enzyme(s), located primarily in the liver, required additional reduced glutathione (GSH, γ-L-glutamyl-L-cysteinylglycine) as a cofactor and acceptor for the catalyzed transfer of the O-alkyl and O-aryl moieties, hence the name GSH S-alkyl or -aryl transferases.[33,34] This enzyme system has been identified now as an important mechanism of detoxification of dimethyl-substituted phosphate and phosphorothioate esters such as dichlorvos, methyl paraoxon, methyl parathion, fenitrothion, guthion, ronnel, etc. yielding S-methylglutathione (GS-CH$_3$) and monodesmethyl products.[33,45,46] The mammalian GSH-alkyltransferases are highly specific for compounds containing methyl substituents, esters containing longer chain alkyl groups being much less readily degraded by this pathway.[47] A number of alkyltransferases with varying substrate specificities exist in animal tissues and their role in insecticide detoxification has not been completely resolved to date.

$$CH_3O\diagdown \underset{CH_3O\diagup}{P}\diagup\overset{S}{\diagdown}Z \; + \; GSH \; \xrightarrow{E} \; HO\diagdown \underset{CH_3O\diagup}{P}\diagup\overset{S}{\diagdown}Z \; + \; GS-CH_3$$

Dimethyl ester   Desmethyl ester

Another group of GSH-dependent transferases, the GSH-aryltransferases, have been identified and can utilize GSH as an acceptor of aryl substituents from organophosphorus triesters to yield dialkyl-phosphoro- or -phosphorothioic acids plus an aryl-glutathione as products.

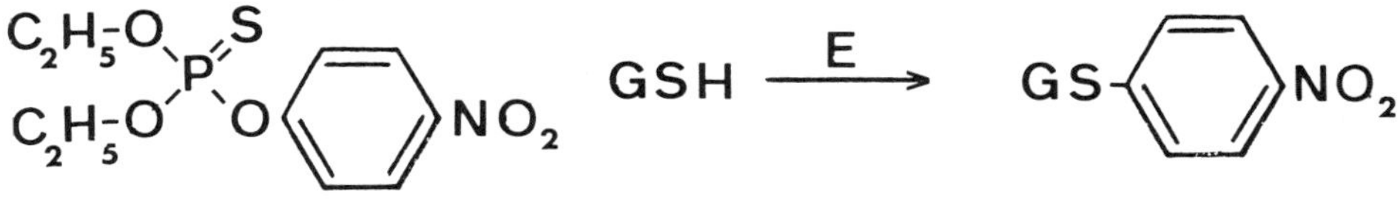

$$\underset{\text{Parathion}}{\begin{array}{c}C_2H_5O\\C_2H_5O\end{array}\!\!\!>\!\!P\!\!\underset{O}{\overset{S}{<}}\!\!-\!\!\bigcirc\!\!-NO_2} \quad GSH \xrightarrow{E} \underset{\text{Dimethylyhsphoro-thioic acid}}{GS-\bigcirc-NO_2}$$

Hollingworth et al.[48] showed that S-P-nitrophenylglutathione was produced from parathion using a postmicrosomal supernatant fortified with additional GSH. The aryltransferase has been partially purified and is distinct from the alkyltransferase. Once again the enzyme system is specific for GSH, other sulfhydryl-containing compounds (cysteine, 2-mercaptoethanol, thioglycolic acid) having no effect on activity.

c. Hydrolases

The organophosphorus insecticides, being esters or anhydrides of a phosphoric acid and some other acid, are susceptible to degradation into biologically inactive products by a variety of hydrolytic enzymes in plant, insect, and animal tissues. Studies to date, and they are by no means complete, have revealed complexity of tissue distribution of activity, overlapping properties, cofactor requirements, and substrate specificities.[49] They can be subdivided into broad classifications, namely phosphorylphosphatases, arylesterases, nonspecific carboxylesterases, carboxyamidases, and phosphotriesterases.

Phosphorylphosphatases have been detected in a variety of microorganisms,[50] fish,[51] insects,[52] and mammalian tissues,[53] both soluble (cytoplasmic) and particulate (organelle-associated) forms being found. Examination of the subcellular localization of these enzymes by differential centrifugation showed the decreasing order of activity to be as follows: supernatant, microsomes, mitochondria, nuclei plus debris.[54] Phosphorylphosphatases are strongly activated by Mg^{2+}, Mn^{2+}, Co^{2+}, and a number of amino acids including histidine and derivatives of pyridine and imidazole.[55,56]

$$\begin{array}{c}RO\\RO\end{array}\!\!\!>\!\!\underset{}{\overset{O}{\underset{\|}{P}}}\!-X \xrightarrow{E} \begin{array}{c}RO\\RO\end{array}\!\!\!>\!\!\underset{}{\overset{O}{\underset{\|}{P}}}\!-OH + HX$$

These enzymes appear to be rather selective for hydrolysis of phosphorus-halide bonds such as found in chemical warfare agents such as bis (1-methylethyl) phosphorofluoridate (DFP), 1-methylethyl methylphosphonofluoridate (Sarin) or 1,2,2-trimethylpropyl methylphosphonofluoridate (Soman) and to be of little use in the detoxification of the wide variety of organophosphorus insecticides in use today.

The arylesterases (aromatic or A-esterases, ArE, EC 3.1.1.2) are a group of enzymes which preferentially hydrolyze aryl esters (phenols, naphthols, indoles) of short-chain aliphatic or phosphoric acids, especially if a double bond is present in the alcohol at a position *alpha* with respect to the ester bond.[57-59]

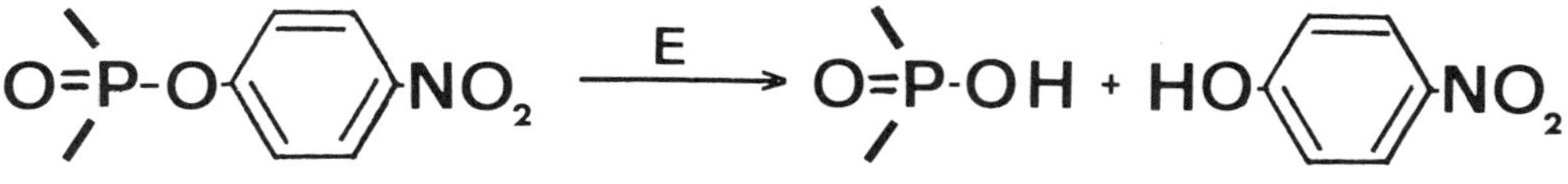

$$O\!\!=\!\!P\!-\!O\!-\!\bigcirc\!\!-NO_2 \xrightarrow{E} O\!\!=\!\!P\!-\!OH + HO\!-\!\bigcirc\!\!-NO_2$$

The enzyme(s) is specifically activated by divalent cations (Co^{2+}, Ca^{2+}, Mn^{2+}) and is inhibited by ethylenediaminetetraacetic acid (EDTA) and mercurial compounds.[60] One can subdivide ArE into two groups according to whether they are activated by calcium (group I) or by cobalt and manganese (group II). Group I enzymes are found only in mammalian tissues and selectively hydrolyze phosphate esters, phosphorothionate esters requiring metabolic desulfuration by microsomal NADPH-dependent monooxygenases prior to hydrolysis by these enzymes. Group II enzymes, found in both mammals and insects, hydrolyze both phosphate and phosphorothioate esters.[61] These enzymes are absent or are present in very low activity in birds, reptiles, amphibians, and fish.[60]

Carboxylesterases (carboxylic acid ester hydrolases, EC 3.1.1.1) are ubiquitously distributed in nature, being found in soil microorganisms as well as in insects, amphibians, piscine, avian, and mammalian tissues.[49] They exist in multiple forms as soluble (cytoplasmic) as well as endoplasmic reticulum-attached molecules.[62,63] They are capable of hydrolyzing a variety of aliphatic and aromatic esters of short chain fatty acids.[64] Carboxylesterase-catalyzed degradation of organophosphorus insecticides is limited to malathion (O,O-dimethyl-S-(1,2-dicarbethoxyethyl phosphorodithioate) where one of the two available carboxylic ethyl ester groups is hydrolyzed to produce malathion and/or malaoxon α-monoacids which are biologically inactive.[65,66] Hydrolysis of the second terminal ester group also occurs, resulting in the formation of malathion diacid.[61] This carboxylesterase-catalyzed reaction is an important feature of resistance to this insecticide in insects and to tolerance in mammals.[67,68] Studies have demonstrated that acethion (O,O-dimethyl-S-carbethoxyethyl phosphorothioate) is degraded in the same manner.[69] Beyond these two insecticides, carboxylesterases play little role in insecticide degradation since most insecticidal organophosphorus esters possess aryl substituents.

Carboxyamidases (acylamide amidohydrolase, EC 3.5.1.4) have been shown to occur in plant, insect, and vertebrate tissues but are of limited interest concerning organophosphorus ester degradation. The highest activities of carboxyamidase are found in mammalian liver, low activities being found in other tissues.[70,71] Considerable species variation has been encountered in activity. Dimethoate (O,O-dimethyl-S-(N-methylcarbamoylmethyl) phosphorodithioate) has been the only amide-containing organophosphorus insecticide shown to be hydrolyzed by mammalian tissue amidases.[72]

$$\begin{array}{c} RO\diagdown\;\;S \\ P \\ RO\diagup\;\;SCH_2\text{-}\overset{O}{\underset{\|}{C}}\text{-NH-CH}_3 \end{array} \xrightarrow{\;\;E\;\;} \begin{array}{c} RO\diagdown\;\;S \\ P \\ RO\diagup\;\;SCH_2\text{-}\overset{O}{\underset{\|}{C}}\text{-OH} \end{array}$$

Other related amide-containing insecticides, Azodrin (3-hydroxy-N-methyl-crotonamide dimethyl phosphate) and Bidrin (3-hydroxy-N,N-dimethylcroton-amide dimethyl phosphate) were hydrolyzed by plant tissues but not by those of mammals.[73] One must question the nature of this enzyme since (1) the mammalian tissue amidase hydrolyzed only the phosphorothioate ester, (2) the toxicity of dimethoate could be potentiated by the coadministration of tri-ortho cresyl phosphate (TOCP) or O-ethyl O-(4-nitrophenyl) phenylphosphonothioate (EPN), (3) the carboxyamidase could be inhibited by incubating it with a variety of organophosphorus and carbamate esters.[74] These results suggest that the enzyme in mammalian tissues may be a carboxylesterase.

O-dealkylation by hydrolases rather than by monooxygenases or glutathione-dependent transferases may be another route of detoxification, desmethylated products being formed by phosphatases, phosphotriesterases, or O-alkylhydrolases. The exact nature of these enzymes has not been extensively investigated in plants and animals, and one may speculate on the involvement of other mechanisms of O-dealkylation more recently identified and studied.

$$\begin{array}{c}R_1O\\R_2O\end{array}\!\!P\!\!\begin{array}{c}{=}O\\Z\end{array} \xrightarrow{E\,+\,Mn^{2+}} \begin{array}{c}R_1O\\HO\end{array}\!\!P\!\!\begin{array}{c}{=}O\\Z\end{array} + R_2OH$$

There appear to be O-alkylhydrolases in insect, plant, and vertebrate tissues, activated by manganese but inhibited by calcium and inorganic and organic mercurials, which can contribute substantially to the degradation of such esters or dichlorvos, trichlorfon, and phosphamidon.[52,75] Desmethyldichlorvos, desmethylphosphamidon, and monomethyl phosphoric acid have been identified following incubation of the parent compounds with tissue extracts containing the enzyme(s).[75] A phosphatase-catalyzed hydrolysis of dimethoate has been studied in insects, vertebrates, and plants, cleavage occurring at both the P—S and S—C bonds (see above) and resulting in the formation of both dimethylphosphorothioic and dimethylphosphorodithioic acids.[76] A similar mechanism of hydrolysis has been suggested for malathion, though it has not been extensively studied.

2. Phase II Detoxifications

In the past, major emphasis has centered on the Phase I mechanisms of detoxification, but it is now being realized that further and often quite complex biotransformations can occur. The Phase II conjugations are biosyntheses in which natural or foreign compounds or their metabolites combine with readily available, endogenous chemicals such as uridine diphosphoglucuronic acid, uridine diphosphoglucose, 3'-phosphoadenosine-5-phosphosulfate, glutathione and amino acids (glycine, ornithine, glutamine) or with reactive groups such as methyl, acetyl, etc. to form a product.[77] Conjugation reactions have been known since the demonstration in 1842 that hippuric acid (N-benzoylglycine) was formed following the administration of benzoic acid to humans.[78] Williams[79] divided the conjugation reactions into two types based on different forms of "activated intermediates". Type I conjugations involved the formation of an activated conjugation agent involving methylation, acetylation, glucuronidation, glucosidation, and sulfation. Type II reactions involved amino acids and usually occurred only in the liver and kidney. Most but not all of these reactions are preceded by Phase I reactions which initiate the attack on the chemical and provide the key reactive groups essential for the conjugation reaction. Conjugation products are generally quite water-soluble, thereby enhancing excretion from the body, reducing the biological half-life of the active agent as well as its toxicity.

While Phase I detoxification mechanisms have been discussed extensively in a recent review,[77] some mention should be made of the diverse nature of these reactions as a warning to those who are not wary. Hutson et al.,[80] studying the biotransformation of chlorfenvinphos (2-chloro-1-[2',4'-dichlorophenyl] vinyl diethylphosphate) in the rat and the dog, identified glucuronide and glycine conjugates of several degradation products. Trichlorfon (O,O-dimethyl 1-hydroxy-2,2,2-trichloroethylphosphonate) was found to be directly conjugated with glucuronic acid in the rat without prior biotransformation.[81] It is important to note that most of the well known glucuronide and glucoside conjugations reported for organophosphorus esters have involved esters structurally similar to trichlorfon which represent only a small proportion of the total organophosphorus esters in use.[77,81] Conjugation pathways are widespread in plant, insect, and mammalian species.[77,82]

III. TOXICITY

A. Acute Toxicity
1. Animal

When observing an animal which has received an acutely toxic dose of an organophosphorus insecticide, one is struck by the sequential appearance of signs of toxicity beginning with marked salivation accompanied by a yawning reaction, rhinorrhea and snuffling, pupillary constriction, lacrimation, piloerection, "bloody tears", or chromodacryorrhea resulting from stimulation of the Harderian gland, bloody exudate around the nostrils and exophthalmos progressing to urination, defecation with loose stools, muscle tremor (fasciculations) and hypersensitivity to auditory stimuli, muscular weakness and ataxia and terminating in incoordination, obvious respiratory difficulty, prostration, clonic (rapid, repetitive) convulsions, and death. Some tonic (limb rigidity) convulsions may be observed. Examination of this animal at necropsy reveals the visceral organs and their blood vessels to be engorged with dark venous blood, peritoneal effusion, a distended right side of the heart, collapsed and ischemic lungs, spasms in the small intestine, and petechial hemorrhages of small vessels in some organs.

Death in mammals due to organophosphorus insecticide poisoning has been generally attributed to asphyxiation. Four mechanisms may be involved, any one of which would hamper adequate respiratory function though they most probably interact. The mechanisms include (1) bronchoconstriction, (2) decreased blood pressure, (3) neuromuscular blockade of the diaphragmatic and intercostal muscles, and (4) depression of the respiratory center in the brain. Many investigators argue that factor 4 is the most important especially when supportive artificial respiration permits animals to survive lethal doses of such agents.[83,84] Other investigators have demonstrated that the importance of the four mechanisms appears to vary with the agent used and the dose administered, reflecting the physicochemical properties of the agent, the rates of uptake, distribution in the body as well as the biotransformation and elimination of the agent. It would seem proper to list the sequence of events leading to death in animals as: (1) inhibition of tissue cholinesterase, (2) the accumulation of acetylcholine, (3) disruption of peripheral and central nerve function, (4) respiratory failure, (5) death by asphyxiation.[85]

All of the observed signs arise from the accumulation of unhydrolyzed acetylcholine and an intense initial stimulation followed by a paralysis of transmission in cholinergic synapses located in the central nervous system, between preganglionic and postganglionic fibers of both sympathetic and parasympathetic neurones at postganglionic nerve endings in the parasympathetic division, and at myoneural junctions of striated muscle (somatic nerves) (Figure 3). By inhibition of the nervous tissue cholinesterase, the organophosphorus insecticides exert a selective influence on certain organs of the body, the effects being divided into muscarinic actions (postganglionic parasympathetic fibers), nicotinic actions at all autonomic (parasympathetic and sympathetic) ganglia, as well as those in the central nervous system and at postganglionic somatic neuromuscular junctions. The nicotinic action on sympathetic ganglia evokes the release of norepinephrine from postganglionic nerve endings and the release of epinephrine from the adrenal medulla. Table 3 lists the key signs and symptoms identified with organophosphorus insecticide poisoning, the nervous tissue and site affected and the nature of the response observed in mammals. The severity of symptoms observed is largely dependent upon the chemical nature of the organophosphorus ester and the duration of exposure.

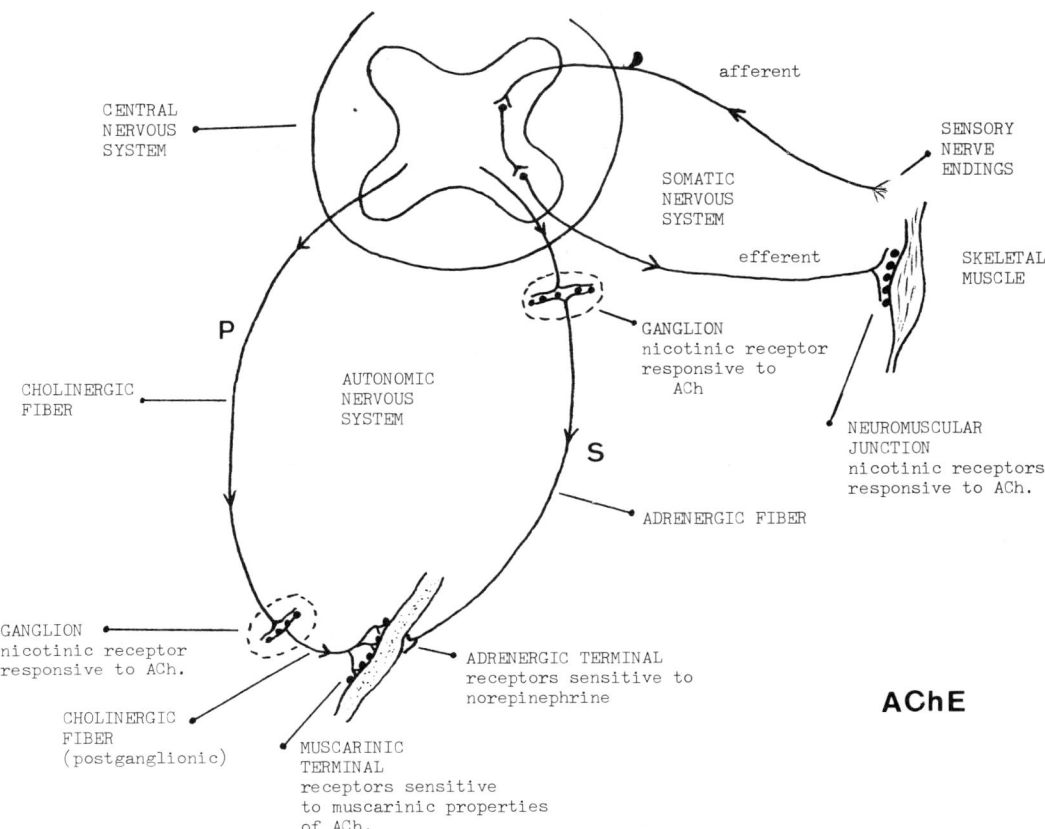

FIGURE 3. A simplified diagram of the peripheral autonomic and somatic nervous systems showing, in the former, the cholinergic parasympathetic (P) fibers and adrenergic sympathetic (S) fibers as well as the ganglia, location and site(s) of action of the neurotransmitter substances and of the enzyme acetylcholinesterase (AChE). A simple reflex of the somatic nervous system is shown along with the neuromuscular junction and the location of the AChE molecules.

2. Human

The termination of World War II brought to light the organophosphorus chemical warfare agents which had been synthesized by the Nazis and others. Such agents as DFP, Sarin, Soman, and Tabun were a fascination to pharmacologists since few had ever seen agents as potent as these and extensive pharmacological and biochemical studies were initiated, even using human volunteers, a practice which certainly would be frowned upon today.[86-88] Excellent descriptions of acute signs and symptoms of poisoning in military personnel purposefully exposed to such agents can be found in the literature between 1946 and 1958.[89-93] As less potent organophosphorus esters came into widespread use as insecticides, acute human poisoning became a relatively common occurrence and the later literature is replete with descriptions of symptoms observed in mild, moderate, and severe cases.[93-95] In general, a regular sequence of events can be discerned. In an extensive study of orchard workers, Sumerford et al.[96] considered the individuals to have been poisoned if they showed any group of three selected symptoms in addition to miosis. Among 38 organophosphorus insecticide-related illnesses, a broad range of symptoms was observed, the different signs appearing in the frequency shown in Table 4.

In mild poisonings, primarily the muscarinic effects will be observed though not all

Table 3
SIGNS AND SYMPTOMS OF ORGANOPHOSPHORUS INSECTICIDE POISONING

Nervous tissue and receptors affected	Site affected	Manifestations
Parasympathetic autonomic (Muscarinic receptors) post-ganglionic nerve fibers	Exocrine glands	Increased salivation, lacrimation, perspiration
	Eyes	Miosis (pinpoint and nonreactive), ptosis, blurring of vision, conjunctival injection, "bloody tears"
	Gastrointestinal tract	Nausea, vomition, abdominal tightness, swelling and cramps, diarrhea, tenesmus, fecal incontinence
	Respiratory tract	Excessive bronchial secretions, rhinorrhea, wheezing, edema, tightness in chest, bronchospasms, bronchoconstriction, cough, bradypnea, dyspnea
	Cardiovascular system	Bradycardia, decrease in blood pressure
	Bladder	Urinary frequency and incontinence
Parasympathetic and sympathetic autonomic fibers (nicotinic receptors)	Cardiovascular system	Tachycardia, pallor, increase in blood pressure
Somatic motor nerve fibers (nicotinic receptors)	Skeletal muscles	Muscle fasciculations (eyelids, fine facial muscles), cramps, diminished tendon reflexes, generalized muscle weakness in peripheral and respiratory muscles, paralysis, flaccid or rigid tone
		Restlessness, generalized motor activity, reaction to acoustic stimuli, tremulousness, emotional lability, ataxia
Brain (acetylcholine receptors)	Central nervous system	Drowsiness, lethargy, fatigue, mental confusion, inability to concentrate, headache, pressure in head, generalized weakness
		Coma with absence of reflexes, tremors, Cheyne-Stokes respiration, dyspnea, convulsions, depression of respiratory centers, cyanosis

may be seen in any one patient. Stimulation of exocrine secretions with marked salivation, lacrimation, rhinorrhea, and excessive bronchial secretions will be observed. Tightness in the chest, bronchoconstriction, wheezing, and coughing may also be observed. If the symptoms do not progress beyond this stage, the condition might be diagnosed as a common cold. An additional complication which might result in a diagnosis of influenza would include the gastrointestinal symptoms of nausea, vomiting, abdominal cramps, and swelling, tenesmus, and diarrhea. These "cold-like" symptoms respond nicely to treatment with atropine. As one can readily appreciate, mild cases of poisoning would not be recognized as such and patients presenting with such symptoms might not receive any specific antidotal treatment. The signs would normally subside in a few days.

In moderate poisonings, the observed signs would be more identifiable and would not be attributed to a cold or influenza. The major symptoms generally include bradycardia accompanied by a decrease in blood pressure, abdominal pain, bradypnea, dyspnea, headache, nausea, vomiting, miosis, diarrhea, generalized weakness, and mild shock.[89] In addition, profuse perspiration, a sensation of numbness (paraesthesia), pallor, tightness, or pain in the chest accompanied by audible coarse crepitations over the lungs may all be observed. Tachycardia, with an increase in blood pressure and pulse rate may be found as well as muscular fasciculation of the fine skeletal muscles of the face, particularly the eyelids.

Table 4
CHARACTERISTIC SIGNS AND SYMPTOMS OF ORGANOPHOSPHORUS INSECTICIDE POISONING IN EXPOSED ORCHARD SPRAYMEN[96]

Symptom reported	Frequency
Headache	29
Nausea	21
Weakness or fatigue	19
Tightness in chest	17
Abdominal pain	13
Vertigo or incoordination	13
Vomition	11
Nervousness, drowsiness, insomnia	9
Perspiration	9
Cough or expectoration	9
Disturbance of vision	9
Loss of appetite	8
Shortness of breath	8
Nasal discharge	6
Miosis	5
Wheezing	5

In severe poisonings which usually involve the ingestion of a quantity of the toxic chemical or the dermal exposure to concentrated material, the patient may be in a semistuporous, preconvulsive state showing a marked general muscle weakness, persistent randomized muscular twitching, and involuntary jerky movements. In severe clinical poisoning, there is a distinct central nervous system component to be observed in the complexity of symptoms and generally manifests itself in the form of tension, anxiety, nervousness, restlessness, giddiness, impairment of memory and of the ability to concentrate on tasks, difficulty in sleeping accompanied by excessive dreaming and nightmares. Apathy, lethargy, fatigue, withdrawal, and depression may be observed in some poisonings.[97]

B. Chronic Toxicity
1. Human
Despite evidence from animal studies, there has been slow recognition of the fact that organophosphorus esters are capable of producing delayed lesions in exposed humans. While the muscarinic and nicotinic effects of these chemicals have been fully documented, delayed effects on the central and peripheral nervous systems are neither as well known nor as extensively investigated. Early data suggested that complete recovery from light or moderate poisoning would occur and authors either ignored or dismissed the question of delayed lesions.[87-89,91,92] The subtle effects produced, the traumatic circumstances of the poisoning, and the latency of appearance of such signs made it difficult to assess the chronic effects of these chemicals. It is a thought worth pondering that most physicians never see the chronic problems arise since the patient is generally discharged following alleviation of the immediate toxic signs without further examination at a later date.

2. Psychopathological Effects
The earliest recorded indications of delayed psychopathological lesions caused by organophosphorus compounds appeared in 1950 in a paper describing the effects of

diisopropylfluorophosphate (DFP) treatment of schizophrenia and manic depressive psychoses, attributing the observed effects to an anticholinesterase action on the central nervous system.[90] Grob et al., in earlier papers, had described electroencephalographic changes in man and also mental symptoms including tremulousness, insomnia, and confusion.[87,88] In 1950, Grob et al.[89] noted that, in patients poisoned by parathion (O,O-diethyl O-4-nitro-phenyl phosphorothioate), unpleasant symptoms including tension, anxiety, depression, crying spells, insomnia, and mild anorexia persisted from 1 to 3 weeks after exposure. In 1953, Bidstrup et al.[98] ascribed the mental disorders (delusions, hallucinations) of two patients exposed to mipafox (N,N'-*bis* (1-methylethyl) phosphorodiamidic fluoride) as resulting from overdoses of the antidote atropine. Their conclusions are equivocal in the light of observations reported in later literature.

In the mid-1950s, Spiegelberg began a comprehensive study of the psychopathological-neurological delayed lesions of workers engaged in the production of chemical warfare agents for the Wehrmacht during World War II. The results, first published in 1961, revealed a psychiatric, delayed-effect syndrome among factory workers and testing site personnel which could be subdivided into two groups of symptoms as quoted from the monographs.[99,100]

Group I — consisting of the majority of those examined, was characterized by: (1) persistently lowered vitality and ambition; (2) defective autonomic regulation leading to cephalalgia, gastrointestinal and cardiovascular symptoms and premature decline in potency and libido; (3) intolerance to alcohol, nicotine and various medicines; (4) impression of premature aging.

Group II — containing fewer of those examined showed, in addition to the above symptoms, one or more of the following: (1) depressive or subdepressive disorders of vital function, (2) cerebral vegetative (syncopal) attack, (3) slight or moderate amnestic or demential defects, (4) slight organoneurological defects (predominantly microsymptoms and singular signs of extrapyramidal character). It would appear that these symptoms developed and persisted for some 5 to 10 years following intermittent exposure to those most toxic of organophosphorus esters during the war years.
phosphorus esters during the war years.

A very controversial paper by Gershon and Shaw[101] in 1961 summarized a study of 16 cases of individuals involved in pesticide spraying who were primarily exposed to organophosphorus ester insecticides. A wide range of signs were observed in the patients including

1. Giddiness, floating sensations
2. Tinnitus, nystagmus, pyrexia
3. Tremor and ataxia
4. Paralysis, paraesthesia, polyneuritis
5. Speech difficulties, including slurring
6. Memory defects, slow recall and of saying what was intended
7. Insomnia, somnambulism and excessive dreaming
8. Drowsiness, lassitude, generalized weakness
9. Difficulty in concentrating, mental confusion, emotional lability
10. Uneasiness, restlessness, anxiety
11. Tremulousness
12. Depression accompanied by weeping spells
13. Dissociation
14. Schizophrenic reactions

They summarized their results by claiming that the 14 men and 2 women, exposed for some 10 to 15 years to organophosphorus insecticides, showed only two forms of psychiatric illness — depressive, and schizophrenic reactions accompanied by some impairment of memory and difficulty in concentration. This report created a furor among scientists because of the imprecise nature of the observations, the data being based upon the biased impressions of physicians and unsupported by blood cholinesterase inhibition data.[102,103] A study quoted for comparison was that of Holmes and Gaon[104] who analyzed 449 cases of poisoning in which blood cholinesterase activities were determined. These authors listed *headache, dreams, poor sleep,* fatigability, nervousness, irritability, *increased perspiration,* dizziness, tremor and twitch, paraesthesia, and cold as distinct signs but those in italics occurred in only 49 severely affected individuals in which the blood cholinesterase was inhibited by 40% or more. Stoller et al.[105] conducted an epidemiological study in Australia to ascertain whether organophosphorus insecticides might be associated with the development of schizophrenia or mental depression. No evidence was produced to link the use of organophosphorus esters with psychiatric disorders. Poisonings were associated with psychological deficit but this became normal during convalescence. No matter how attractive the signs reported by Gershon and Shaw[101] might be in supporting a hypothesis of central nervous system intoxification, proper experimental design is essential, and investigators must be aware of the inherent difficulties in objectively studying psychopathological phenomena.

The literature on potential, suspected, and established sequelae of phosphorus ester poisoning does not confirm the often read statement that clinical recovery from nonfatal poisoning is always complete in a few days.[106] Easily recognized, serious, or permanent sequelae have never been observed frequently enough to establish a recognizable pattern. As has been stated by West,[106] somewhere between the two extremes there are a variety of suspected or actual sequelae which may occur in a sizable minority of cases. The literature contains just sufficient anecdotal information to suggest that there are persistent and serious complaints lasting from 6 months to several years and it is worthwhile examining these data.

In assessing the few extensive epidemiological studies of the possible chronic effects of organophosphorus insecticides on man, it has been difficult to relate effects to specific causal agents since occupationally exposed workers have encountered a variety of insecticides, both organophosphorus esters and chlorinated hydrocarbons. It is well known that different effects can be caused by interactions of several chemicals. Invariably, few effects of clinical significance have been observed in occupationally exposed individuals and the differences which have been observed are of questionable significance to health.[107] Sumerford et al.,[96] Kay et al.,[108] and Hayes et al.[109] have carried out extensive studies on population groups exposed to organophosphorus insecticides and have demonstrated that, during the spraying season, there was a lowering of blood cholinesterase levels accompanied by symptoms of toxicity among individuals in close contact with these chemicals, but the cholinesterase levels among people living near sprayed orchards did not differ significantly from those people living in nonagricultural regions. These observations suggested that the chronic effects of insecticides in man would not be identified easily. Davignon et al.,[110] in a study of 441 apple growers, 170 individuals living in the same environment and 162 other persons having no contact with insecticides, demonstrated a higher incidence of leukopenia and neurological manifestations among insecticide handlers, suggesting that insecticides may have some chronic effects in man. There was also a good positive correlation between the incidence of neurological abnormalities (loss of reflexes, tremors, disturbances of equilibrium) and the number of years of exposure to insecticides. An epidemiological study of 114 pesticide workers, located and examined 3 years after incidents of organophosphorus ester poisoning, demonstrated that complaints categorized as optic, gastrointestinal, cephalagia, cardiorespiratory, and neuropsychiatric persisted for up to 3 years.[111] It was concluded that such studies would detect serious after-effects of high incidence, those which followed severe poisoning, those from exposure to specific compounds, and those occurring in hypersensitive individuals but not those effects which were minor.[106]

Following acute poisoning by parathion, most patients recovered from the acute effects within 25 hr though some signs including anorexia, giddiness, anxiety, uneasiness, insomnia, and headache persisted in seriously poisoned patients for up to 72 hr.[89] A few patients have complained of moderate tension, anxiety, depression with crying spells, insomnia, generalized weakness, and anorexia for periods of 1 to 3 weeks after exposure.[89,112] Electroencephalographic changes have been noted in poisoned individuals which persisted for several weeks. In three severe cases, the patients showed definite electroencephalographic changes, interpreted, or hypothalmic spikes, still present 3 years after exposure and accompanied by mental disturbances.[113] Some patients have complained of irritability, nervousness, fatigue, lethargy, and impairment of memory for weeks or months following exposure.[89,112,114] The ability to perform functional tests designed to assess mental alertness was disturbed in pesticide sprayers but was only found in those who already showed physical signs of acute poisoning.[97] Three out of 10 Russian workers who experienced organophosphorus poisoning were affected 2 to 4 years afterward by general weakness, fatigue, and insomnia which reduced their work capacity.[115] In a recent study of individuals accidentally exposed to toxic concentrations of chemical warfare agents, Sarin and Soman, two patients showed persistent psychiatric sequelae.[116] One patient did poorly on a series of psychological tests (visual retention test, word association, proverbs, ink blot, and the Minnesota Multiphasic Personality Inventory) though his performance gradually improved within 6 months. The other patient demonstrated fatigability, dyspnea, restlessness, anxiety, and depression up to 4 months after poisoning but some of these effects might be attributed to concern about a possible cardiac problem apparently unassociated with the pesticide exposure.

Cases involving aircraft pilots are of particular interest since they may be exposed to considerable contamination. Minor poisoning of a pilot may have more serious consequences when his life depends upon the skillful manipulation of a complex machine. In 1964 and 1965, agricultural pilots were involved in 402 and 366 accidents, respectively, while engaged in spraying operations.[117] The toxic effects of insecticides may frequently contribute to accidents of the pilot-error category. Stories of spills when loading insecticides into aircraft, of ruptured pressurized lines flooding cockpits, etc. can be found in the literature. Confusion or slowed reflex activity, being rather commonly observed in those poisoned by organophosphorus esters, would be of great hazard to anyone experiencing these symptoms while operating an aircraft. There are many reports in the literature linking altered mental states to aircraft accidents.[117,118]

A report by Quinby et al.[119] describes a case history of a fatal crash of a pilot who was observed to have been flying erratically following exposure over a period of days to methyl parathion (O,O-dimethyl O-[4-nitrophenyl] phosphorothioate). In another case, a pilot, experiencing severe visual disturbances following a lengthy exposure to methyl parathion, found that his double vision was corrected when he shut one eye during the landing of his aircraft.[119] Other pilots have reported difficulty in landing aircraft following extensive spray routines. Loss of judgment and automatism has contributed to accidents, the former evidenced by indifference to obvious danger, the latter resembling sleepwalking. Durham et al.[97] mention one pilot, the survivor of a crash, who reported that he had seen the obstacle which he had struck but just did not care. Usually cautious pilots have been known to attempt maneuvers which would not have been tried by a rational person. Another report, which discusses psychiatric symptoms in pilots following chronic exposure to organophosphorus insecticides, described neurotic reactions, obsessive-compulsive, phobic, and depressive symptoms as well as despondency with crying spells and a fear of being alone.[114] The chief complaint of another pilot was of recurring episodes of acute anxiety. It was also observed that, over a 2-year interval, his usual genial nature gave way to sudden outbursts of temper. Marked changes in personality have been reported in other studies as well. Once again, the contributions of pesticide exposure to such incidents are often difficult to assess since pilots are exposed to a potpourri of different chemicals and often work 16 to 18 hr in a day, thereby suffering from fatigue and lack of sleep. Durham et al.[97] attempted to answer the question of whether or not the mental effects observed in organophosphorus poisoning cases preceded or took place in the complete absence of the other more traumatic signs of intoxication. Their studies involved two tests designed to assess mental alertness. In the study of 187 cases of suspected organophosphorus poisoning, there were no cases in which mental effects were noted in the complete absence of physical signs or symptoms of illness. In administering the mental alertness tests to 11 individuals who had shown symptoms of recent poisoning, 9 subjects scored poorly in the acute or early-recovery stage and showed progressive improvement during and after convalescence. The results suggested that mental alertness and reflex response could be impaired in aircraft pilots continually exposed to organophosphorus insecticides but that recovery should be adequate if they were removed from the source of contamination for a period of time. This has been confirmed by anecdotal reports in the literature. The major difficulty would concern exposure to those organophosphorus esters which are extensively stored in adipose tissue, being released slowly at a rate sufficient to maintain a constant inhibition of the cholinesterase with the individual subsequently being acutely and perhaps mildly poisoned by another agent causing further depression of brain cholinesterase activity.

Is there a critical level of acetylcholinesterase inhibition at which central effects can be expected to occur? The results in the literature are equivocal as was examined in an earlier section dealing with problem-solving in control and organophosphorus ester-treated rats. In one study, rats having brain acetylcholinesterase activity comparable to 59% of control activity made fewer errors than control animals, but those groups with brain acetylcholinesterases 41, 30, and 25% of control values made significantly more mistakes.[120] In another study, poisoned rats having brain acetylcholinesterase 41, 33, and 24% of controls made significantly fewer errors than control animals.[121] The divergent results could be the result of one group of rats having fluctuating levels of brain enzyme activity whereas in the other study the inhibition was relatively stable.[122] Such studies render the "critical level" concept very unlikely.

The problem has not been put to rest yet, for a recent review of the literature has reexamined the issue and has come to the conclusion that, despite methodologic shortcomings in many of the published studies, there are several behavioral sequelae of organophosphorus insecticide poisoning. These effects include

1. Impaired vigilance and reduced concentration
2. Slowing of information processing and psychomotor speed
3. Memory deficit
4. Linguistic disturbance
5. Depression
6. Anxiety and irritability[123]

Equivocal results have been obtained concerning the presence of less severe or latent forms of behavioral abnormalities in asymptomatic workers repeatedly exposed to organophosphorus insecticides. In a recent study in which exposed workers were treated for abnormalities in memory, signal processing, vigilance, language, and propioceptive feedback performance, none of the individuals was deficient in performance. The results suggested that higher nervous system functions were relatively resistant to mild chronic intoxication.[124]

IV. NEUROLOGICAL LESIONS

Much more clearly defined and recognized than the psychopathological effects caused by organophosphorus insecticides are the neurotoxic effects in man and animals. These lesions, both unpleasant and of a long-lasting nature, consist of a polyneuritis with a flaccid paralysis of the distal skeletal muscles of the lower and upper extremities accompanied by degeneration of the myelin sheaths and axons of peripheral nerves, spinal cord, and medulla.

A. Historical

The earliest reported incident of organophosphorus ester induced lesions occurred in 1899 when 6 cases of polyneuritis were reported among 41 cases of pulmonary tuberculosis treated with phospho-creosote later shown to contain 15% TOCP.[125] Some 53 additional cases were recorded in the next 35 years as this bizarre treatment for tuberculosis enjoyed a certain degree of popularity.[126]

The initiation of prohibition in the U.S. resulted in an outbreak of paralysis of epidemic proportions in an estimated 20,000 individuals who had consumed a certain brand of an alcoholic extract of Jamaican ginger as a substitute for more palatable but unobtainable spiritous liquors. The ethanol content, 60 to 90% was, of course, the attraction to those who used it as a beverage either "neat" or diluted with soft drinks.[127,128] This medical phenomenon, commonly referred to as "Ginger Jake Paralysis" or "Jake Leg" was intensively studied by Maurice Smith and colleagues at the National Institute of Health, U.S. Public Health Service, the results appearing in a series of now classical papers and identifying the harmful contaminants as tri-cresyl phosphate esters and confirming by animal experiments in calves, monkeys, dogs, and rabbits that one isomer, tri-ortho-cresyl phosphate (TOCP) was the specific potent neurotoxin.[129-131] As was stated by Aring in 1942, this tragedy furnished a ramifying medical and social problem for 12 years after the inception of "that noble experiment commonly known as prohibition".[132] It was the subject of a large number of studies in different regions of the U.S. and, perhaps, has been one of the most thoroughly investigated and documented medical disasters on record.[133-135] It even became an integral part of American folklore music between 1928 and 1934, revealing a preepidemic cultural familiarity with this condition.[136] This finding confirms a statement by Kidd and Langworthy[135] that consumption of Jamaican ginger extract went on in not inconsiderable quantities even before the advent of prohibition.

While many of the Jamaican ginger paralysis victims appeared to recover slowly from the exposure, few long-term follow-up studies have been conducted. Studies done in 1937 to 1938, some 6 or more years after exposure revealed that many of the more severely poisoned individuals still suffered serious impairment.[132] The characteristic flaccidity and muscle weakness were replaced by a spasticity, hypertonicity, hyperreflexia, clonus, and abnormal reflexes, indicative of damage to the pyramidal tracts, and a permanent upper motor neuron syndrome. In many patients, recovery was limited to the arms and hands and abnormalities of the lower limbs remained. In a recently published follow-up study of 11 survivors of the original "Ginger Jake" epidemic 47 years earlier, Morgan and Penovich[147] found evidence of persistent damage related principally to an upper motor neuron syndrome with spasticity of the lower extremities accompanied by hyperactive reflexes in the lower extremities. These observations signify that the effects were quite permanent and evidence suggested damage to the spinal cord rather than the peripheral nerves.

A bizarre series of polyneuritic cases occurred in various European countries (Holland, Germany, France, Yugoslavia, Switzerland) in the years 1931 and 1932 as a result of using an alcoholic extract of the oleoresins of the parsley plant called Apiol as an abortifacient.[137,138] It was established that this extract was adulterated with TOCP, present to the extent of 28 to 50%. No one, to date, has fathomed why this substance was chosen as an adulterant since color, odor, and taste of the two preparations are quite dissimilar.[138]

The years 1937 to 1945 saw the introduction of a new hazard in the form of adulterated cooking and salad oils and fat substitutes. Some 68 individuals in Natal developed lower motor neuronal paralysis of the feet and hands as a consequence of eating a soya-bean cooking oil which contained 0.4% TOCP.[140] The deliberate use of mineral oils and fat substitutes for cooking in Germany during World War II resulted in many instances of ataxia and paralysis since these lubricant oils contained significant amounts of crude cresyl phosphate esters.[139,141] A severe outbreak of poisoning occurred in the Swiss army in 1940 when some 80 men ate food cooked by mistake in a TOCP-containing oil.[142] A similar outbreak involving TOCP-contaminated cottonseed oil and 8 individuals occurred in Liverpool in 1946.[143] In 1957, 11 individuals were poisoned by drinking water stored in metal drums which earlier had contained TOCP.[144] The most recent outbreak of epidemic proportions occurred in Morocco, eventually afflicting some 10,000 people who ingested an olive oil which had been mixed with aircraft lubricating and hydraulic fluids containing TOCP by unscrupulous merchants.[145] Similar outbreaks of this affliction occurred in Quang Tri Province, Viet Nam, during 1970 to 1972 following the use of "black market" cooking oil which, on investigation, proved to be a TOCP-containing U.S. military aviation lubricant supplied to South Vietnamese helicopter units.[146]

Occupation-related poisonings with TOCP have occurred infrequently even though this chemical has been used extensively as a plasticizing agent in polyvinyl chloride material and also in a process to recover phenol residues from gas-plant effluents.[139,148] In the first case reported, the three seriously affected individuals probably inhaled cresyl phosphate vapors from wash tanks in a poorly ventilated factory.[139] The ortho-isomer content of the fluid was approximately 60% though the men were also exposed to triphenyl phosphate as well. The workers in question made a slow recovery and were able to return to work within 12 to 42 months of being afflicted.

It would appear that some organophosphorus ester insecticides have the ability to cause a TOCP-like syndrome of muscle weakness, polyneuritis, and paralysis. The first, well-documented incident involved a research chemist and a process worker at a plant manufacturing the pesticide mipafox (isopestox, *bis*-(monoisopropylamino)-fluoro-phosphine oxide).[98] A progressive clinical syndrome closely resembling TOCP-poisoning began following a latent interval of 2 weeks after exposure and resulted in an obvious lower motor neuron lesion with paralysis from which the patients slowly recovered within a period of a year. These poisonings bore a marked similarity to a case in Germany involving parathion which was sprayed in enclosed hot-houses 10 to 12 times in a period of 4 weeks by a single worker.[149] Petty described two cases of permanent nerve damage with attendant muscle, auditory, and vestibular dysfunction, weakness, and easy fatigability as a consequence of multiple exposures to parathion, EPN, and some chlorinated hydrocarbon insecticides in one case and to malathion (O,O-dimethyl-S-1,2-dicarboethoxyethyl phosphorothioate) in the second case.[150] In a study of workers engaged in the manufacture and formulation of insecticides, exposure to two dimethyl phosphate esters, Bidrin (*cis*-isomer of 3-(dimethoxy phosphinyloxy)-N,N-dimethyl crotonamide) and Gardona® (β-isomer of 2-chloro-1-[2,4,5-trichlorophenyl] vinyl dimethyl phosphate), was associated with a high incidence (50%) of electromyographic signs of impaired nerve and muscle function.[151] A severe acute poisoning with another organophosphorus insecticide, fenitrothion (O,O-dimethyl O-[4-nitro-m-tolyl] phosphorothioate), resulted in prolonged muscular weakness and easy fatigability of some 6 months duration in the patient even though electromyographic conduction studies were within normal limits.[112] Recently, Hierons and Johnson[152] reported a progressive neuropathy characteristic of classical delayed neurotoxicity in a man 2 to 8 weeks following acute poisoning by trichlorfon (O,O-dimethyl-1-hydroxy-2-trichloroethyl phosphate). Electrophysiological studies demonstrated widespread denervation in the limbs accompanied by a progressive developing muscle weakness in the lower limbs. Morgan, in a recent report of a study of workers involved in manufacturing leptophos (O-[4-bromo-2,5-dichlorophenyl] 0-methyl phenylphosphonothioate) in a plant in Texas, linked this agent with the spasticity, hyperreflexia of lower limbs, and extensor plantar responses measured.[147,153] It has been demonstrated that leptophos causes a delayed neurotoxicity of the classical TOCP type in chickens.[154]

B. Signs and Symptoms

There are any number of excellent descriptions of the symptomatology of aryl phosphate ester poisoning in the literature and the following description is a distillation of the signs shown by a few examples.[127-131,134,135] In general, the time to onset ranged from 5 to 21 days after ingesting the TOCP and while it should be noted that not all patients showed all symptoms, there was a remarkable constancy in the reported sequence of events.

> The patient would notice a sensation of coldness, numbness, tingling dull aching and soreness of the calf muscles of the legs. The tenderness of the muscles was enhanced by walking or exercise. The peculiar cramp-like sensation in the leg muscles would give way to a weakness of the anterior tibial group of muscles accompanied by difficulty in moving the toes. This progressed rapidly to a flaccid paralysis with a bilateral and symmetrical "foot-drop" at the ankle making it necessary to raise the foot higher in order to avoid dragging it. This exaggerated stepping gate with the feet slapping the ground would be accompanied by an inability to walk properly or to stand still without moving backward and forward in an attempt to stabilize the stance and prevent falling. The thigh muscles would be affected only in severe cases and cyanosis and edema of the lower extremities might be seen.
>
> In most instances, approximately 1 week after onset of the leg paralysis, involvement of the upper extremities would be noted. A numbness of the fingers progressing to weakness of the hand muscles and an inability to accomplish such simple tasks as buttoning up a shirt, lighting a match or holding a heavy object would be experienced. Wrist drop would occur approximately 7 to 10 days after the foot drop, the intrinsic muscles of the hand (the enterossei and those situated at the thenar and hypothenar eminences) becoming weak and paralyzed and then atrophic. Paralysis in the upper extremities usually would not occur beyond the elbow. The severity of the poisoning could be judged by the extent of involvement of the upper limbs.
>
> The patient would experience some sensory disturbances of the glove and stocking type though he/she would usually be able to distinguish between hot and cold, would detect pain and pressure. Superficial reflexes would be normal except for the knee-jerk which would be hyperactive and exaggerated in most cases, hypoactive or absent in severe poisonings. The ankle jerk reflex would be absent and the plantar response would normally be flexor. Sphincter muscles would show normal tone and there would be no visual disturbances and no evidence of impairment of cranial nerves or ganglia. In severe cases, wrist jerks would be impaired.
>
> Recovery from the paralysis would occur in the muscle groups in exactly the reverse order of the appearance of symptoms; i.e., recovery would begin in the extensor muscles of the forearm and in about 5 months after the onset, the wrist drop would have recovered. Tone would return to the interossei and pollicis muscles would recover by 12 to 14 months. In the legs, the thigh muscles would recover fast, the calf muscles functioning sufficiently to raise the toes with effort. Rehabilitation would take anywhere up to 3 or 4 years.

In brief, the clinical picture of TOCP-related poisoning is uniformly that of a rapidly developing, flaccid, bilateral paralysis of the distal muscles of the lower and upper extremities, pointing to an involvement of the distal ends of the lower motor neurons localized in the lower lumbar and lower cervical regions of the spinal cord.[130] In the mildest cases, only foot drop occurs. In moderately severe poisonings, there is weakness of the hand muscles and at least, partial paralysis of the legs; whereas, in severe cases, wrist drop is complete, severe paralysis of the legs occurs, and the muscles of the hand become paralyzed with atrophy and subsequent wasting.

By comparison, few complete and detailed descriptions of poisoning by alkyl phosphate esters exist in the literature beyond coverage of the characteristic acute toxicity accompanying the inhibition of vital tissue cholinesterases. As shall be explained in a later section, not all akyl phosphate esters cause delayed neurological effects and patients showing such effects generally have been exposed to undetermined but high concentrations of the particular agent in question.[98,112,152] One particular case, that of a research chemist engaged in the manufacture of mipafox in a pilot plant, demonstrates the similarity of the development of paralysis to that observed following exposure to TOCP.[98]

The patient, a 28-year-old woman, had experienced acute signs of anticholinesterase poisoning and had received atropine continuously over a 4 day period. Her erythrocyte and plasma cholinesterase were markedly inhibited and on the basis that she was apparently well by the 14th day after the poisoning, she was discharged from hospital.

During the third week after the exposure, she noted weakness of the legs which became worse after exertion. She was readmitted to hospital 25 days after the onset of the acute symptoms and was found to be suffering from a flaccid paralysis of both legs with tenderness and weakness of the thigh muscles. The knee jerks were diminished and ankle jerks were absent. Cutaneous sensation to pinprick or light touch was normal. Five days later, the paralysis of the lower limbs was complete, both knee and ankle jerks were absent and no plantar response could be elicited. Fasciculations of the leg muscles were observed and they were tender to palpation.

Tone and power in the forearms and hands were greatly reduced. The biceps jerk was just present but neither the supination nor the triceps jerk could be elicited. There was weakness of the trunk muscles and muscle twitchings were observed in the deltoid and facial muscles. There was weakness of the long extensor muscles of the fingers and the small muscles of the hand were wasted and completely paralyzed at 6 months after poisoning. The cranial nerves were normal and no signs of central nervous system involvement or disturbance of cutaneous sensation could be demonstrated.

Power returned slowly over a period of weeks to the muscles of the thighs, arms, and forearms. Power and muscle tone were normal in the upper arm and the biceps and triceps jerks were present 65 days after the acute episode. Power had returned to the thigh muscles, the knee jerks were exaggerated, and patellar clonus could be elicited but there was no spasticity. The leg muscles and the small muscles of the feet were completely paralyzed and wasted and ankle jerks were absent. Muscle fasciculations were still present and the patient complained of cramp-like pains of the muscles of the lower limbs. Alternating sensations of burning and coldness in the hands and feet were accompanied by flushing, pallor, and cyanosis.

A slow improvement occurred in those groups of muscles which had begun to regain their normal function by 4 to 7 months after the poisoning episode but the muscles of the legs and feet remained paralyzed. The wasted small muscles of the hands began to recover by 9 months after the initial incident. Power returned slowly to the lumbrials, interossei and opponens pollicis and, by 2 years after poisoning, function was normal. After 9 months, the ankle jerks were first elicited and quite quickly, within a few days, there was a return of movement in the toes. Progress was steady but very slow.*

The other case reported in the study obviously received a more limited exposure to the mipofox since his symptomatology was not as severe as that of the research chemist.[98] Indeed, his signs were more characteristic of the TOCP-related condition as will be observed below.

Nine days after the signs of acute poisoning which were treated with atropine, the patient noted weakness of the legs, associated with cramp-like pain in the muscles of the calves and feet. There was, on being readmitted to hospital, a loss of tone in the muscles of the lower limbs, particularly below the knee. The patient was unable to dorsiflex either foot, the power of plantar flexion was absent in the right foot and weak in the left, ankle jerks were diminished and the plantar reflexes could not be elicited. While his condition improved gradually, the normal function in the right leg was more impaired than that in the left leg. There was a bilateral foot-drop, more pronounced on the right than on the left side. There was wasting of the small muscles of the feet, particularly on the right side.

Six months after the initial poisoning, the patient complained that he still became tired easily and could walk only 200 or 300 yards before his feet, especially the right, became "floppy". He resumed working 10

---

* From Bidstrup, P. L., Bonnell, J. A., and Beckett, A. G., *Br. Med. J.*, I, 1068, 1953. With permission.

months after poisoning but found that standing caused low back pain and a dull aching pain in his right foot and calf muscles which occurs most frequently at night. It was necessary for him to have a more sedentary job.*

Since few people died as a result of exposure to TOCP, it was difficult to carry out an extensive pathological examination of tissues. The few cases which were examined did demonstrate what had happened but not the mechanisms by which it occurred.

As might be expected from the clinical signs, the lesions were localized chiefly in the peripheral (radial, anterior tibial, sciatic) nerves and consisted mainly of scattered, patchy degenerative changes in the myelin sheaths with occasional fusiform enlargements of the axis cylinders.[155,156] The myelin sheaths of all axons were not injured nor did the degeneration necessarily extend throughout the entire section. The degenerated foci were characterized by a widening of the sheaths and the formation within of dark blue or black-staining balls, circles, and irregular broken fragments of myelin material. It is worth noting that myelin sheaths outside of the foci of definitive degeneration did not have the clear cut herring-bone pattern of a normal neuron, suggesting evidence of an early stage of degeneration.[134]

Jeter described the essential pathological changes in the brain of a fatal case of TOCP poisoning as consisting of thickening and edema of the meninges, perineural exudate, thickening, and fibrosis of the perineurium of the peripheral nerves and endarteritis of the vessels of the central nervous system.[157] Aring described myelin degeneration in the spinal cord, being most severe in the lower segments though stopping short of the medulla and not being seen in the cerebrum.[132] The anterior horn cells of the lumbar, upper thoracic, and lower cervical regions of the spinal cord were affected.[158] The changes were characterized by occasional absence of the nucleolus, eccentric nuclei being found in the periphery of the cell, a swollen appearance, chromatolysis, and the formation of a granular material in the cytoplasm.[155] Sections of spinal cords taken at various levels showed the presence of five reddish-brown granules in the center and the periphery of the cytoplasm of practically all anterior horn cells.[134,156] Cells in the grey matter of the anterior horns were also affected, being markedly reduced in number with those remaining being pyknotic with an ill-defined outline.[132] Sections of the medulla and pons showed some changes in many of the ganglion cells, suggesting different stages of degeneration. No changes were observed in the nerve cells of the cerebral and cerebellar cortex.[134]

In concluding his study of fatal cases, Vonderahe[156] made an uncanny prediction of the mechanism of action of the toxin. Remembering that this was in 1931, he stated that

the pathological observations suggest a toxin which reaches the peripheral nerves in the extremities through the circulation but which subsequently travels in the central nervous system along the nerve fibers and anterior roots. There is an obvious tendency of the toxin to exert its pathological effect most severely on the motor cells and fibers.**

C. Mechanisms

The mechanisms by which organophosphorus esters cause their psychopharmacological and neurological trauma have been studied at great length though it would be fair to state that only the relationship between chemical structure and neurotoxicity has proven to be a fruitful field of investigation. That most of the neurotoxic agents have been anticholinesterase agents led to the obvious hypothesis involving nervous tissue cholinesterases.

---

*   From Bidstrup, P. L., Bonnell, J. A., and Beckett, A. G., *Br. Med. J.*, I, 1068, 1953. With permission.
** From Vonderache, A. R., *Arch Neurol. Psychiat.*, 25, 29, 1931. With permission.

Studying TOCP, Bloch[159] suggested that inactivation of motor end-plate cholinesterase was the possible cause of paralysis. This hypothesis proved difficult to support with experimental evidence since TOCP is almost totally inactive toward nervous tissue AChE. There did not appear to be any relationship between AChE inhibition and paralysis.[160] Other neurotoxic agents such as DFP and mipafox were good inhibitors of nervous tissue AChE.[161] It was discovered that these three agents (DFP, mipafox, TOCP) were potent inhibitors of nervous tissue butyrylcholinesterases (BuChE), generally associated with white matter in the brain and spinal cord where demyelination occurred.[162,163] This observation promoted the hypothesis that the inactivation of this enzyme was responsible for the demyelination though it was realized that, while the plasma, brain, and spinal cord BuChE of hens was markedly inhibited, the tissue AChE was not affected.[163] Marked species selectivity was also observed in that while the BuChE of human nervous tissue was markedly inhibited, TOCP was less effective toward this enzyme in hen tissues and totally inactive toward BuChE in rat nervous tissue.[161] While TOCP readily causd demyelination in humans and chickens, it was ineffective in the rat, observations which appeared to support the BuChE-related hypothesis. The defect in this hypothesis is, of course, that any potent inhibitor of BuChE should be neurotoxic if inactivation of this enzyme is a necessity for demyelination. This hypothesis was disproven by an extensive study of a number of phosphate esters which showed that all, with the exception of TOCP, caused a rapid (within 2 hr) depression of central nervous system BuChE, followed by a recovery of enzyme activity in 6 to 8 days.[164] With TOCP, the rate of inhibition was much slower and the BuChE remained inhibited for 10 days or more. Continual daily administration of potent BuChE inhibitors for 10 to 14 days did not result in neurotoxic effects.[165] Later research demonstrated that the paralysis produced in chickens treated with EPN or malathion differed from that noted with TOCP, DFP, and mipafox in both onset and duration.[166] Whereas, the "classical" paralysis developed in 10 to 14 days, the syndrome was observed within 1 to 2 days of receiving malathion or EPN.

With the development of potent fluorine-containing alkyl phosphate esters, a new hypothesis was developed which centered around the release of the fluorine *in situ* with an interaction occurring in some unknown metabolic pathway and resulting in a biochemical lesion.[167] While this was a useful theory and contributed to the overall investigation of the phenomenon, it did not explain the potent neurotoxic actions of triaryl phosphate esters.

In a review of the topic in 1963, Davies[168] described some typical features which must be considered when formulating any complete theory of the neurotoxic actions of organophosphorus esters. These features, as valid today as when they were written, include

1. The characteristic delay in the onset of clinical signs.
2. The marked species differences in response to such compounds.
3. The highly specific sites in which the histological lesions are found.
4. Why, although all the active neurotoxic agents with but one exception are anticholinesterase agents, there are many anticholinesterase agents which are inactive.

No hypothesis has been formulated to date which will support all of these features, but, in the following section, we shall examine evidence, both animal and human, which demonstrates that we are closer to the answers concerning the mechanism(s) of action.

1. Psychopharmacological

Changes in behavior and neurological function have given rise to a new field of occupational health research termed "behavioral toxicology" in which a key concept is that toxic effects may manifest themselves as subtle disturbances of behavior (and function) long before any classic symptoms of poisoning become apparent.[169] One might also add that behavioral signs may persist long after apparent recovery of the more acute signs of poisoning. Behavioral and neurophysiological methodology may be more relevant and more sensitive to nervous tissue dysfunction resulting from long-term, low-dose exposure to toxic agents and may be more useful in predicting irreversible tissue damage.[170] Advances have been made in the study of behavioral responses following exposure to organophosphorus esters.

The importance of the cholinesterase (acetylcholine-acetylcholinesterase) system in the central nervous tissue can scarcely be questioned but its precise role is not well defined even though it has been studied longer (and in greater depth) than any other neurochemical mechanism in the central nervous system.[171] For a better understanding of the interpretation of the vast number of experiments conducted with cholinergic agents and anticholinesterase agents upon the CNS, the reader is referred to Holmstedt,[94] Diamant,[172] Machne and Unna,[173] and Karczmar.[174] It seems, therefore, that an excess of free endogenous acetylcholine as produced by anticholinesterase administration leads to altered states in the central nervous tissue. There is little doubt that acetylcholine does have direct central pharmacological effects but the full importance of this neurotransmitter in central nervous system physiology has yet to be realized. Some insight can be obtained by an examination of the biological effects of anticholinesterases.

Organophosphorus anticholinesterase agents can induce "excitatory" epileptiform alterations in the electroencephalograms (EEGs) of experimental animals, resulting in an alerting effect accompanied by desynchronization. An initial effect, desynchronization implies that the EEG is composed of low voltage, fast activity over the entire surface of the brain, the hypothetical implication being that the neural units are firing out of phase with one another.[94] In contrast, synchronization implies that the EEG record exhibits large slow waves and spindle bursts as are found during drowsiness and sleep.[94]

Bowers et al.,[171] in a study of military volunteers receiving dermal applications of an organophosphorus nerve gas (EA-1701) similar to Sarin, noted a good correlation between CNS signs and erythrocyte AChE depression. Psychological tests were employed to assess subject capability in addition to recording descriptions elicited from the individuals. The duration of psychological impairment occurred over a 24-hr period following administration though it was maximal in the first 8 to 10 hr. The signs included a feeling of fatigue, lethargy, irritability, and tenseness, a "slowed down all over feeling" accompanied by sluggish though appropriate responses to questions, and loss of ability to read with comprehension. It is of interest to note that the subjects were cognizant of the fact that they were performing poorly on the tests but consciously could do nothing about it except apologize. Subjects could be placed in one of three groups, associating psychopharmacological effects with whole-blood cholinesterase activities. Group I, those with activities not less than 81% of control levels, showed little effects. Psychological symptoms (anxiety, psychomotor depression, intellectual impairment, and unusual dreams) were observed in those with cholinesterase activities between 41 to 80% of normal whereas the same symptoms became quite severe in individuals with activities 10 to 40% of normal. While the preceding was an acute study, Metcalf and Holmes[175] demonstrated a slowing down of performance in psychological tests in individuals chronically exposed to organophosphorus pesticides. As has been mentioned in an earlier section, several scientists have commented on the altered physical and mental capabilities of individuals acutely or chronically exposed to organophosphorus esters.[97,104,117,122]

Behavioral studies in experimental animals exposed to organophosphorus esters have been confined to three research groups, the results being reviewed by Russell[176] and by Clark.[122] The central theme developed by Russell was that inhibition of brain AChE markedly affected extinction of a response without influencing other functions. The dose-response relationship was not linear in that, above a "critical level" of 40 to 50% of normal activity, responses occurred but, below this level, the rate of extinction became slower as AChE decreased. Richardson and Glow,[177,178] in a study of discrimination behavior in rats poisoned with DFP, concluded that the major effect of inhibited cholinesterase activity was to reduce the stability of performance. Banks and Russell[179] showed that as the inhibition of the brain AChE increased, animals (rats) in problem solving situations tended to make more mistakes. In contrast, the experiments conducted by the Pearson group demonstrated that disulfoton (O,O-dimethyl S-[2 ethylmercaptoethyl] phosphorodithioate) poisoned rats in problem-solving and maze situations made significantly fewer errors and had shorter running times than did control rats.[122,180] According to Clark,[122] the only decrement in behavior of experimental animals poisoned with organophosphorus insecticides that appears to be established is an increase in the number of trials needed for extinction of a response. In a mouse study involving the acute administration of parathion and effect on memory in a one-trial, passive avoidance task, learning was disrupted when the organophosphorus ester was given within an hour of the trial.[181] Subacute treatment had no effect upon learning and it appeared that compensation occurred with the prolonged depression of brain AChE activity. In a more recent study of the effects of malathion on conditioned avoidance performance in the rat, avoidance was significantly impaired 1 hr after injection (50 mg/kg bw) even though there was little inhibition of brain AChE by this ester.[182] Higher doses produced more significant enzyme inhibition as well as disruption of avoidance performances though the authors found that the behavioral and enzymatic decrements did not necessarily coincide. Possibly, the important feature of this study may be that behavior can be affected without significantly altering enzymatic activities. The subacute administration of fenitrothion to rats markedly altered conditioned avoidance responses, diminishing the number of responses, lengthening the latency period, and slowing the tendency toward extinction.[183]

In an earlier section, the persistent central effects of organophosphorus esters resulting from intermittent exposure of pesticide handlers to rather high doses were described. In most cases, as the tissue cholinesterase activities returned to normal at a rate of approximately 1.0%/day, recovery would be complete within a period of 3 to 4 months of the last exposure.[93] However, one is reminded of the study of Spiegelberg who showed persistent symptoms in military personnel some 10 years after exposure to nerve gases.[99] Recent studies in primates (Rhesus monkeys) exposed to single large doses of Sarin revealed increases in the relative amount of beta voltage (15 to 50 Hz) of electroencephalograms (EEGs) which persisted for 1 year.[184] Similar effects on the frequency spectrum were elicited following a series of subclinical exposures to sarin. In their study, Metcalf and Holmes[175] demonstrated a high incidence of low-to-medium voltage, slow wave-activity in the theta range of frequencies (4 to 7 Hz) occurring in bursts of 2 to 4 sec duration. Korsak and Sato,[185] in a study of workers occupationally exposed to organophosphorus esters, found maximal effects in the frontal areas of the brain, the EEG results being similar to those reported by Metcalf and Holmes with the exception of a mixed pattern of slower alpha range of frequencies (9 to 13 Hz), theta and beta (22 to 25 Hz) range frequencies. A recent paper, studying the waking and sleep EEGs of a population of industrial workers with histories of accidental exposure to Sarin, revealed statistically significant increased beta activity, increased delta and theta slowing, decreased alpha activity, and increased amounts of rapid eye movement sleep.[186] The authors of this paper suggested that their EEG findings represent an unexpected persistence (more than 1 year) of known short-term organophosphate actions resulting in long-term changes in brain function. While sarin is a particularly potent agent, one must consider the results of these studies in the context of those observations described for pilots and other spray personnel exposed to the more commonly used insecticidal organophosphorus esters. Evidence of central nervous system abnormalities attributable to organophosphorus compounds may be found in workers at a time when tissue cholinesterase activity is probably within normal limits.

The pharmacological mechanisms underlying this persistent effect have not been elucidated, but Duffy et al.[186] have speculated on at least three possible explanations. The first suggested that the low tissue cholinesterase activity and resultant acetylcholine excess at the time of exposure may have induced chronic changes in synaptic morphological and biochemical organization which are exceedingly slow to reverse. A second explanation suggested that the high acetylcholine concentration at the time of exposure altered post-synaptic receptors in the central nervous system, making them chronically more sensitive to endogenous acetylcholine. A third mechanism raised the possibility that exposure to organophosphorus esters may result in alterations of cerebral blood flow resulting in hypoxia, a mechanism which was suggested by evidence that cardiovascular regulatory mechanisms in the brain are under partial cholinergic control.[187] A fourth explanation for the chronic effects suggested that organophosphorus esters may have actions in addition to that of potent anticholinesterases, a subject which will be presented in the next section.

2. Demyelination

Subsequent to a detailed study of the epidemiological and experimental evidence associated with the episode of "Ginger Jake" paralysis in the early 1930s, Smith and Lillie[131] concluded that the neurotoxic action of TOCP was associated with a degeneration of the myelin sheaths at the distal ends of peripheral nerves, with varying amounts of degenerative changes affecting the anterior horn cells throughout the lumbar and cervical regions of the spinal cord. In both animal experiments and affected humans, it was observed that myelin degeneration was matched by an equal amount of axonal destruction. In most species studied, there was a delay of 8 to 12 days (up to 21 days in humans) after a single dose before structural and functional changes occurred in the nervous system. It has never been possible to experimentally shorten this time interval. What is happening in the nerve cells during this "silent" period has been regarded as being the most important question relating to the underlying metabolic lesion.[188] In 1954, Cavanagh[189] stated that little attention had been paid to the mode of onset of neurological changes in organophosphorus ester-induced neuropathies, whether changes in the myelin were primary or secondary to the degeneration of the neural axis cylinder.[189] If myelin degeneration was a primary event, then investigations should be directed toward myelin synthesis. If the myelin effect was secondary, studies should be directed toward biochemical effects of TOCP on the neuron itself.

In the species (man, cat, and the chicken) which are sensitive to those organophosphorus esters which induce neurotoxicity, the morphological changes follow a stereotyped pattern.[189-193] The earliest microscopically visible changes occur after the onset of neurological signs and appear first at the distal ends of small myelinated branches of large-diameter fibers innervating peripheral skeletal muscle.[189,193] Axons begin to balloon, show nodularity and fragmentation. Different fibers do not show the same stages of degeneration at any one time.[189] Nerve cells show no morphological changes other than a chromatolytic reaction to the degeneration of the distal axon. Short nerve fibers with proximally lying terminals are less affected than longer fibers.[189] Changes are observed in large diameter fibers in nerve trunks at a later stage and, still later, neuromuscular junctions show swelling and morphological changes including lamellated inclusions or "myelin figures".[193] While distal nerve endings appeared normal, intramuscular nerve bundles showed extensive degeneration, the most affected end-organ being the annulospiral formation of the muscle spindle which is supplied by-large-diameter fibers.[194] Ultrastructural studies have provided the most conclusive evidence that axonal changes are well advanced before the myelin begins to disrupt.[192,193,195] The nerve fibers continue to degenerate for several weeks following observance of the first visible damage. It has been concluded that the process by which the neuron is affected is a distal atrophy or "dying back" phenomenon.[192,193] This process is a relatively common neuropathological change in motor neuron disease, Friedreich's ataxia and other spinocerebellar abiotrophies, and in amyotrophic lateral sclerosis.[188]

In the central nervous system, a similar degeneration has been observed, starting at the distal ends of small myelinated long fibers in both ascending (dorsal columns and spinocerebellar tracts) and descending (corticospinal and tectospinal) tracts of the gray matter. The ascending tracts showed most damage rostrally, whereas, caudal regions of descending tracts showed the greatest change.[189] With the onset of neurological signs, the boutons terminaux in the spinal gray matter become enlarged, take on an irregular shape and a granular appearance.[193,196] At a later stage, degenerative changes appear in the preterminal axons.[197] In contrast to the observations of Cavanagh and colleagues, Prineas[193] found that the TOCP-induced changes in the central nervous system closely resembled those observed in peripheral axons and neuromuscular junctions.

The morphological changes in nerve fibers are indistinguishable from those of Wallerian degeneration which occurs when a nerve fiber is transected from the cell body, thereby interfering with nutrient transport. The axis cylinder is destroyed apace with the degeneration of the sheath, no evidence being observed which would suggest that demyelination occurs at a different rate than axonal fragmentation.[189] Segmental destruction of the myelin sheath may be dependent upon some local metabolic inadequacy in Schwann cells which may control adjacent internodal stretches of sheath but such distinct focal lesions have not been observed in the early degeneration phase. However, it is improbable that myelin degeneration alone could produce such a dramatic effect on the physiological function of the axon since it has been shown that, in starvation, the myelin in rat nerves disintegrated though the nerve fiber remained intact and functional.[198] Manjo and Karnovsky,[199] comparing the sciatic nerves of mipafox-treated rats with those in which a surgical transection of the nerve had been done, demonstrated a breakdown of the "blood-nerve barrier" in both groups, the trypan blue dye leaking into the swollen nerves. Dye uptake was much more pronounced in the transected nerve, coloration being observed up to the point of transection. In contrast, dye penetrated only the distal portion of the sciatic nerve in the organophosphate-treated animals.

Cavanagh[188] suggested that the metabolism in all nerve cells may be affected following exposure to organophosphorus esters but that the degree of morphological damage may be severest in those fibers which make the greatest demands for cellular nutrients. There is very little biochemical evidence to suggest that interference in a metabolic pathway, resulting in faulty or inadequate synthesis of essential nutrients, may be the cause of the lesion. In mipafox-treated rats and hens, oxygen uptake and lipid biosynthesis from acetate was markedly depressed in the proximal portion of morphologically normal sciatic nerves but was substantially increased in the distal degenerating branches.[199] The same functions were depressed in slices of brain and spinal cord of treated rats and hens, suggesting that this is not necessarily a specific toxicity-related effect. The observed morphological changes may suggest that the transport of essential macromolecules along the length of the axon may be impaired, the nutrients not being replenished in the damaged distal axon, thereby aggravating the degenerative situation.[193] This concept is supported by recent experiments demonstrating that neurotoxic organophosphorus esters inhibited axoplasmic transport of tritiated proline in the rat optic nerve.[200] In contrast, the importance of microtubular transport mechanisms in neural axons cannot be quantitated since the intraneural injection of colchicine which depolymerizes microtubular proteins causes Wallerian degeneration backward from the injection site and not distally to it.[192-194]

A recent reexamination of organophosphate-induced neuropathy using a teased fiber preparation of the recurrent laryngeal nerve of DFP-treated cats has revealed that axonal degeneration was initially focal, not associated with the nerve ending and that it subsequently spread in a somatofugal direction and eventually affected the entire distal axon.[201,202] Varicosities in the nerve fiber and paranodal demyelination preceded axonal degeneration. The varicosities were associated with intra-axonal and intramyelinic vacuoles.[202] It is now suggested that neurotoxic organophosphorus compounds induce a focal "chemical transection" of distal but not terminal portions of the axon, this effect precipitating Wallerian degeneration of the more distal portion. Bouldrin and Cavanagh[202] now claim that the traditional hypothesis of the retrograde axonal degeneration observed in dying-back neuropathies is not valid for organophosphorus ester-induced neuropathies.

V. STRUCTURE — ACTIVITY RELATIONSHIPS AND NEUROTOXICITY

While the clinical and morphological effects of organophosphorus-induced neuropathy were widely known by the early 1940s, the process which initiated the degeneration of the peripheral nerves and spinal cord tracts was poorly understood. In 1941, Bloch demonstrated that TOCP inhibited the cholinesterase of horse serum and was the first to suggest that the inactivation of the cholinesterase at the motor end-plates of peripheral nerves might be the cause of the paralysis.[203] While this was an interesting and challenging hypothesis, it was also untenable since TOCP had little or no effect on the nervous tissue acetylcholinesterase.[204] As the pharmacology of other organophosphorus esters such as DFP became known and the importance of the nervous tissue acetylcholinesterase was recognized in association with DFP-induced paralysis, interest was renewed in the cholinesterase-TOCP relationship.[86,205-207] At a somewhat later date, it was discovered that the areas of white matter in the brain and spinal cord contained a cholinesterase, a butyrylcholinesterase, or pseudocholinesterase, similar to that detected in horse serum.[208,209] Some 15% of the brain cholinesterase activity could be attributed to this enzyme. This raised the possibility of the butyrylcholinesterase being involved in myelin metabolism, the argument being supported by (1) the histochemical localization of the enzyme in the Schwann and supportive glial cells around the nerve axon in close proximity to the myelin and (2) the fact that the only known neurotoxic agents at that time, TOCP, DFP, and mipafox, were potent inhibitors of this enzyme.[210,211] Earl and Thompson[212] concluded that the inactivation of butyrylcholinesterase was a prelude to demyelination. This hypothesis, too, was overthrown by several investigators who demonstrated either that organophosphorus esters were potent inhibitors of butyrylcholinesterase but had no effect on myelin or that, when the nervous tissue butyrylcholinesterase activity was inhibited, there was no appreciable change in the turn-over of lipid phosphorus.[164,213,214] Other experiments showed that the brain butyrylcholinesterase could be kept inhibited for 10 to 14 days by repeated administration of TOCP, the length of the "silent period", without causing neurotoxicity.[165,215] Both Barnes and Denz[190] and Cavanagh[194] found that delayed neurotoxicity in the cat or hen by either TOCP or mipafox was not associated with the level of inhibition of nervous tissue butyrylcholinesterase. These observations initiated an intense study into the structural properties of organophosphorus esters necessary to elicit delayed neurotoxicity.

The original detailed pharmacological studies by Smith et al.[129,131] left no doubt that the active ingredient in the adulterated ginger extracts was the phosphoric acid ester of ortho-cresol, this particular ester being far more biologically active than the esters of para- or meta-cresol.[129,131] Despite a number of studies to ascertain the nature of the biochemical lesion, it was not until the mid 1950s that Aldridge demonstrated that homogenized liver fortified with reduced diphosphopyridine nucleotide enhanced the toxicity of TOCP.[216] Myers and colleagues found that the material extracted from rat tissues was much more toxic than the TOCP administered.[217] Casida et al.[218,219] published a short paper demonstrating that metabolites possessed the potent anticholinesterase activity rather than the parent chemical. In fact, it appeared to be one particular metabolite. Subsequent to the in vivo hydroxylation of the O-methyl by hepatic microsomal enzymes, spontaneous cyclization resulted in a product which has lost one of the cresol groups. This reactive compound appeared to actively react with serine proteases by opening the cyclic phosphate structure at the P-O-aryl bond.[220] The cyclized metabolite (2-[O-cresyl]-4-H-1:3:2-benzodioxaphosphoran-2-one or CBDP) has been synthesized and found to induce ataxia in hens and cats, at a dose of 4 to 8 mg/kg

bw, to be a potent inhibitor of chymotrypsin and to cause a neurotoxic syndrome in hens identical to that produced by TOCP.[220-222] Interestingly, while CBDP was originally isolated from the liver, intestine, and feces of rats, the rat is resistant to the typical neurotoxic effects. The administration of TOCP to cats, a susceptible species, resulted in the isolation of CBDP from the intestine some 15 to 36 hr after dosing.[223] Eto et al.[220] demonstrated that while spontaneous cyclization of TOCP could occur, this reaction was catalyzed by plasma albumin. Sharma and Watanabe[224] demonstrated that TOCP administered to chickens was extensively concentrated in the liver, the amount of CBDP representing 71 and 74% of the total concentration at 48 and 72 hr, respectively. Extensively bound in the liver, CBDP may be responsible for the gradual increase in levels detected in various visceral organs and nervous tissue for several days following a single oral dose.[224] The delayed neurotoxic effect observed with TOCP may be due to the slow accumulation of this biologically active metabolite in the nervous tissue.

A recent report suggests that TOCP may exert a direct effect on the nerve axon by altering some structural components, particularly the membranes of synaptic vesicles and axons, causing a breakdown of the permeability barrier.[225] The TOCP, a rather inert molecule, may be incorporated into structural lipid elements of the nervous tissue.

Table 5
STRUCTURE-ACTIVITY RELATIONSHIPS OF NEUROTOXIC AND NONNEUROTOXIC ORGANOPHOSPHORUS ESTERS[230]

Neurotoxic	Nonneurotoxic
$(2\text{-}CH_3\text{-}C_6H_4\text{-}O\text{-})_3 P=O$	$(4\text{-}CH_3\text{-}C_6H_4\text{-}O\text{-})_3 P=O$
$(2\text{-}C_2H_5\text{-}C_6H_4\text{-}O\text{-})_3 P=O$	$(4\text{-}C_2H_5\text{-}C_6H_4\text{-}O\text{-})_3 P=O$
$(C_2H_5O)(C_6H_5)P(=O)\text{-}O\text{-}C_6H_4\text{-}NO_2$	$(C_2H_5O)(CH_3(CH_2)_4)P(=O)\text{-}O\text{-}C_6H_4\text{-}NO_2$
$(Cl\text{-}CH_2CH_2O\text{-})_2 P(=O)\text{-}O\text{-}C_6H_4\text{-}NO_2$	$(C_2H_5O\text{-})_2 P(=O)\text{-}O\text{-}C_6H_4\text{-}NO_2$
$(CH_3O)(Cl\text{-}CH_2CH_2O)P(=O)\text{-}O\text{-}CH=CCl_2$	$(CH_3O\text{-})_2 P(=O)\text{-}O\text{-}CH=CCl_2$
$(iC_3H_7NH)_2 P(=O)F$	$((CH_3)_2N)_2 P(=O)F$
$(iC_3H_7O\text{-})_2 P(=O)F$	$(iC_3H_7O\text{-})_2 P(=O)\text{-}O\text{-}P(=O)(\text{-}O\text{-}iC_3H_7)_2$

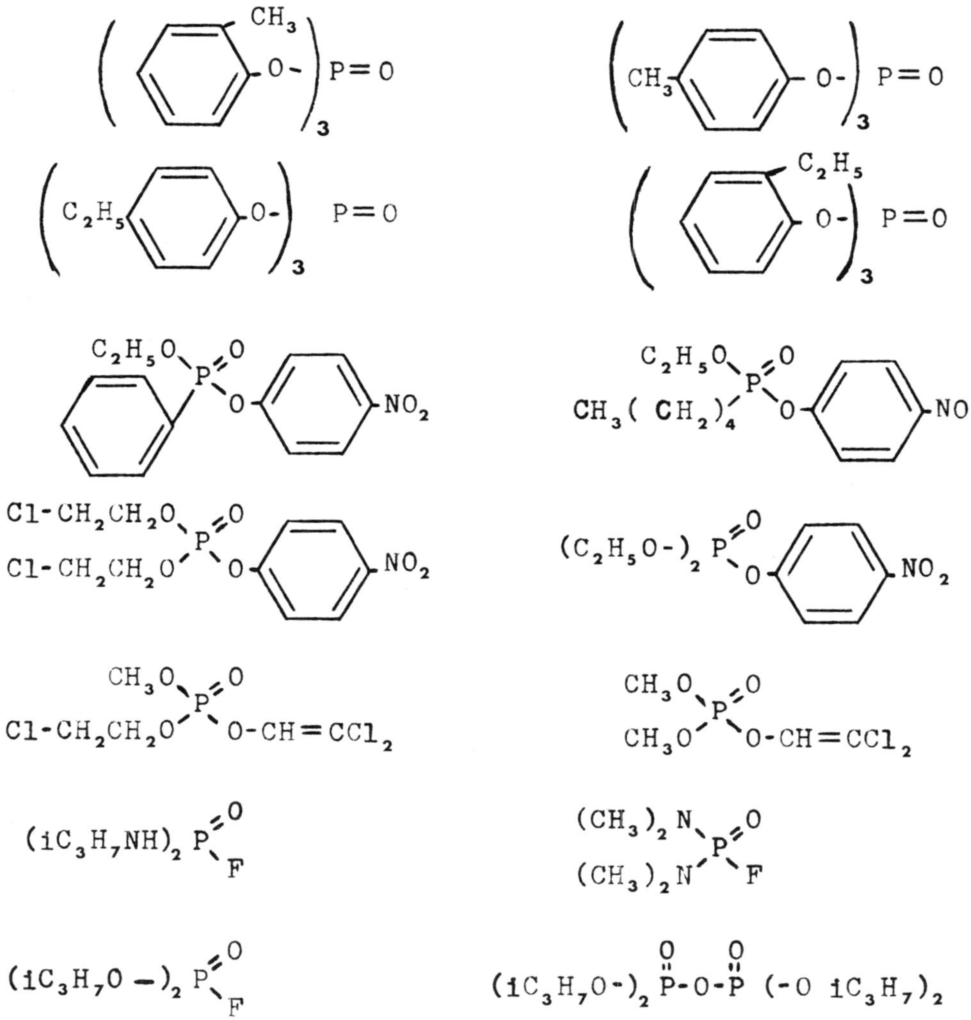

The importance of these findings await further experimentation but it is of interest to note that the interpretation of Bischoff finds support in the observations of Morazain and Rosenberg[226] who were unable to detect the presence of the potent antiesterase metabolites in spinal cords or sciatic nerves of TOCP-poisoned chickens.

While it was recognized that many organophosphorus esters could inhibit nervous tissue cholinesterases without causing neurotoxicity, no correlation was evident between chemical structure and this unique property.[227-230] A group of structurally similar but biologically unrelated agents is shown in Table 5.[230] There was also no relationship between the prolonged inhibition of brain cholinesterase and the paralytic syndrome.[231] The prolonged inhibition of cholinesterases by organophosphorus esters may be due to the persistence of these lipophilic inhibitors in nervous tissue. It was also suggested that administration of "neurotoxic" organophosphorus esters resulted in a selective inhibition rather than a prolonged inhibition of nerve esterases.[232]

In an extensive study of the potential of a series of organophosphorus esters to cause neurotoxicity in the chicken, Davies and colleagues[233] postulated that a compound having the following chemical structure would be neurotoxic.

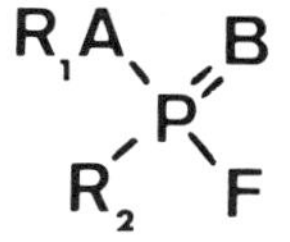

$$\begin{array}{c} R_1A \diagdown \quad B \\ \quad P \\ R_2 \diagup \quad F \end{array}$$

F = fluorine
R_1 = alkyl
R_2 = alkyl, alkoxy or alkylamino
A = oxygen or 2° amine
B = oxygen or sulfur

All of the phosphorofluoridates tested were neurotoxic. At least one alkoxy group was necessary, alkylphosphorofluoridates being neurotoxic, whereas dialkylphosphinic fluorides were not. The ester oxygen (A group) could be replaced by an amino group as in mipafox, an analog of DFP, as is shown below. The N,N'-di-n-butyl analog of mipafox was also markedly neurotoxic.[234]

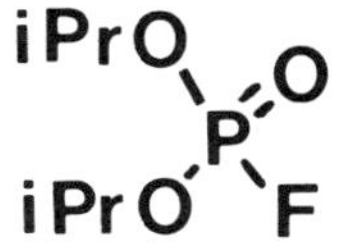

$$\begin{array}{c} iPrO \diagdown \quad O \\ \quad P \\ iPrO \diagup \quad F \end{array} \qquad \begin{array}{c} iC_3H_7NH \diagdown \quad O \\ \quad P \\ iC_3H_7NH \diagup \quad F \end{array}$$

DFP 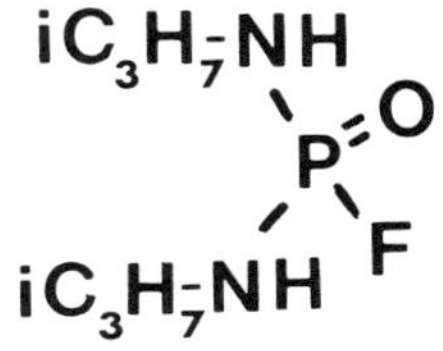   Mipafox

The significance of the fluorine atom suggested that it played some direct role in producing the lesion. It is just possible that the potency of phosphorofluoridates may be related to their physicochemical properties. These acyl halide esters are hydrolytically more stable and are more lipophilic and, therefore, possess superior penetration and partition properties.[230] Steric limitations are minimized because of the small size of the fluorine atom and it is found that an ester such as DFP will react at the esteratic sites of all serine proteases including all esterases, chymotrypsin, trypsin, cathepsin, etc. Changes in structure from phosphoro- to phosphono-fluoridates result in increased activity since the latter are more stable and hence more reactive.

Extensive studies have been carried out to assess the neurotoxic potential of many triaryl phosphates, the results being summarized efficiently in Table 6, taken from Eto.[235] The crucial step for any of the active esters appears to involve the "lethal synthesis", as with TOCP, of hydroxylated intermediates which in turn can cyclize to form neurotoxic products. Ortho-substituted phenyl phosphates are transformed to cyclic saligenin phosphates which are neurotoxic. The p-substituted phenyl phosphates are oxidized to phosphates of p-hydroxybenzaldehyde which may be destroyed rapidly by other mechanisms. The O-ethylphenyl phosphates are unable to form saligenin type compounds after hydroxylation. The p-ethylphenyl phosphates may be dehydrogenated to stable and neurotoxic phosphates of p-hydroxyacetophenone.[236] Meta-alkyl phenyl phosphates do not produce neurotoxic products, the hydroxylation followed by dehydrogenation being degradative steps. The chemical reactivity of triaryl phosphates as phosphorylating agents is dependent upon the cyclic ester structure and the electron attractive substituent group while the steric property of the nonleaving group is important for the selectivity of the ester. As insecticides continue to be developed, routine assessment of their neurotoxic potential will be carried out and several potent esters have been identified including S-seven (O-ethyl O-2,4-dichlorophenyl phenylphosphonothioate).[237-239] Many of these agents conform to the structural requirements essential for causing delayed neurotoxicity while a few do not.

Table 6
COMPARISON OF THE STRUCTURES AND NEUROTOXIC POTENTIAL OF TRIARYL PHOSPHATE ESTERS

A	B	C	Dose mg/kg	Neurotoxicity
H	H	H	1,000	−
2-Me	H	H	50	+
2-Me	2-Me	2-Me	25—200	+
2-Me	2-Me	3-Me	250	+
2-Me	2-Me	4-Me	25	+
2-Me	3-Me	3-Me	100	+
2-Me	3-Me	4-Me	100	+
2-Me	4-Me	4-Me	50	+
2-Me	3,5-diMe	3,5-diMe	1,000	+
2-Et	H	H	1,000	+
2-Et	4-Me	4-Me	200	+
2-Et	2-Et	H	250	+
2-Et	2-Et	4-Me	500	+
2-Et	2-Et	2-Et	500 × 5	+
2-Et	3,5-diMe	3,5-diMe	500	+
2-n-Pr	4-Et	4-Et	200	+
2-n-Pr	2-n-Pr	4-Me	1,000	+
2-n-Pr	2-n-Pr	2-n-Pr	1,000	−
2-n-Pr	3,5-diMe	3,5-diMe	1,000	−
2,3-diMe	2,3-diMe	2,3-diMe	4,250	+
2,4-diMe	2,4-diMe	2,4-diMe	2,000	+
2,4-diMe	2,3-diMe	3,5-diMe	10,400	+
2,5-diMe	2,5-diMe	2,5-diMe	1,000	−
2,6-diMe	2,6-diMe	2,6-diMe	1,000	−
2,6-diMe	3,5-diMe	3,5-diMe	24,000	+
2-Me-4-Et	2-Me-4-Et	2-Me-4-Et	500	−
3-Me	3-Me	3-Me	5,000	−
3-Et	3-Et	3-Et	1,000	−
4-Me	4-Me	4-Me	5,000	−
4-Et	4-Et	H	2,000	−
4-Et	4-Et	4-Et	200—1,000	+
4-Et	4-Et	4-Me	500	−
4-Et	4-Me	4-Me	500 × 5	−
4-Et	H	H	1,000	+
2-Cl	H	H	1,000	−
2-Cl	2-Cl	2-Cl	1,000	−

From Eto, M., *Organophosphorus Pesticides: Organic and Biological Chemistry*, CRC Press, Boca Raton, Fla., 1974, 219. With permission.

The kinetics of the phosphorylation of cholinesterases has been thoroughly studied, the organophosphorus group attaching to a serine hydroxyl group which serves as an important part of the enzyme's active site.[240,241] Inhibited enzymes change gradually with time into a nonreactivatable form by a process known as "aging" which is generally accepted as being due to the dealkylation of the dialkoxyphosphorylated enzyme with the formation of a stable, irreversibly inhibited enzyme.[240] For phosphorylated serine proteases, two aging mechanisms are proposed.

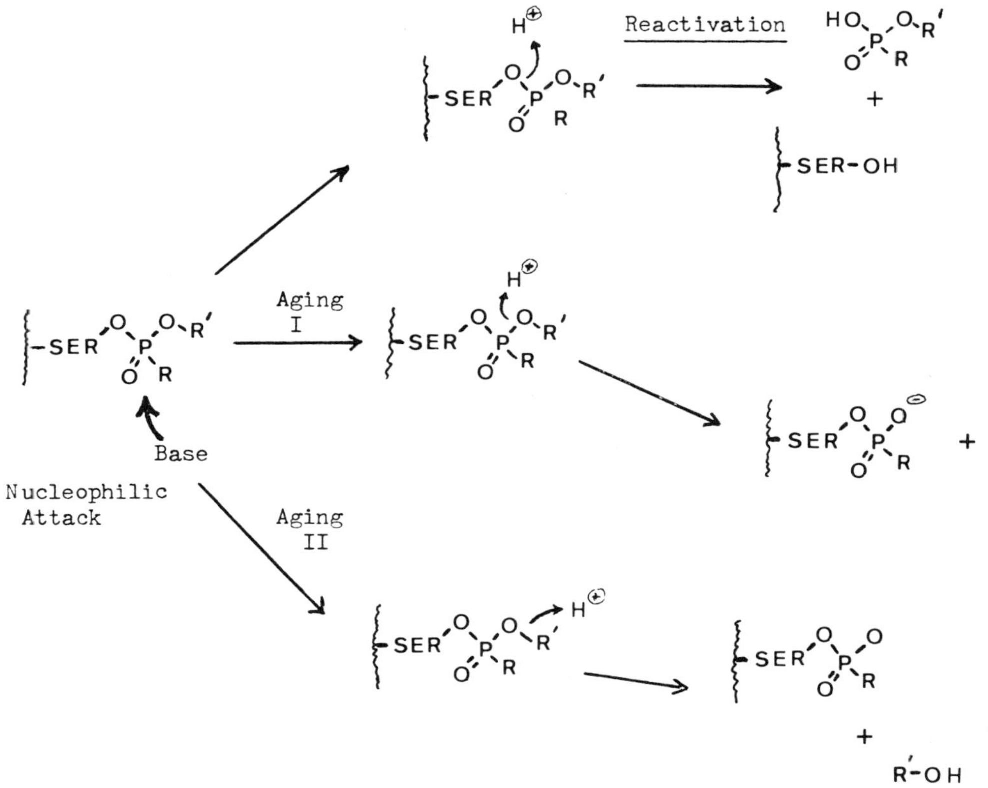

The first mechanism (I) involves the hydrolysis of the P—O bond following a nucleophilic attack on the P atom while the second mechanism (II) involves O—C cleavage by an acid-type catalysis resulting in the formation of a carbonium ion as the leaving group.[242] A recent paper has demonstrated the phosphorylation, aging, and possible alkylation reactions of saligenin cyclic phosphorus esters with alpha-chymotrypsin.[243] The evidence points to the possibility of two stabilized forms of the "aged" enzyme, both of which have utilized the imidazole group of a neighboring histidine as is shown below.

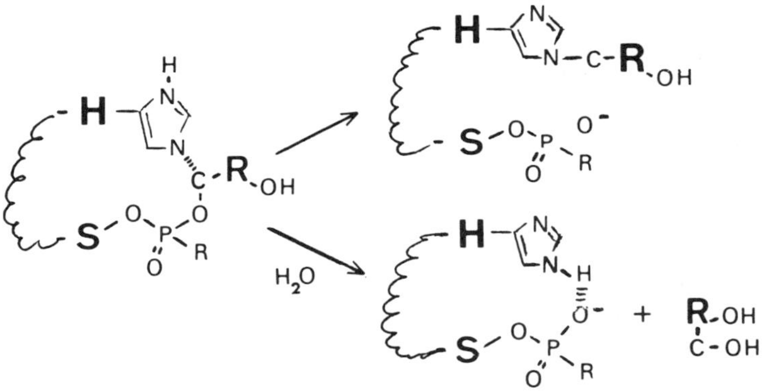

In one reaction, an hydroxylated phenol is released and the phenylphosphoryl-chymotrypsin is stabilized by a hydrogen on the imidazole group. In the second reaction, the released benzyl carbonium ion is attached to the imidazole to yield a N-(O-hydroxybenzyl) derivative of the phenylphosphorylated chymotrypsin. These highly selective reactions may be responsible for the inhibition of the specific enzyme associated with the delayed neurological lesion but confirmation of the role must await the isolation and purification of the enzyme involved.

A recently introduced insecticide, leptophos (O-[4-bromo-2,5-dichloro-phenyl] O-methyl phenylphosphonothioate, Velsicol VCS-506, Phosvel®) caused widespread paralysis and the death of some 1300 water buffaloes when it was used extensively in the Nile Delta.[244] No further reports on this particular incident have been seen in the literature. There also have been reports of poisoning by leptophos of chemical plant workers in Bayport, Texas in 1974 to 1975 though the causative agent(s) has not been identified conclusively since these workers had also been exposed to organic solvents such as n-hexane and toluene which can cause peripheral neuropathies.[245-248] Some of these workers were examined months after their initial illness and were found to have stigmata of an upper motor neuron syndrome with spasticity and hyperflexia of the lower extremities and extensor plantar responses.[247]

Leptophos

Despite the recognizable structural similarity of the agent to TOCP, neurotoxicity was apparently not detected during the routine toxological assessment of the chemical. In 1974, it was demonstrated conclusively that leptophos, administered as a single oral dose (180 to 3000 mg/kg bw) to hens, resulted in a slow-developing typical ataxia, loss of appetite and body weight, progressing to paralysis and death in a manner generally associated with other neurotoxic esters of phosphoric acid.[154] Subsequent studies of the peripheral nerves and spinal cords of hens receiving single large (200 to 800 mg/kg bw) doses or multiple (1,5,10 and 20 mg/kg/day) doses of leptophos revealed ataxia in these birds associated with posterior and lateral degeneration of the anterior descending tract of the spinal cord and degeneration of the smaller branches of the sciatic nerve.[249-253] Leptophos appeared to exert a greater effect on the spinal cord, the earliest changes being noted 4 days after treatment and consisting of a swelling of axons within the gray matter of the cervical cord. Following this, axonal swelling was observed in the posterior and lateral columns of the cervical cord followed by similar changes in the anterior column of the thoracic and lumbar cord.[252,253] The axonal swelling was followed by disintegration of both axons and myelin in these areas. The changes observed were comparable to those described for TOCP by Cavanagh.[194] With larger or multiple doses, leptophos was capable of damaging peripheral nerves.[250] An interesting point raised by Abou-Donia[253] in his comparative studies was that TOCP and leptophos had more profound effects in the spinal cord in the chicken than in peripheral nerves. In contrast, TOCP, in both human and nonhuman primates, initiated more severe peripheral nerve lesions at low doses.

Leptophos is quite persistent in the environment, being retained by pasture grasses, forage corn, leaves of tomato plants, and grapes for some length of time following treatment.[254-256] Leptophos is rapidly cleared from the bloodstream of treated animals, the biological half-life being of the order of 3 to 5 hr.[257,259] This, of course, does not mean that it is eliminated from the animal's body and, as has been demonstrated, it has a long biological half-life in the adipose tissue of the order of 5 to 7 days following a single oral dose.[257,259] It was found that erythrocyte AChE was initially inhibited following treatment but subsequently recovered. The plasma pseudocholinesterase followed a similar pattern though it became more severely inhibited as signs of neurotoxicity progressed following treatment with a single large oral dose.[249] These results suggested that the absorbed insecticide was extensively stored in adipose tissue and was released slowly, exerting an effect on this enzyme and the nervous tissue. A recent paper has demonstrated that two metabolites of leptophos, desbromoleptophos and leptophos oxon were approximately 3- and two-fold, respectively, more effective neurotoxins than was leptophos in young pullets.[260] This observation, when considered with those of Abou-Donia et al. would suggest that the slow release of low levels of leptophos from adipose tissue accompanied by biotransformation to considerably more toxic products could account for the marked inhibition of plasma pseudocholinesterase and the delayed onset of neurotoxicity.

A. Neurotoxic Esterase

Over the years, many well-defined esterases have been candidates for possible involvement in the genesis of the lesion associated with neurotoxic esters only to have the hypothesis destroyed again when comparisons of in vivo and in vitro studies revealed no correlation of neurotoxicity with the inhibitory potency of these organophosphorus esters toward these esterases.[216,227-229] As was stated by Johnson,[261] all of these enzymes were chosen for study by what might be called the "guess and test" approach. However, it appeared that phosphorylation of protein in the central nervous system was an essential prerequisite for the development of the lesion. Despite the large amount of protein which appeared to be phosphorylated, considerable success was achieved when this hypothesis was tested.

Using $DF^{32}P$ as a known neurotoxic agent, hen brain homogenates were exposed simultaneously to this agent plus other suspected neurotoxic esters. Those which were neurotoxic in the hen in vivo markedly reduced the phosphorylation of brain protein by $DF^{32}P$ whereas nonneurotoxic esters had little effect.[262] On the basis of such experiments to examine the competition between two organophosphorus esters for the same site(s), the concept of a "neurotoxic protein" was established. Neurotoxic esters phosphorylated this brain protein, reducing by some 70% the available sites for covalently binding $DF^{32}P$ whereas nonneurotoxic compounds had little effect as is shown in Table 7. Similar effects were also observed in the spinal cord and the sciatic nerve. For neurotoxic effects to be seen, more than 70% of this "neurotoxic protein" must be phosphorylated shortly after dosing.[261]

The "neurotoxic protein" was identified as an esterase following the observation that phenyl phenylacetate (PPA) which was structurally similar to TOCP could block the phosphorylation of this protein and that the protein could hydrolyze this ester.[263] Only a small proportion, less than 4%, of the total activity toward PPA found in hen brain homogenate was contributed by the "neurotoxic esterase". More than 80% of the PPA-hydrolyzing activity was attributable to two distinct enzymes, both very sensitive to inhibition by paraoxon.[264] The remaining, paraoxon-insensitive activity could be divided into two components, one large fraction which was partially inhibited by DFP and a second smaller fraction which was sensitive to mipafox and which appeared to be associated with delayed neurotoxicity.[261] The subcellular distribution of "neurotoxic esterase" has been studied by differential and sucrose gradient centrifugation, the results indicating that up to 70% of the activity of this enzyme was localized in a microsomal fraction believed to be enriched with synaptosomal membranes.[265] These

Table 7
THE RELATIONSHIP BETWEEN THE AVAILABILITY OF PHOSPHORYLATION SITES, NEUROTOXIC ESTERASE INHIBITION, AND NEUROTOXIC ESTERS[a]

Compound	Phosphorylation site	Esterase activity
tri-o-tolyl phosphate (TOTP)	8	—
tri-o-ethylphenyl phosphate	13	—
Mipafox	8, 11	0
DFP	8	5, 10
Trichlorfon (dimethyl[2,2,2-trichloro-1-hydroxyethyl] phosphate)	—	32
2,2-dichlorovinyl diethyl phosphate	—	26
diphenyl-o-tolyl phosphate	5, 13	7
ethyl p-nitrophenyl 2-phenethyl phosphonate	35	14

[a] Data from Aldridge and Johnson.[261] Values represent percent of control measurements.

results have been confirmed by others for hen brain as well as the sciatic nerve though the activity was much lower in the latter tissue.[266] Myelin fractions were devoid of neurotoxic esterase activity. The association of this esterase with synaptosomal membranes suggested that early events in the organophosphorus ester-induced neuropathy may well occur at the axonal or the synaptic surface.[265] The evidence appears conclusive enough to suggest that this "neurotoxic esterase" may be the primary site of action of the neurotoxic organophosphorus esters but just how the phosphorylation of this enzyme is related to the demyelination of nerve tissue is not known.[263,267] This enzyme appears to play an important and as yet unidentified role in cellular metabolism which may not be associated with its esteratic activity.

Having isolated the possible biochemical factor involved in the neuropathy, extensive studies were conducted to compare the inhibitory and neurotoxic esters.[268-270] The structures shown below have been identified as being those which cause neurotoxicity and which inhibit the "neurotoxic esterase".

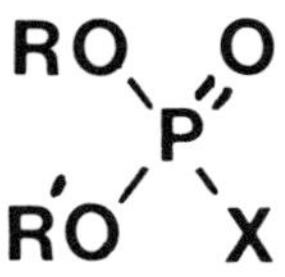

Phosphate 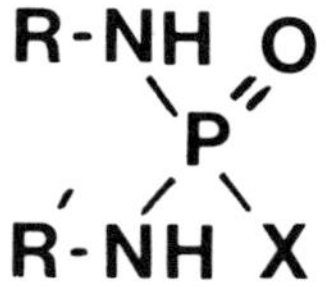   Phosphoramidate / Phosphordiamidate   Phosphonate

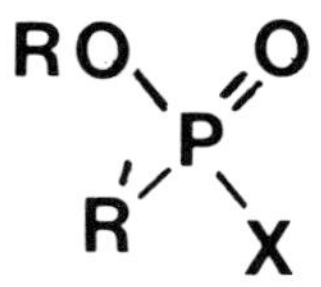

It has been proposed that the phosphorylation of the esterase by neurotoxic compounds is followed by hydrolysis of one of the ester (R-O-P) or amide (R-NH-P) bonds of the bound residue resulting in an ionized acidic group of the phosphorus which upsets normal physiological processes leading to myelin degeneration and ataxia.[269,270] The genesis of the lesion appears to depend upon the nature of the acyl group covalently bound to the active site of the enzyme and not just on the fact that the enzyme has been inhibited.[261] A recent paper presented evidence that there is an aging of the "neurotoxic esterase" following inhibition by an agent such as DFP, with the formation of an irreversibly inhibited mono-isopropylphosphorylated enzyme from the original di-isopropylphosphorylated enzyme.[271] The kinetics of this reaction suggest that it is different from that observed for the aging of DFP-inhibited cholinesterases.[271]

If the initial biochemical event of the neuropathy is the inhibition of the neurotoxic esterases, can this enzyme be used as a means of predicting the ability of new insecticides to induce delayed neurotoxicity? Lotti and Johnson[272] carried out a series of experiments to estimate the "neurotoxic esterase" inhibitory power of a number of esters, matching these values with previous in vivo toxicity data. When inhibition data for "neurotoxic esterase" of human brain was compared with that of hen brain, the results suggested that such experiments have predictive value for neurotoxicity in the hen and might be of use in predicting relative toxicities of these compounds in man. With a few exceptions, there was good predictive agreement between the toxic properties in hens and humans. The two exceptions, dichlorvos and trichlorfon, both dimethyl phosphate esters, have never caused neuropathies in hens following the administration of single doses far above the unprotected LD_{50} but recent reports have demonstrated that they can cause neuropathies in humans following the ingestion of large amounts of these agents in suicide attempts.[152,272,273] When the anomalous behavior of dichlorvos and some analogs was tested in hens, it was found that high doses of these agents caused marked inhibition of the brain "neurotoxic esterase" but caused no ataxia.[274] When repeated doses of dimethyl phosphates were administered, the inhibition of the "neurotoxic esterase" in spinal cord increased and the birds became ataxic, results which suggested that the ataxia arose from the inhibition of the spinal cord enzyme rather than that of the brain. The results also suggested that, for a few organophosphorus esters, the in vitro measurement of brain "neurotoxic esterase" inhibition was an oversensitive monitoring parameter. Johnson also believes that one should not consider dichlorvos (and other dimethyl phosphate esters) in the same class as those agents which cause a neurotoxicity following the administration of single doses comparable to the LD_{50} value.[274]

VI. SKELETAL MUSCLE NECROSIS

One of the features of organophosphorus ester poisoning is a peripheral muscular weakness which may be persistent for some months following exposure. While this weakness has been attributed to damage to the nerve axons and myelin, evidence in the literature suggests that a more direct action of certain organophosphorus esters may be focused on the skeletal muscle.

A single injection of paraoxon, the oxygen analog of parathion, has been shown to produce skeletal muscle lesions, the early evidence of such lesions occurring near the region of the motor end-plate.[275] These lesions were reversible. Fenichal et al.[276] demonstrated that the chronic administration of paraoxon in doses approximately 66% of the LD_{50} value resulted in a progressive myopathy in the soleus, gastrocnemius and quadriceps muscles of the rat. This myopathy could be prevented by denervation, and modified to some extent by the administration of hemicholinium. A mechanism by which the lesion was initiated was proposed as being due to excessive acetylcholine stimulation at the neuromuscular junction due to chronic cholinesterase inhibition. Similar studies in which parathion was administered for 2 weeks showed that this ester induced histological evidence of skeletal muscle lesions as well as weakness comparable to those induced by paraoxon.[277] Within 30 min of receiving paraoxon (0 to 24 mg/kg subcutaneously), the diaphragmatic muscle beneath the subsynaptic folds had a distorted appearance and contained large vacuoles, dilated mitochondria, and expanded sarcoplasmic reticulum in contrast to the normal-appearing muscle some distance from the motor end-plate.[278,279] The average distance that the pathological changes extended from the nerve terminal membrane was 8:0 μm. After 2 days of paraoxon treatment, the ultrastructural appearance of nerve terminals revealed swollen mitochondria and coated vesicles.[280] The most severely affected end-plates were associated with necrotic fibers and phagocytosis.[278,281] The results from studies to date have demonstrated marked differences in the susceptibility of muscle types to paraoxon and parathion. The myopathy was much more severe in the diaphragm, predominantly a slow-contracting, tonic, or red muscle, than in the soleus and gastrocnemius muscles which are fast-contracting, phasic, or white muscles.[281] One possible explanation for the effect on the diaphragm may be the quality and quantity of use, in that slow muscles receive a larger number of impulses of acetylcholine than do fast muscles and, therefore, have greater activity.[281]

Following exposure, there may be a critical time-period of cholinesterase inhibition necessary to initiate the neurally mediated neuropathy. Such an action on nervous tissue acetylcholinesterase would be dependent upon the quantity of organophosphorus ester absorbed into the body and subsequently reaching the site of action. Denervation some 2 weeks prior to paraoxon administration effectively prevented the myopathy.[282] The myopathy was shown to be dose-dependent and 85% inhibition of the nervous tissue acetylcholinesterase during the first hour after injection was necessary for maximal skeletal muscle fiber breakdown.[283] Following a single dose of paraoxon (0.23 mg/kg bw) which caused prolonged muscle fasciculations for 1 to 2 hr, a grouped skeletal muscle (diaphragm) fiber necrosis was evident at 24 hr. The administration of 2-PAM (pyridine-2-aldoxine methochloride) between 10 and 120 min after administering paraoxon completely reactivated the phosphorylated acetylcholinesterase in the junctional and nonjunctional regions of the diaphragmatic muscle and the time interval between the injection of the organophosphorus ester and reactivator was positively correlated with the severity of the myopathy.[283] The importance of the acetylcholinesterase in the initiation of this lesion was emphasized by the fact that repeated injections of reversible anticholinesterase agents such as physostigmine and neostigmine designed to prolong skeletal muscle hyperactivity also produced myopathies which were qualitatively similar to the paraoxon-induced myopathy.[283]

VII. TREATMENT

The origin of the symptoms and physical signs identified with acute organophosphorus ester poisoning have been presented in Table 3. The regimen of treatment has been adequately discussed by many authors and the reader is referred to the papers by Durham and Hayes,[284] Ellin and Wills,[285] Wills,[286] Namba et al.,[287] and Milby[288] for a complete presentation. The regimen of therapy for acute poisoning based on the analysis of plasma pseudocholinesterase activity is presented in brief form in Table 8.[112]

Frequent small doses of atropine are indicated following brief intensive exposure to control the initial muscarinic symptoms. Relatively large cumulative doses of atropine, up to 50 mg daily, may be necessary to control severe muscarinic symptoms. Atropine is of particular importance in acute, life-threatening intoxication. The subsequent involvement of the neuromuscular junctions and the central nervous system may be alleviated by the use of specific oxime antidotes (2-PAM, pralidoxime chloride, P2S, 2-hydroxyimino-methyl-N-methylpyridinium methyl methane sulfonate) which reactivate the phosphorylated acetylcholinesterase, other drugs including tranquilizers and antidepressant agents and other supportive symptomatic therapy.

Protection against the lethal effects of an organophosphorus ester does not offer

Table 8
CLASSIFICATION AND TREATMENT OF ORGANOPHOSPHORUS
INSECTICIDE POISONING BASED ON PLASMA
PSEUDOCHOLINESTERASE ACTIVITY[a]

Classification of poisoning	Enzyme activity (% of normal)	Treatment — Atropine	Treatment — Pralidoxime
Mild	20—50	1.0 mg subcutaneously	1.0 g IV over 20—30 min.
Moderate	10—20	1.0 mg every 20—30 min until sweating and salivation disappear and slight flush and mydriasis appear	1.0 g IV over 20—30 min.
Severe	10	5.0 mg IV every 20—30 min until sweating and salivation disappear and slight flush and mydriasis appear	1.0 g IV as above. If no improvement, administer another 1.0 g IV. If no improvement, start IV infusion at 0.5 g/hr.

[a] Modified from Ecobichon et al.[112]

protection against the possibility of delayed neurotoxicity as was described in the TOCP-induced neuropathies earlier in this chapter. Davies and Holland[289] found that the oximes P2S and PAD (the dodecyl iodide salt of P2S) with atropine were without effect upon the development of delayed neurotoxicity in chickens following poisoning by DFP. These results have been observed by others.[290,291] Pretreatment with phenyl benzylcarbamate or phenyl methanesulfonyl fluoride protected hens from DFP-induced delayed neurotoxicity to some extent, the protection with the carbamate lasting a few hours while that with the sulfonyl fluoride persisted for several days.[267] The administration of these chemicals even 1 hr after treatment with DFP did not afford any protection. In contrast, Wecker et al.[283] have shown that the administration of the oxime 2-PAM between 10 and 120 min of administering paraoxon did protect against the skeletal muscle myopathy. These oximes appear to have no effect on the delayed neurotoxic effects caused by TOCP or other organophosphorus esters.[290,291]

REFERENCES

1. von Rumker, R., Lawless, E. W., and Meiners, A. F., Production, Distribution, Use and Environmental Impact Potential of Selected Insecticides, *EPA Report* 540-1-74-001, U.S. Environmental Protection Agency, Washington, D.C., 1975.
2. O'Brien, R. D., *Toxic Phosphorus Esters. Chemistry, Metabolism and Biological Effects,* Academic Press, New York, 1960, chap. 1.
3. Melnikov, N. N., Chemistry of pesticides, in *Residue Reviews,* Vol. 36, Gunther, F. A. and Gunther, J. D., Eds., Springer-Verlag, New York, 1971, chap. 26.
4. Fest, C. and Schmidt, K.-J., *The Chemistry of Organophosphorus Pesticides. Reactivity, Synthesis, Mode of Action, Toxicology,* Springer-Verlag, New York, 1973, chap. 2.
5. Eto, M., *Organophosphorus Pesticides: Organic and Biological Chemistry,* CRC Press, Boca Raton, Fla., 1974, chap. 1.
6. Aldridge, W. N. and Davison, A. H., The inhibition of erythrocyte cholinesterase by tri-esters of phosphoric acid, *Biochem. J.,*51, 62, 1952.
7. Schrader, G., *Die Entwicklung Neuer Insektizider Phosphorsaureester,* 3rd ed., Verlag Chemie., Weinheim, 1963.

8. Clark, V. M., Hutchinson, D. W., Kirby, A. I., and Warren, S. G., Phosphorylation: principle structures and reaction mechanism, *Angew. Chem.*, 76, 704, 1964.
9. Hammett, L. P., *Physical Organic Chemistry*, McGraw-Hill, New York, 1940.
10. Jaffe, H. H., Reexamination of the Hammett equation, *Chem. Rev.*, 53, 191, 1953.
11. Fukuto, T. R. and Metcalf, R. L., Structure and insecticidal activity of some diethyl substituted phenyl phosphates, *J. Agric. Food Chem.*, 4, 930, 1956.
12. Fukuto, T. R., Physico-organic chemical approach to the mode of action of organophosphorus insecticides, in *Residue Reviews*, Vol. 25, Gunther, F. A. and Gunther, J. D., Eds., Springer-Verlag, New York, 1969, 327.
13. Fukuto, T. R. and Metcalf, R. L., The effect of structure on the reactivity of alkylphosphonate esters, *J. Am. Chem. Soc.*, 81, 372, 1959.
14. Hansch, C. and Deutsch, E. W., The use of substituent constants in the study of structure-activity relationships in cholinesterase inhibitors, *Biochim. Biophys. Acta*, 126, 117, 1966.
15. Meyer, H. H., The theory of alcohol narcosis, *Arch. Exp. Pathol. Pharmakol.*, 42, 109, 1899.
16. Overton, E., *Studien Über die Narkose, Zugleich Ein Beintragzur Allgemeinen Pharkologie*, Fischer, Gustaf Gena, Germany, 1901, 101.
17. Meyer, K. H. and Hemmi, H., The theory of narcosis, *Biochem. Z.*, 277, 39, 1935.
18. Hansch, C., A quantitative approach to biochemical structure-activity relationships, *Accounts Chem. Res.*, 2, 232, 1969.
19. Hansch, C. and Fujita, T., ϱ-σ-π analysis. A method for the correlation of biological activity and chemical structure, *J. Am. Chem. Soc.*, 86, 1616, 1964.
20. Fujita, T., Iwasa, J., and Hansch, C., A new substituent constant, π, derived from partition coefficients, *J. Am. Chem. Soc.*, 86, 5175, 1964.
21. Hansch, C., The use of physicochemical parameters and regression analysis in pesticide design, in *Biochemical Toxicology of Insecticides*, O'Brien, R. D. and Yamamoto, I., Eds., Academic Press, New York, 1970, 33.
22. Matsumura, F., *Toxicology of Insecticides*, Plenum Press, New York, 1975.
23. Nakatsugawa, T. and Morelli, M. A., Microsomal oxidation and insecticide metabolism, in *Insecticide Biochemistry and Physiology*, Wilkinson, C. F., Ed., Plenum Press, New York, 1976, chap. 2.
24. Ecobichon, D. J., Hydrolytic transformation of environmental pollutants, in *Handbook of Physiology-Reactions to Environmental Agents*, Lee, D. H. K., Ed., Am. Physiol. Soc., Bethesda, Md., 1977, chap. 27.
25. Hodgson, E., *The Enzymic Oxidation of Toxicants*, North Carolina State University, Raleigh, 1968, 5.
26. Hollingworth, R. M., Comparative metabolism and selectivity of organophosphate and carbamate insecticides, *Bull. W.H.O.*, 44, 155, 1971.
27. Kulkarni, A. P. and Hodgson, E., Metabolism of insecticides by mixed function oxidase systems, *Pharmac. Ther.*, 8, 379, 1980.
28. Diggle, W. M. and Gage, J. C., Cholinesterase inhibition *in vivo* by O,O-diethyl-O-p-nitrophenyl thiophosphate (parathion, E605), *Biochem. J.*, 49, 491, 1951.
29. Gage, J. C., A cholinesterase inhibitor derived from O,O-diethyl-O-p-nitrophenyl thiophosphate *in vivo*, *Biochem. J.*, 54, 426, 1953.
30. Davison, A. N., The conversion of Schradan (OMPA) and parathion into inhibitors of cholinesterase by mammalian liver, *Biochem. J.*, 61, 203, 1955.
31. Neal, R. A. and DuBois, K. P., Studies on the mechanism of detoxification of cholinergic phosphorothioates, *J. Pharmacol. Exp. Therap.*, 148, 185, 1965.
32. Kojima, K. and O'Brien, R. D., Paraoxon hydrolyzing enzymes in rat liver, *J. Agric. Food Chem.*, 16, 574, 1968.
33. Hollingworth, R. M., The dealkylation of organophosphorus triesters by liver enzymes, in *Biochemical Toxicology of Insecticides*, O'Brien, R. D. and Yamamoto, I., Eds., Academic Press, New York, 170, 75.
34. Bowman, J. S. and Casida, J. E., Further studies on the metabolism of thimet by plants, insects and mammals, *J. Econ. Entomol.*, 51, 838, 1958.
35. Bull, D. A., Metabolism of Disyston by insects, isolated cotton leaves and rats, *J. Econ. Entomol.*, 58, 249, 1965.
36. Bull, D. A., Metabolism of organophosphorus insecticides in animals and plants, in *Residue Reviews*, Vol. 43, Gunther, F. A. and Gunther, J. D., Eds., Springer-Verlag, New York, 1972, 1.
37. Metcalf, R. L., Fukuto, T. R., and March, R. B., Plant metabolism of dithiosystox and thimet, *J. Econ. Entomol.*, 50, 338, 1957.
38. Eto, M., *Organophosphorus Pesticides: Organic and Biological Chemistry*, CRC Press, Boca Raton, Fla., 1974, 164.
39. Nakatsugawa, T. and Dahm, P. A., Microsomal metabolism of parathion, *Biochem. Pharmacol.*, 16, 25, 1967.

40. Nakatsugawa, T., Tolman, N. M., and Dahm, P. A., Degradation and activation of parathion analogs by microsomal enzymes, *Biochem. Pharmacol.*, 17, 1517, 1968.
41. Yang, R. S. H., Hodgson, E., and Dauterman, W. C., Metabolism *in vitro* of diazinon and diazoxon in rat liver, *J. Agric. Food Chem.*, 19, 10, 1971.
42. Motoyama, N. and Dauterman, W. C., The *in vitro* metabolism of azinophosmethyl by mouse liver, *Pestic. Biochem. Physiol.*, 2, 170, 1972.
43. Eto, M., *Organophosphorus Pesticides: Organic and Biological Chemistry*, CRC Press, Boca Raton, Fla., 1974, 173.
44. Ahmad, S. and Forgash, A. J., Nonoxidative enzymes in the metabolism of insecticides, *Drug Metab. Rev.*, 5, 141, 1976.
45. Fukami, J. and Shishido, T., Nature of the soluble, glutathione dependent enzyme system active in cleavage of methyl parathion to desmethyl parathion, *J. Econ. Entomol.*, 59, 1338, 1966.
46. Fukunaga, K., Fukami, J., and Shishido, T., The *in vitro* metabolism of organophosphorus insecticides by tissue homogenates from mammal and insect, in *Residue Reviews*, Vol. 25, Gunther, F. A. and Gunther, J. D., Eds., Springer-Verlag, New York, 1969, 223.
47. Appleton, H. T. and Nakatsugawa, T., Paraoxon deethylation in the metabolism of parathion, *Pestic. Biochem. Physiol.*, 2, 286, 1972.
48. Hollingworth, R. M., Alstott, R. L., and Litzenberg, R. D., Glutathione S-aryl transferase in the metabolism of parathion and its analogs, *Life Sci.*, 13, 191, 1973.
49. Ecobichon, D. J., Hydrolytic mechanisms of pesticide degradation, in *Advances in Pesticide Science*, Part 3, Geissbuhler, H., Ed., Pergamon Press, New York, 1979, 516.
50. Mounter, L. A., Baxter, R. F., and Chanutin, A., Dialkylfluorophosphatases in microorganisms, *J. Biol. Chem.*, 215, 699, 1955.
51. Hogan, J. W. and Knowles, C. O., Degradation of organophosphates by fish liver phosphatases, *J. Fisheries Res. Bd. Can.*, 25, 1571, 1968.
52. Krueger, H. R. and Casida, J. E., Hydrolysis of certain organophosphate insecticides by housefly enzymes, *J. Econ. Entomol.*, 54, 239, 1961.
53. Mounter, L. A., Dien, L. T. H., and Chanutin, A., The distribution of dialkylfluorophosphatases in the tissue of various species, *J. Biol. Chem.*, 215, 691, 1955.
54. Ramachandran, B. V. and Agren, G., Esterases of rat liver cell fractions. Correlation of $DF^{32}P$-binding capacity to esterase activity, *Biochem. Pharmacol.*, 12, 981, 1963.
55. Mounter, L. A. and Chanutin, A., Dialkylfluorophosphatase of kidney II. studies of activation and inhibition by metals, *J. Biol. Chem.*, 204, 837, 1953.
56. Mounter, L. A. and Chanutin, A., Dialkylfluorophosphatase of kidney III. studies of activation and inhibition by cofactors, *J. Biol. Chem.*, 210, 219, 1954.
57. Mounter, L. A. and Whittaker, V. P., The hydrolysis of esters of phenol by cholinesterases and other esterases, *Biochem. J.*, 54, 551, 1953.
58. Augustinsson, K.-B. and Ekedahl, G., On the specificity of arylesterases, *Acta Chem. Scand.*, 16, 240, 1962.
59. Augustinsson, K.-B., Arylesterases, *J. Histochem. Cytochem.*, 12, 744, 1964.
60. Augustinsson, K.-B., Multiple forms of esterases in vertebrate blood plasma, *Ann. N.Y. Acad. Sci.*, 94, 844, 1961.
61. Dauterman, W. C., Extramicrosomal metabolism of insecticides, in *Insecticide Biochemistry and Physiology*, Wilkinson, C. F., Ed., Plenum Press, New York, 1976, 149.
62. Schwark, W. S. and Ecobichon, D. J., Subcellular localization and drug-induced changes of rat liver and kidney esterases, *Can. J. Physiol. Pharmacol.*, 46, 207, 1968.
63. Chow, A. Y. K. and Ecobichon, D. J., Characterization of the esterases of guinea pig liver and kidney, *Biochem. Pharmacol.*, 22, 689, 1973.
64. Ecobichon, D. J. and Kalow, W., Properties and classification of the soluble esterases of human liver, *Biochem. Pharmacol.*, 11, 573, 1962.
65. Main, A. R. and Braid, P. E., Hydrolysis of malathion by aliesterase *in vitro* and *in vivo*, *Biochem. J.*, 84, 255, 1962.
66. Dauterman, W. C. and Main, A. R., Relationship between acute toxicity and *in vitro* inhibition and hydrolysis of a series of carbolboxy homologs of malathion, *Toxicol. Appl. Pharmacol.*, 9, 408, 1966.
67. Matsumura, F. and Brown, A. W. A., Biochemistry of malathion resistance in *Culex tarsalis*, *J. Econ. Entomol.*, 54, 1176, 1961.
68. Matsumura, F. and Ward, C. T., Degradation of insecticides by the human and rat liver, *Arch. Environ. Health*, 13, 257, 1966.
69. O'Brien, R. D., Thorn, G. C., and Fisher, R. W., New organophosphate insecticides developed on rational principles, *J. Econ. Entomol.*, 51, 714, 1958.
70. Uchida, T., Dauterman, W. C., and O'Brien, R. D., The metabolism of dimethoate by vertebrate tissue, *J. Agric. Food Chem.*, 12, 48, 1964.

71. Chen, P. R. S. and Dauterman, W. C., Studies on the toxicity of dimethoate analogs and their hydrolysis by sheep liver amidase, *Pestic. Biochem. Physiol.*, 1, 340, 1971.
72. Fukuto, T. R. and Metcalf, R. L., Metabolism of insecticides in plants and animals, *Ann. N.Y. Acad. Sci.*, 160, 97, 1969.
73. Bull, D. L. and Lindquist, D. A., Metabolism of 3-hydroxy-N,N-dimethyl crotonamide dimethyl phosphate by cotton plants, insects and rats, *J. Agric. Food Chem.*, 12, 310, 1964.
74. Seume, F. W. and O'Brien, R. D., Potentiation of the toxicity to insects and mice of phosphorothionates containing carboxyester and carboxyamide groups, *Toxicol. Appl. Pharmacol.*, 2, 495, 1960.
75. Bull, D. L., Lindquist, D. A., and Grabbe, R.R., Comparative fate of the geometric isomers of phosphamidon in plants and animals, *J. Econ. Entomol.*, 60, 332, 1967.
76. Uchida, T., Zchintzsch, J., and O'Brien, R. D., Relation between synergism and metabolism of dimethoate in mammals and insects, *Toxicol. Appl. Pharmacol.*, 8, 259, 1966.
77. Yang, R. S. H., Enzymatic conjugation and insecticide metabolism, in *Insecticide Biochemistry and Physiology*, Wilkinson, C. F., Ed., Plenum Press, New York, 1976, 177.
78. Williams, R. T., *Detoxification Mechanisms*, 2nd ed., Chapman and Hall, London, 1959.
79. Williams, R. T., The biogenesis of conjugation and detoxication products, in *Biogenesis of Natural Compounds*, Bernfeld, P., Ed., Pergamon Press, Oxford, 1967, 427.
80. Hutson, D. H., Akintowa, D. A. A., and Hathway, D. E., The metabolism of 2-chloro-1-(2', 4'-dichlorophenyl) vinyl diethylphosphate in the dog and rat, *Biochem. J.*, 102, 133, 1967.
81. Bull, D. L. and Ridgway, R. L., Metabolism of trichlorfon in animals and plants, *J. Agric. Food Chem.*, 17, 837, 1969.
82. Bull, D. L., Metabolism of organophosphorus insecticides in animals and plants, in *Residue Rev.*, Vol. 43, Gunther, F. A. and Gunther, J. D., Eds., Springer-Verlag, New York, 1972, 1.
83. Barnes, J. M., The reactions of the rabbit to poisoning by p-nitro-phenyldiethylphosphate (E600), *Br. J. Pharmacol.*, 8, 208, 1953.
84. DeCandole, C. A., Douglas, W. W., Evans, C. L., Holmes, R., Spencer, K. E. V., Torrence, R. W., and Wilson, K. M., The failure of respiration in death by anticholinesterase poisoning, *Br. J. Pharmacol.*, 8, 466, 1953.
85. O'Brien, R. D., *Toxic Phosphorus Esters. Chemistry, Metabolism and Biological Effects*, Academic Press, New York, 1960, 176.
86. Koelle, G. B. and Gilman, A., The chronic toxicity of di-isopropyl fluorophosphate (DFP) in dogs, monkeys, and rats, *J. Pharmacol. Exp. Therap.*, 87, 435, 1946.
87. Grob, D., Lilienthal, J. L., Harvey, A. M., and Jones, B. F., The administration of diisopropylfluorophosphate (DFP) to man: I. Effects on plasma and erythrocte cholinesterase; general systemic effects: use in study of hepatic function and erythropoiesis; and some properties of plasma cholinesterase, *Bull. Johns Hopkins Hosp.*, 81, 217, 1947.
88. Grob, D., Harvey, A. M., Langworthy, O. R., and Lilienthal, J. L., The administration of diisopropylfluorophosphate (DFP) to man: III. Effect on the central nervous system with special reference to the electrical activity of the brain, *Bull. Johns Hopkins Hosp.*, 81, 257, 1947.
89. Grob, D., Garlick, W. L., and Harvey, A. M., The toxic effects in man of the anticholinesterase insecticide parathion (p-nitrophenyl diethyl thionophosphate), *Bull. Johns Hopkins Hosp.*, 87, 106, 1950.
90. Rowntree, D. W., Nevin, S., and Wilson, A., The effects of diisopropylfluorophosphate in schizophrenia and manic depressive psychoses, *J. Neurol. Neurosurg. Psychiatry*, 13, 47, 1950.
91. Grob, D. and Harvey, A. M., Effects in man of the anticholinesterase compound sarin (isopropyl methyl phosphonofluoridate), *J. Clin. Invest.*, 37, 350, 1958.
92. Grob, D., The manifestations and treatment of poisoning due to nerve gas and other organic phosphate anticholinesterase compounds, *A.M.A. Arch. Internal Med.*, 98, 221, 1956.
93. Grob, D. and Harvey, A. M., The effects and treatment of nerve gas poisoning, *Am. J. Med.*, 14, 52, 1953.
94. Holmstedt, B., Pharmacology of organophosphorus cholinesterase inhibitors, *Pharmacol. Rev.*, 11, 567, 1959.
95. Karczmar, A. G., Anticholinesterase agents, in *International Encyclopedia of Pharmacology and Therapeutics*, Vol. 1, Pergamon Press, New York, 1970.
96. Sumerford, W. T., Hayes, W. J., Jr., Johnston, J. M., Walker, K., and Spillane, J., Cholinesterase response and symptomatology from exposure to organic phosphorus insecticides, *A.M.A. Arch Ind. Hyg. Occup. Med.*, 7, 383, 1953.
97. Durham, W. F., Wolfe, H. R., and Quinby, G. E., Organophosphorus insecticides and mental alertness, *Arch. Environ. Health*, 10, 55, 1965.
98. Bidstrup, P. L., Bonnell, J. A., and Beckett, A. G., Paralysis following poisoning by a new organic phosphorus insecticide (Mipafox), *Br. Med. J.*, I, 1068, 1953.

99. Spiegelberg, U., *Psychopathologisch-neurologische Schaden nach Einwirkung synthetischer Gifte*, Wehrdienst und Gesundheit, Vol. III, Darmstaedt, Wehr und Wissen, Verlagsgesellschaft mbH, 1961.
100. *SIPRI Monograph*, Delayed toxic effects of chemical warfare agents, Stockholm International Peace Research Institute, Almquist and Wiskell International, Stockholm and New York, 1975.
101. Gershon, S. and Shaw, F. B., Psychiatric sequelae of chronic exposure to organophosphorus insecticides, *Lancet*, I, 1371, 1961.
102. Barnes, J. M., Psychiatric sequelae of chronic exposure to organophosphorus insecticides, *Lancet*, II, 102, 1961.
103. Bidstrup, P. L., Psychiatric sequelae of chronic exposure to organophosphorus insecticides, *Lancet*, II, 103, 1961.
104. Holmes, J. H. and Gaon, M. D., Observations on acute and multiple exposures to anticholinesterase agents, *Trans Am. Clin. Climatol. Assoc.*, 68, 86, 1956.
105. Stoller, A., Krupinski, J., Christophers, A. J., and Blanks, G. K., Organophosphorus insecticides and major mental illness, *Lancet*, I, 1387, 1965.
106. West, I., Sequelae of poisoning from phosphate ester pesticides, *Ind. Med. Surg.*, 37, 832, 1968.
107. Warnick, S. L. and Carter, J. E., Some findings in a study of workers occupationally exposed to pesticides, *Arch. Environ. Health*, 25, 265, 1972.
108. Kay, K., Monkman, L., Windish, J. P., Doherty, T., Paré, J., and Racicot, C., Parathion exposure and cholinesterase response of Quebec apple growers, *A.M.A. Arch. Ind. Hyg. Occup. Med.*, 6, 252, 1952.
109. Hayes, W. J., Jr., Dixon, E. M., Batchelor, G. S., and Upholt, W. M., Exposure to organic phosphorus sprays and occurrence of selected symptoms, *Public Health Rep.*, 72, 787, 1959.
110. Davignon, L. F., St-Pierre, J., Charest, G., and Tourangeau, F. J., A study of the chronic effects of insecticides in man, *Can. Med. Assoc. J.*, 92, 597, 1965.
111. Tabershaw, I. R. and Cooper, W. C., Sequelae of acute organic phosphate poisoning, *J. Occup. Med.*, 8, 5, 1966.
112. Ecobichon, D. J., Ozere, R. L., Reid, E., and Crocker, J. F. S., Acute fenitrothion poisoning, *Can. Med. Assoc. J.*, 116, 377, 1977.
113. Holmes, J. H., Organophosphorus insecticides in Colorado, *Arch. Environ. Health*, 9, 445, 1964.
114. Dille, J. R. and Smith, P. W., Central nervous system effects of chronic exposure to organophosphate insecticides, *Aerosp. Med.*, 35, 475, 1964.
115. Faerman, I. S., After effects of acute poisoning with organophosphorus insecticides, *Gig. Tr. Prof. Zabol.*, 4, 39, 1967.
116. Sidell, F. R., Soman and sarin. Clinical manifestations and treatment of accidental poisoning by organophosphates, *Clin. Toxicol.*, 7, 1, 1974.
117. Smith, P. W., Stavinoha, W. B., and Ryan, L. C., Cholinesterase inhibition in relation to fitness to fly, *Aerosp. Med.*, 39, 754, 1968.
118. Woods, W., Gavica, J., Brown, W., Watson, M., and Benson, W. W., Implications of organophosphate pesticide poisoning in the plane crash of a duster pilot, *Aerosp. Med.*, 42, 1111, 1971.
119. Quinby, G. E., Walker, K. C., and Durham, W. F., Public health hazard involved in the use of organic phosphorus insecticides in cotton culture in the delta area of Mississippi, *J. Econ. Entomol.*, 51, 831, 1958.
120. Banks, A. and Russell, R. W., Effects of chronic reductions in acetylcholinesterase activity on serial problem-solving behavior, *J. Comp. Physiol. Psychol.*, 64, 262, 1967.
121. Pearson, D. W., Clark, G., and Moore, C. M., Comparison of behavioral effects of various levels of Di-syston poisoning, *Proc. 77th Ann. Conven. A.P.A.*, 871, 1969.
122. Clark, G., Organophosphate insecticides and behavior. A review, *Aerosp. Med.*, 42, 735, 1971.
123. Levin, H. S. and Rodnitzky, R. L., Behavioral aspects of organophosphate pesticides in man, *Clin. Toxicol.*, 9, 391, 1976.
124. Rodnitzky, R. L., Levin, H. S., and Mick, D. L., Occupational exposure to organophosphorus pesticides. A neurobehavioral study, *Arch. Environ. Health*, 30, 98, 1975.
125. Lorot, C., Les combinaisons de la créosote dans le traitement de la tuberculose pulmonaire, Thèse de Paris, 1899.
126. Roger, H. and Recordier, M., Les polynévrites phosphocréosotiques (phosphate) de créosote, ginger paralysis, apiol, *Ann. Med.*, 35, 44, 1934.
127. Bennett, C. R., A group of patients suffering from paralysis due to drinking Jamaica ginger, *South. Med. J.*, 23, 371, 1930.
128. Harris, S., Jr., Jamaica ginger paralysis (a peripheral polyneuritis), *South. Med. J.*, 23, 375, 1930.
129. Smith, M. I., Elvove, E., Valaer, P. J., Frazier, W. H., and Mallory, G. E., Pharmacological and chemical studies of the cause of so-called ginger paralysis, *Public Health Rep.*, 45, 1703, 1930.

130. Smith, M. I., Elvove, E., and Frazier, W. H., The pharmacological action of certain phenol esters, with special reference to the etiology of so-called ginger paralysis, *Public Health Rep.*, 45, 2509, 1930.
131. Smith, M. I. and Lillie, R. D., The histopathology of triorthocresyl phosphate poisoning. The etiology of so-called ginger paralysis (third report), *Arch. Neurol. Psychiatry*, 26, 976, 1931.
132. Aring, C. D., The systemic nervous affinity of triorthocresyl phosphate (Jamaica ginger palsy), *Brain*, 65, 34, 1942.
133. Smith, M. I. and Elvove, E., The epidemic of so-called ginger paralysis in southern California in 1930-31, *Public Health Rep.*, 46, 1227, 1931.
134. Burley, B. T., Polyneuritis from tricresyl phosphate, *J.A.M.A.*, 98, 298, 1932.
135. Kidd, J. G. and Langworthy, O. R., Jake paralysis. Paralysis following the ingestion of Jamaica ginger extract adulterated with tri-ortho-cresyl phosphate, *Bull. Johns Hopkins Hosp.*, 52, 39, 1933.
136. Morgan, J. P. and Tulloss, T. C., The Jake Walk Blues. A toxicologic tragedy mirrored in American popular music, *Ann. Intern. Med.*, 85, 804, 1976.
137. Ter Braak, J. W. C., A polyneuritis epidemic of peculiar origin, *Ned. T. Geneesk*, 7, 2329, 1931.
138. Germon, C., Intoxication mortelle par apiol, *Thèse de Paris*, 1932.
139. Hunter, D., Perry, K. M. A., and Evans, R. B., Toxic polyneuritis arising during the manufacture of tricresyl phosphate, *Br. J. Ind. Med.*, 1, 227, 1944.
140. Sampson, B. F., The strange Durban epidemic of 1937, *S. Afr. Med. J.*, 16, 1, 1942.
141. Davies, D. R., Neurotoxicity of organophosphorus compounds, in *The Anticholinesterases. Handbuch der Exp. Pharmacol.*, Vol. 15, Koelle, G. B., Ed., Springer-Verlag, Berlin, 1963, chap. 19.
142. Walthard, K. M., Aperçu des resultats obtenus lors les derniers examens des malades intoxiques en 1940 par le phosphate triorthocrésilique, *Achweiz. Arch. Neurol. Psychiat.*, 58, 189, 1946.
143. Hotston, R. D., Outbreak of polyneuritis due to orthotricresyl phosphate poisoning, *Lancet*, I, 207, 1946.
144. Susser, M. and Stein, Z., An outbreak of tri-ortho-cresyl phosphate (TOCP) poisoning in Durban, *Br. J. Ind. Med.*, 14, 111, 1957.
145. Smith, H. V. and Spalding, J. M. K., Outbreak of paralysis in Morocco due to ortho-cresyl phosphate poisoning, *Lancet*, II, 1019, 1959.
146. Dennis, D. T., Jake walk in Viet Nam, *Ann. Intern. Med.*, 86, 665, 1977.
147. Morgan, J. P. and Penovich, P., Jamaica ginger paralysis. Forty-seven-year follow-up, *Arch. Neurol.*, 35, 530, 1978.
148. Bondy, H. F., Field, E. J., Worden, A. N., and Hughes, J. P. W., A study on the acute toxicity of the tri-aryl phosphates used as plasticizers, *Br. J. Ind. Med.*, 17, 190, 1960.
149. Petry, H., Polyneuritis through E-605, *Zentralbl. Arbeitsmed. u. Arbeitsschutz.*, 1, 86, 1951.
150. Petty, C. S., Organic phosphate insecticide poisoning, *Am. J. Med.*, 24, 467, 1958.
151. Jager, K. W., Roberts, D. V., and Wilson, A., Neuromuscular function in pesticide workers, *Br. J. Ind. Med.*, 27, 273, 1970.
152. Hierons, R. and Johnson, M. K., Clinical and toxicologic investigations of a case of delayed neuropathy in man after acute poisoning by an organophosphorus pesticide, *Arch. Toxicol.*, 40, 279, 1978.
153. Morgan, J. P., Apparent delayed neurotoxicity in humans caused by leptophos, an organophosphate pesticide, *Int. U. of Pharmacology Congress*, Abstract 645, Paris, 1978.
154. Abou-Donia, M. B., Othman, M. A., Tantawy, G., Zaki Khalil, A., and Shawer, M. F., Neurotoxic effect of Leptophos, *Experientia*, 30, 63, 1974.
155. Goodale, R. H. and Humphreys, M. B., Ginger jake paralysis. Autopsy observation, *J.A.M.A.*, 96, 14, 1931.
156. Vonderahe, A. R., Pathologic changes in paralysis caused by drinking Jamaica ginger, *Arch. Neurol. Psychiatry*, 25, 29, 1931.
157. Jeter, H., Autopsy report of a case of so-called jake paralysis, *J.A.M.A.*, 95, 112, 1930.
158. Bowden, D. T., Turley, L. A., and Shoemaker, H. A., The incidence of "jake" paralysis in Oklahoma, *Am. J. Publ. Health*, 20, 1179, 1930.
159. Bloch, H., Specificity of inhibitors in esters inhibited by triortho-cresyl phosphate, *Helv. Chim. Acta.*, 26, 733, 1943.
160. Hine, C. H., Dunlap, M. K., Rice, E. G., Coursey, M. M., Gross, R. M., and Anderson, H. H., The neurotoxicity and anticholinesterase properties of some substituted phenyl phosphates, *J. Pharmacol. Exp. Ther.*, 116, 227, 1956.
161. Earl, C. J. and Thompson, R. H. S., The inhibitory action of tri-ortho-cresyl phosphate on cholinesterase, *Br. J. Pharmacol.*, 7, 261, 1952.
162. Ord, M. G. and Thompson, R. H. S., Pseudocholinesterase activity in the central nervous system, *Biochem. J.*, 51, 245, 1952.
163. Earl, C. J. and Thompson, R. H. S., Cholinesterase levels in the nervous system of tri-ortho-cresyl phosphate poisoning, *Br. J. Pharmacol.*, 7, 685, 1952.

164. Davison, A. N., Some observations on the cholinesterases of the central nervous system after administration of organophosphorus compounds, *Br. J. Pharmacol.*, 8, 212, 1953.
165. Austin, L. and Davies, D. R., The part played by inhibition of cholinesterase of the CNS in producing paralysis in chickens, *Br. J. Pharmacol.*, 9, 145, 1954.
166. Durham, W. F., Gaines, T. B., and Hayes, W. J., Jr., Paralytic and related effects of certain organic phosphorus compounds, *Am. Med. Assoc. Arch. Ind. Health*, 13, 326, 1956.
167. Davies, D. R., Holland, P., and Rumens, M. J., The relationship between the chemical structure and neurotoxicity of alkyl organophosphorus compounds, *Br. J. Pharmacol.*, 15, 271, 1960.
168. Davies, D. R., Neurotoxicity of organophosphorus compounds, in *The Anticholinesterases. Handbuch der Exp. Pharmacol.*, Vol. 15, Koelle, G. B., Ed., Springer-Verlag, Berlin, 1963, 878.
169. Xintaras, C. and Johnson, B. L., Behavioral toxicology: Early warning and worker safety and health, in *Essays in Toxicology*, Vol. 7, Hayes, W. J., Jr., Ed., Academic Press, New York, 1976, 155.
170. Teichner, W. H., Methodology in behavioral toxicology research, in *Behavioral Toxicology*, Xintaras, C., Johnson, B. L., and de Groot, I., Eds., National Institute of Occupational Safety and Health, Cincinnati, Ohio, 1974, 441.
171. Bowers, M. B., Goodman, E., and Sim, V. M., Some behavioral changes in man following anticholinesterase administration, *J. Nerv. and Ment. Dis.*, 138, 383, 1964.
172. Diamant, H., Cholinesterase inhibitors and vestibular function. A study of vestibular syndrome in guinea pigs caused by intracarotid centripetal injection of cholinesterase inhibitors and cholinesters, *Acta Oto-Laryngol. Suppl.*, 111, 1954.
173. Machne, X. and Unna, K. R. W., Actions on the central nervous system, in *The Anticholinesterases. Handbuch der Exp. Pharmakol.*, Vol. 15, Koelle, G. B., Ed., Springer-Verlag, Berlin, 1963, chap. 14.
174. Karczmar, A. G., Central cholinergic pathways and their behavioral implications, in *Principles of Psychopharmacology*, Clark, W. G., Dittman, K. S., Leake, C. D., and Freedman, D. X., Eds., Academic Press, New York, 1969, 155.
175. Metcalf, D. R. and Holmes, J. H., EEG, psychological and neurological alterations in humans with organophosphorus exposure, *Ann. N.Y. Acad. Sci.*, 160, 357, 1969.
176. Russell, R. W., Behavioral aspects of cholinergic transmission, *Fed. Proc.*, 28, 121, 1969.
177. Richardson, A. J. and Glow, P. H., Discrimination behavior in rats with reduced cholinesterase activity, *J. Comp. Physiol. Psychol.*, 63, 240, 1967.
178. Richardson, A. J. and Glow, P. H., Post-criterion discrimination in rats with reduced cholinesterase activity, *Psychopharmacologia (Berlin)*, 11, 435, 1967.
179. Banks, A. and Russell, R. W., Effects of chronic reductions in acetylcholinesterase activity in serial problem-solving behavior, *J. Comp. Physiol.*, 64, 262, 1967.
180. Pearson, D. W., Clark, G., and Moore, C. M., Comparison of behavioral effects of various levels of chronic di-syston poisoning, *Proc. 77th Ann. Convent. A.P.A.*, 871, 969.
181. Reiter, L., Talens, G., and Woolley, D., Acute and subacute parathion treatment: effect on cholinesterase activities and learning in mice, *Toxicol. Appl. Pharmacol.*, 25, 582, 1973.
182. Kurtz, P. J., Dissociated behavioral and cholinesterase decrements following malathion exposure, *Toxicol. Appl. Pharmacol.*, 42, 589, 1977.
183. Lehotsky, K. and Ungvary, Gy., Experimental data on the neurotoxicity of fenitrothion, *Acta Pharmacol. Toxicol.*, 39, 374, 1976.
184. Burchfiel, J. L., Duffy, F. N., and Sim, V. M., Persistent effects of sarin and dieldrin upon the primate electroencephalogram, *Toxicol. Appl. Pharmacol.*, 35, 365, 1976.
185. Korsak, R. J. and Sato, M. M., Effects of chronic organophosphate pesticide exposure on the central nervous system, *Clin. Toxicol.*, 11, 83, 1977.
186. Duffy, F. H., Burchfiel, J. L., Bartels, P. H., Gaon, M., and Sim, V. M., Long-term effects of an organophosphate upon the human electroencephalogram, *Toxicol. Appl. Pharmacol.*, 47, 161, 1979.
187. Aoyagi, M., Meyer, J. S., and Deshmukh, V. D., Central cholinergic control of cerebral blood flow in the baboon. Effect of cholinesterase inhibition with neostigmine on auto-regulation and CO_2 responsiveness, *J. Neurosurg.*, 43, 689, 1975.
188. Cavanagh, J. B., Toxic substances and the nervous system, *Br. Med. Bull.*, 25, 268, 1969.
189. Cavanagh, J. B., The toxic effects of tri-ortho-cresyl phosphate on the nervous system, *J. Neurol. Neurosurg. Psychiatry*, 17, 163, 1954.
190. Barnes, J.M. and Denz, F. A., Experimental demyelination with organophosphorus compounds, *J. Pathol. Bacteriol.*, 65, 597, 1953.
191. Wills, J. H., Toxicity of anticholinesterases and treatment of poisoning, in *Anticholinesterase Agents*, Vol. I, Karczmar, A. G., Usdin, E., and Wills, J. H., Eds., Pergamon Press, New York, 1970, 357.
192. Cavanagh, J. B., Peripheral nerve changes in ortho-cresyl phosphate poisoning in the cat, *J. Pathol. Bacteriol.*, 87, 365, 1964.

193. **Prineas, J.**, The pathogenesis of dying-back polyneuropathies Part I. An ultrastructural study of experimental tri-ortho-cresyl phosphate intoxication in the cat, *J. Neuropathol. Exp. Neurol.*, 28, 571, 1969.
194. **Cavanagh, J. B.**, Peripheral neuropathy caused by chemical agents, *CRC Crit. Rev. Toxicol.*, 2, 365, 1973.
195. **Bischoff, A.**, The ultrastructure of tri-ortho-cresyl phosphate poisoning. 1. Studies on myelin and oxonal alterations in the sciatic nerve, *Acta Neuropathol. (Berlin)*, 9, 158, 1967.
196. **Illis, L., Patangia, G. N., and Cavanagh, J. B.**, Boutons terminaux and tri-ortho-cresyl phosphate neurotoxicity, *Exp. Neurol.*, 14, 160, 1966.
197. **Cavanagh, J. B. and Patangia, G. N.**, Changes in the central nervous system in the cat as the result of tri-ortho-cresyl phosphate poisoning, *Brain*, 88, 165, 1965.
198. **Swank, R. L.**, Avian thiamin deficiency. A correlation of the pathology and clinical behavior, *J. Exp. Med.*, 71, 683, 1940.
199. **Manjo, G. and Karnovsky, M. L.**, A biochemical and morphologic study of myelination and demyelination. II. Effect of an organophosphorus compound (mipafox) on the biosynthesis of lipid by nervous tissue of rats and hens, *J. Neurochem.*, 8, 1, 1961.
200. **Reichert, B. L. and Abou-Dohia, M. B.**, Inhibition of axoplasmic transport by delayed neurotoxic organophosphorus esters: A possible mode of action, Abstract #392, *Toxicol. Appl. Pharmacol.*, 48, A196, 1979.
201. **Bouldin, T. W. and Cavanagh, J. B.**, Organophosphorus neuropathy. 1. A teased-fiber study of the spatio-temporal spread of axonal degeneration, *Am. J. Pathol.*, 94, 241, 1979.
202. **Bouldin, T. W. and Cavanagh, J. B.**, Organophosphorus neuropathy. 2. A fine-structural study of the early stages of axonal degeneration, *Am. J. Pathol.*, 94, 253, 1979.
203. **Bloch, H.**, Der Einfluss von Trikresylphosphat auf die Aktivitat der Cholinesterase, *Helv. Med. Acta.*, 8(Suppl. 7), 15, 1941.
204. **Mendel, B. and Rudney, H.**, The cholinesterases in the light of recent findings, *Science*, 100, 499, 1944.
205. **Comroe, J. H., Jr., Todd, J., and Koelle, G. B.**, The pharmacology of di-isopropyl fluorophosphate (DFP) in man, *J. Pharmacol. Exp. Ther.*, 87, 281, 1946.
206. **Modell, W., Krop, S., Hitchcock, P., and Riker, W. F., Jr.**, General systemic actions of di-isopropyl fluorophosphate (DFP) in cats, *J. Pharmacol. Exp. Ther.*, 87, 400, 1946.
207. **Horton, R. G., Koelle, G. B., McNamara, B. P., and Pratt, H. J.**, The acute toxicity of di-isopropyl fluorophosphate, *J. Pharmacol. Exp. Therap.*, 87, 414, 1946.
208. **Burgen, A. S. V. and Chipman, L. M.**, Cholinesterase and succinic dehydrogenase in the central nervous system of the dog, *J. Physiol.*, 114, 296, 1951.
209. **Ord, M. G. and Thompson, R. H. S.**, Pseudo-cholinesterase activity in the central nervous system, *Biochem. J.*, 51, 245, 1952.
210. **Koelle, G. B.**, The histochemical differentiation of types of cholinesterases and their localization in tissues of the cat, *J. Pharmacol. Exp. Ther.*, 100, 158, 1950.
211. **Earl, C. J. and Thompson, R. H. S.**, The inhibitory action of tri-ortho-cresyl phosphate on cholinesterases, *Br. J. Pharmacol.*, 7, 261, 1952a.
212. **Earl, C. J. and Thompson, R. H. S.**, Cholinesterase levels in the nervous system in tri-ortho-cresyl phosphate poisoning, *Br. J. Pharmacol.*, 7, 685, 1952b.
213. **Earl, C. J., Thompson, R. H. S., and Webster, G. R.**, Observations on the specificity of the inhibition of cholinesterases by tri-ortho-cresyl phosphate, *Br. J. Pharmacol.*, 8, 110, 1953.
214. **Webster, G. R.**, The distribution and metabolism of phosphorus compounds in normal and demyelinating nervous tissue of the chicken, *Biochem. J.*, 57, 153, 1954.
215. **Mendel, B. and Myers, D. K.**, Pseudo-cholinesterase of brain, *Nature (London)*, 170, 928, 1952.
216. **Aldridge, W. N.**, Tricresyl phosphates and cholinesterases, *Biochem. J.*, 56, 185, 1954.
217. **Myers, D. K., Rekel, J. B. J., Veeger, C., Kemp, A., and Simons, E. G. L.**, Metabolism of triaryl phosphates in rodents, *Nature (London)*, 176, 259, 1955.
218. **Casida, J. E., Eto, M., and Baron, R. L.**, Biological activity of a tri-o-cresyl phosphate metabolite, *Nature (London)*, 191, 1396, 1961.
219. **Casida, J. E.**, Specificity of substituted phenyl phosphorus compounds for esterase inhibition in mice, *Biochem. Pharmacol.*, 5, 332, 1961.
220. **Eto, M., Casida, J. E., and Eto, T.**, Hydroxylation and cyclization reactions involved in the metabolism of tri-o-cresyl phosphate, *Biochem. Pharmacol.*, 11, 337, 1902.
221. **Baron, R. L., Bennett, D. R., and Casida, J. E.**, Neurotoxic syndrome produced in chickens by a cyclic phosphate metabolite of tri-o-cresyl phosphate — a clinical and pathological study, *Br. J. Pharmacol. Chemother.*, 18, 405, 1962.
222. **Taylor, J. D.**, A neurotoxic syndrome produced in cats by a cyclic phosphate metabolite of tri-o-cresyl phosphate, *Toxicol. Appl. Pharmacol.*, 11, 538, 1967a.

223. Taylor, J. D. and Buttar, H. S., Evidence for the presence of 2-(o-cresyl)4H-1:3:2-benzodioxaphosphoran-2-one in cat intestine following tri-o-cresyl phosphate administration, *Toxicol. Appl. Pharmacol.*, 11, 529, 1967b.
224. Sharma, R. P. and Watanabe, P. G., Time related disposition of tri-o-tolylphosphate (TOTP) and metabolites in chicken, *Pharmacol. Res. Commun.*, 6, 475, 1974.
225. Bischoff, A., Tri-ortho-cresyl phosphate neurotoxicity, in *Neurotoxicity*, Vol. 1, Roizin, L., Shiraki, H., and Gorcevic, N., Eds., Raven Press, New York, 1977, 431.
226. Morazain, R. and Rosenberg, P., Lipid changes in tri-o-cresyl phosphate-induced neuropathy, *Toxicol. Appl. Pharmacol.*, 16, 461, 1970.
227. Aldridge, W. N. and Barnes, J. M., Neurotoxic and biochemical properties of some triaryl phosphates, *Biochem. Pharmacol.*, 6, 177, 1961.
228. Aldridge, W. N. and Barnes, J. M., Further observations on the neurotoxicity of organophosphorus compounds, *Biochem. Pharmacol.*, 15, 541, 1966.
229. Aldridge, W. N. and Barnes, J. M., Esterases and neurotoxicity of some organophosphorus compounds, *Biochem. Pharmacol.*, 15, 549, 1966.
230. Aldridge, W. N. and Barnes, J. M., Neurotoxicity of drugs, in *Proc. of the European Society for the Study of Drug Toxicity*, Vol. VIII, Intern. Cong. Ser. #118, Excerpta Medica Foundation, Amsterdam, 1967, 162.
231. Witter, R. F. and Gaines, T. B., Relationship between depression brain or plasma cholinesterase and paralysis in chickens caused by certain organic phosphorus compounds, *Biochem. Pharmacol.*, 12, 1377, 1963.
232. Baron, R. L. and Casida, J. E., Enzymatic and antidotal studies on the neurotoxic effect of certain organophosphates, *Biochem. Pharmacol.*, 11, 1129, 1962.
233. Davies, D. R., Holland, P., and Rumens, M. J., The relationship between the chemical structure and neurotoxicity of alkyl organophosphorus compounds, *Brit. J. Pharmacol.*, 15, 271, 1960.
234. Davies, D. R., Holland, P., and Rumens, M. J., The delayed neurotoxicity of phosphorodiamidic fluorides, *Biochem. Pharmacol.*, 15, 1783, 1966.
235. Eto, M., *Organophosphorus Pesticides: Organic and Biological Chemistry*, CRC Press, Boca Raton, Fla., 1974, 219.
236. Eto, M. and Abe, M., Metabolic activation of alkylphenyl phosphates, *Biochem. Pharmacol.*, 20, 967, 1971.
237. Gaines, T. B., Acute toxicity of pesticides, *Toxicol. Appl. Pharmacol.*, 14, 515, 1969.
238. Johannsen, F. R., Wright, P. L., Gordon, D. E., Levinskas, G. J., Radue, R. W., and Graham, P. R., Evaluation of delayed neurotoxicity and dose-response relationships of phosphate esters in the adult hen, *Toxicol. Appl. Pharmacol.*, 41, 291, 1977.
239. Abou-Donia, M. B., Graham, D. G., and Komeil, A. A., Delayed neurotoxicity of O-ethyl O-2,4-dichlorophenyl phenylphosphonothioate: effects of a single oral dose on hens, *Toxicol. Appl. Pharmacol.*, 49, 293, 1979.
240. O'Brien, R. D., *Toxic Phosphorus Esters. Chemistry, Metabolism and Biological Effects*, Academic Press, New York, 1960, 106.
241. Eto, M., *Organophosphorus Pesticides: Organic and Biological Chemistry*, CRC Press, Boca Raton, Fla., 1974, 142.
242. Aldridge, W. N. and Reiner, E., *Enzyme Inhibitors as Substrates. Interaction of Esterases with Esters of Organophosphorus and Carbamic Acids*, American Elsevier, New York, 1972, 79.
243. Toia, R. F. and Casida, J. E., Phosphorylation, "aging" and possible alkylation reactions of saligenin cyclic phosphorus esters with α-chymotrypsin, *Biochem. Pharmacol.*, 28, 211, 1979.
244. Anon., *Near East News Roundup*, FAO, RNEA, Cairo, Nov. 22, 1971.
245. Pesticide, disease connection feared, *San Antonio Express News*, p. 9E, Thursday, Dec. 2, 1976a.
246. Deadly pesticide was handled in 11 plants, *San Antonio Express News*, p. 18A, Saturday, Dec. 4, 1976b.
247. Morgan, J. P., Apparent delayed neurotoxicity in humans by leptophos, an organophosphate pesticide, *IUPHAR*, Paris, Abstract #645, 1978.
248. National Institute of Occupational Safety and Health Technical Report, *Occupational exposure to leptophos and other chemicals*, U.S. Dept. Health, Education and Welfare, Publ. 78-136, 1978.
249. Abou-Donia, M. B. and Preissig, S. H., Delayed neurotoxicity of leptophos: toxic effects on the nervous system of hens, *Toxicol. Appl. Pharmacol.*, 35, 209, 1976a.
250. Abou-Donia, M. B. and Preissig, S. H., Delayed neurotoxicity from continuous low-dose oral administration of leptophos to hens, *Toxicol. Appl. Pharmacol.*, 38, 595, 1976b.
251. Preissig, S. H. and Abou-Donia, M. B., The chronologic effects of leptophos on the spinal cord and sciatic nerve of hens, *J. Neuropathol. Exp. Neurol.*, 35, 303, 1976.
252. Abou-Donia, M. B. and Graham, D. G., Delayed neurotoxicity from long-term, low-level topical administration of leptophos to the comb of hens, *Toxicol. Appl. Pharmacol.*, 46, 199, 1978a.

253. Preissig, S. H. and Abou-Donia, M. B., The neuropathy of leptophos in the hen: a chronologic study, *Environ. Res.*, 17, 242, 1978b.
254. Leuck, C. B., Bowman, M. C., and McWilliams, J. M., Persistence of Velsicol VCS-506 (0-(4-bromo-2,5-dichlorophenyl)-0-methyl phenyl-phosphonothioate), its oxygen analogue and its phenol in coastal Bermuda grass pasture, *J. Econ. Entomol.*, 63, 1346, 1970.
255. Johnson, J. C., Bowman, M. C., Leuck, D. B., and Knox, F. E., Persistence of Phosvel in corn silage and effects of feeding dairy cows the treated silage, *J. Dairy Sci.*, 54, 1840, 1971.
256. Aharonson, N. and Ben-Aziz, A., Persistence of residues of Velsicol VCS-506 and two of its metabolites in tomatoes and grapes, *J. Agric. Food Chem.*, 22, 704, 1974.
257. Abou-Donia, M. B., Pharmacokinetics of neurotoxic oral dose of leptophos in hens, *Arch. Toxicol.*, 36, 103, 1976.
258. Whitacre, D. M., Badie, M., Schwemmer, B. A., and Diaz, L. I., Metabolism of ^{14}C-leptophos and ^{14}C-4-bromo-2,5-dichlorophenol in rats: A multiple dosing study, *Bull. Environ. Contam. Toxicol.*, 16, 689, 1976.
259. Konno, N. and Kinebuchi, H., Residues of Phosvel in plasma and adipose tissue of hens after single oral administration, *Toxicol. Appl. Pharmacol.*, 45, 541, 1978.
260. Sanborn, J. R., Metcalf, R. L., and Hansen, L. G., The neurotoxicity of 0-(2,5-dichlorophenyl) 0-methyl phenylphosphonothioate, an impurity and photoproduct of leptophos (Phosvel) insecticide, *Pestic. Biochem. Physiol.*, 7, 142, 1977.
261. Aldridge, W. N. and Johnson, M. K., Side effects of organophosphorus compounds: delayed neurotoxicity, *Bull. W.H.O.*, 44, 259, 1971.
262. Johnson, M. K., A phosphorylation site in brain and the delayed neurotoxic effects of some organophosphorus compounds, *Biochem. J.*, 111, 487, 1969a.
263. Johnson, M. K., The delayed neurotoxic effect of some organophosphorus compounds. Identification of the phosphorylation site as an esterase, *Biochem. J.*, 114, 711, 1969b.
264. Poulsen, E. and Aldridge, W. N., Studies on esterases in the chicken central nervous system, *Biochem. J.*, 90, 182, 1964.
265. Richardson, R. J., Davis, C. S., and Johnson, M. K., Subcellular distribution of marker enzymes and of neurotoxic esterase in adult hen brain, *J. Neurochem.*, 32, 607, 1979.
266. Olajos, E. J. and Rosenblum, I., Measurement of neurotoxic esterase activity in various subcellular fractions in hen brain and sciatic nerve homogenates and the effect of diisopropylfluorophosphate (DFP) administration, *Ecotoxicology and Environ. Safety*, 3, 18, 1979.
267. Johnson, M. K., Organophosphorus and other inhibitors of brain "neurotoxic esterase" and the development of delayed neurotoxicity in hens, *Biochem. J.*, 120, 523, 1970.
268. Johnson, M. K., The primary biochemical lesion leading to the delayed neurotoxic effects of some organophosphorus esters, *J. Neurochem.*, 23, 785, 1974.
269. Johnson, M. K., Structure-activity relationships for substrates and inhibitors of hen brain neurotoxic esterase, *Biochem. Pharmacol.*, 24, 797, 1975a.
270. Johnson, M. K., The delayed neuropathy caused by some organophosphorus esters: mechanism and challenge, *CRC Crit. Rev. Toxicol.*, 3, 289, 1975b.
271. Clothier, B. and Johnson, M. K., Rapid aging of neurotoxic esterase after inhibition by di-isopropylphosphorofluoridate, *Biochem. J.*, 177, 549, 1979.
272. Lotti, M. and Johnson, M. K., Neurotoxicity of organophosphorus pesticides: predictions can be based on *in vitro* studies with hen and human enzymes, *Arch. Toxicol.*, 41, 215, 1978.
273. Johnson, M. K., Organophosphorus esters causing delayed neurotoxic effects: mechanism of action and structure/activity studies, *Arch. Toxicol.*, 34, 259, 1975c.
274. Johnson, M. K., The anomalous behaviour of dimethyl phosphates in the biochemical test for delayed neurotoxicity, *Arch. Toxicol.*, 41, 107, 1978.
275. Ariens, A. Th., Meeter, E., Wolthius, O. L., and Van Benthem, R. M. J., Reversible necrosis at the end-plate region in streated muscle of the rat poisoned with cholinesterase inhibitors, *Experientia*, 25, 57, 1969.
276. Fenichal, G. M., Kibler, W. B., Olsen, W. H., and Dettbarn, W.-D., Chronic inhibition of cholinesterase as a cause of myopathy, *Neurology*, 22, 1026, 1972.
277. Kibler, W. B., Skeletal muscle necrosis secondary to parathion, *Toxicol. Appl. Pharmacol.*, 25, 117, 1973.
278. Laskowski, M. B., Olsen, W. H., and Dettbarn, W.-D., Initial ultrastructural abnormalities at the motor end-plate produced by a cholinesterase inhibitor, *Exp. Neurol.*, 57, 13, 1977.
279. Wecker, L., Laskowski, M. B., and Dettbarn, W.-D., Neuromuscular dysfunction induced by acetylcholinesterase inhibition, *Fed. Proc.*, 37, 2818, 1978.
280. Laskowski, M. B., Olsen, W. H., and Dettbarn, W.-D., Ultrastructural changes at the motor end-plate by an irreversible cholinesterase inhibitor, *Exp. Neurol.*, 47, 290, 1975.
281. Wecker, L. and Dettbarn, W.-D., Paraoxon-induced myopathy: muscle specificity and acetylcholine involvement, *Exp. Neurol.*, 51, 281, 1976.

282. Wecker, L. and Dettbarn, W.-D., Effect of denervation on the production of an experimental neuropathy, *Exp. Neurol.,* 57, 94, 1977.
283. Wecker, L., Kiauta, T., and Dettbarn, W.-D., Relationship between acetylcholinesterase inhibition and the development of a myopathy, *J. Pharmacol. Exp. Ther.,* 206, 97, 1978.
284. Durham, W. F. and Hayes, W. J., Jr., Organic phosphorus poisoning and its theory, *Arch. Environ. Health,* 5, 27, 1962.
285. Ellin, R. I. and Wills, J. H., Oximes antagonistic to inhibitors of cholinesterase, Parts I and II, *J. Pharmaceut. Sci.,* 53, 995, 1143, 1964.
286. Wills, J. H., Toxicity of anticholinesterases and its treatment, in *Anticholinesterase Agents,* Vol. 1, Karczmar, A. G., Ed., Pergamon Press, New York, 1970, 400.
287. Namba, T., Nolte, C. T., Jackrel, J., and Grob, D., Poisoning due to organophosphate insecticides. Acute and chronic manifestations, *Am. J. Med.,* 50, 475, 1971.
288. Milby, T. H., Prevention and management of organophosphate poisoning, *J.A.M.A.,* 216, 2131, 1971.
289. Davies, D. R. and Holland, P., Effect of oximes and atropine upon the development of delayed neurotoxic signs in chickens following poisoning by DFP and sarin, *Biochem. Pharmacol.,* 21, 3145, 1972.
290. Baron, R. L. and Casida, J. E., Enzymatic and antidotal studies on the neurotoxic effect of certain organophosphorus compounds, *Biochem. Pharmacol.,* 11, 1129, 1962.
291. Casida, J. E., Baron, R. L., Eto, M., and Engel, J. L., Potentiation and neurotoxicity induced by certain organophosphates, *Biochem. Pharmacol.,* 12, 73, 1963.

Chapter 6

CARBAMIC ACID ESTER PESTICIDES

Donald J. Ecobichon

I. INTRODUCTION

The synthesis of aliphatic esters of carbamic acid led to the development of a new class of pesticides in the early 1930s which were introduced and marketed as fungicides. Interest in this group of chemicals, poor insecticides in comparison with the chlorinated hydrocarbons and organophosphorus esters, lagged until the mid-1950s when renewed interest in insecticides having anti-cholinesterase activity led to the synthesis of several potent aryl esters of methyl-carbamic acid. The carbamates became the insecticides of the 1960s following the restriction of chlorinated hydrocarbon usage and the recognized toxicity of many of the organophosphorus esters. While a large number of carbamates have been developed to the experimental level, relatively few progressed past this stage. Carbaryl and carbofuran have been the most widely used commercial insecticides. Despite widespread use, the carbamates represent only a small proportion of the pesticide market, some 21 agents being listed in 1969.[1] Their use is increasing slowly (see Chapter 2, Figure 3).

II. CHEMISTRY

A. Nomenclature

Compared to the chlorinated hydrocarbon, organophosphorus ester, and pyrethroid insecticides, the nomenclature of the carbamate esters is quite straightforward. As is shown in Figure 1, the basic structure of all carbamate insecticides is carbamic acid (I), the monoamide of carbon dioxide, a highly unstable molecule which decomposes readily into carbon dioxide and ammonia. Carbamic acid may be stabilized by forming salts such as ammonium carbamate or by synthesizing alkyl (II) or aryl [phenolic (III), naphtholic (IV), benzofuranyl (V)] esters. The replacement of one of the protons associated with the nitrogen by a methyl group results in the formation of N-monomethylcarbamic acids (VI) which, when combined with esterified aryl groups, possess insecticidal properties. The majority of carbamate insecticides in use are N-monomethylcarbamates, frequently just called N-methylcarbamates or methylcarbamates. N-dimethylcarbamate (VII) insecticides are also found and substituted heterocyclic N,N-dimethylcarbamates (VII) (dimetilan, Isolan®, Pyramat®) have been marketed for many years as fungicides. The addition of suitable substituents to certain positions on the aryl or heterocyclic rings shown in Figure 1 yield a variety of chemical compounds with a wide range of insecticidal potency and selectivity. The structures and the numbering of substituent positions on the aryl rings provide the key to the nomenclature of the carbamic acid esters. By way of example, Table 1 shows the structures, the common and/or "generic" and chemical names of some commonly used carbamates.

In addition to the above several thiocarbamates and dithiocarbamates have been marketed for years as herbicides and fungicides, the structures of some being shown in Table 1. The newest group of methylcarbamate insecticides to be developed are derivatives of aliphatic oximes which resemble aldehydes or ketones in structure. The structures of two commercially used oxime derivatives, aldicarb and methomyl, are shown in Table 1. In the present chapter, the insecticides will be referred to by the trade name or common name frequently used in the literature, since this will simplify discussion.

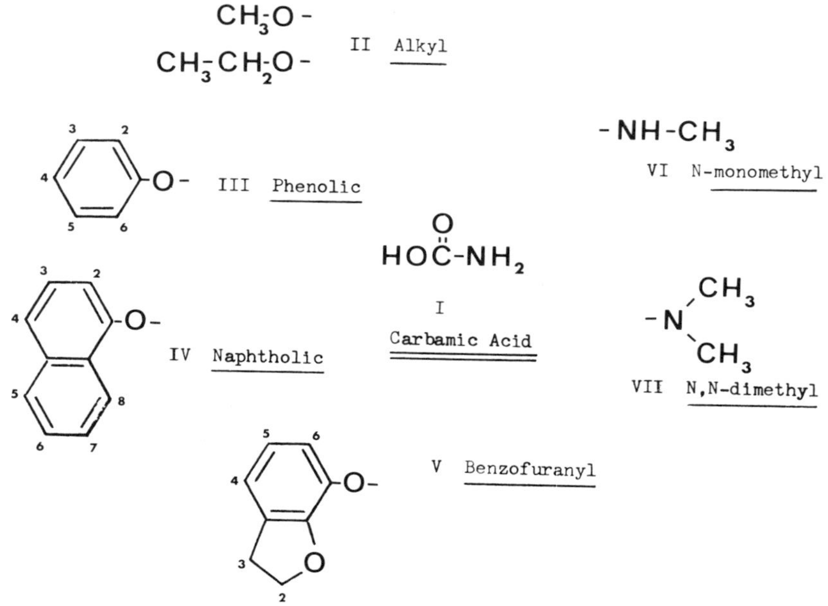

FIGURE 1. Structural components of carbamic acid insecticides showing substituent groups at the amino and at the carbonyl positions.

B. Physicochemical Properties

As can be gathered from the previous section, the nature of the substituent groups not only alters the biological properties of the carbamates but exerts considerable influence on the physicochemical properties of the insecticides. Prior to discussing the significance of the substituents, it should be stated that carbamic acid esters dissolve readily in organic solvents but are only slightly soluble in water. The exceptions are the methylcarbamoyloximes (aldicarb and methomyl) (Table 1). A wide range of melting points (50 to 150°C) can be found, determined largely by the size of the substituent aryl group and vapor pressures range from 1×10^{-3} to 5×10^{-2} mm Hg.[2]

Alkyl and aryl esters may be formed from carbamic acid. It was found that alkyl esters were relatively stable whereas aryl esters (phenolic, naphtholic, etc.) tended to be unstable. Stability was enhanced by attaching substituents to either the aryl structure or to the carbamoyl nitrogen. Thus, N-monomethylcarbamates were considerably more stable than unsubstituted esters.[3] Considering a few of the structurally different N-methylcarbamates shown in Table 1 and incubating them at 25°C in an alkaline solution (0.01 M barbital buffer, pH 9.3), the biological half-lives of Banol®, Mesurol®, carbaryl, Zectran®, Baygon®, and Matacil® were 0.1, 0.4, 0.5, 2.3, 3.1, and 4.0 hr, respectively.[4] It can be seen, on examining the structures of these agents, that the larger the aliphatic substituent, i.e., dimethylamino, isopropoxy, the more stable was the chemical in alkaline solution. Small substituents such as a nitro group had little effect and the rate of hydrolysis was very rapid, the second order rate constant being 3.5×10^6 mol/min.[5] Substituting one chloro group on the aryl ring decreased the rate constant of hydrolysis to 1.0 to 2.0×10^3 mol/min while substitution with a dimethylamino or isopropoxy group reduced the constant of hydrolysis by another 100-fold to approximately 1.0×10^1 mol/min.[5]

Being esters, carbamates decompose slowly in water at an acidic pH but alkalinity is a serious problem since substituent groups which tend to withdraw electrons from the phenyl ring reduce the electronegativity around the ester group, thereby weakening it and accelerating the hydrolysis by hydroxide ions. Mono- or dimethyl substitution on the carbamoyl nitrogen results in the stabilization of the ester bond. N-monomethylcarbamates degrade slowly, i.e., carbaryl at pH 7.0 has a half-life of 10 days.

Table 1
STRUCTURES, COMMON OR TRADE NAMES AND CHEMICAL NAMES OF
SOME COMMERCIALLY IMPORTANT CARBAMATE ESTER INSECTICIDES

Structure	Common or trade name	Chemical name
	Carbaryl (Sevin®)	1-naphthyl N-methyl carbamate
	Carbofuran	2,3-dihydro2,2-dimethyl-7-benzofuranyl N-methyl-carbamate
	Banol®	2-chloro-4,5-dimethylphenyl N-methylcarbamate
	Baygon®	2-isopropoxyphenyl N-methylcarbamate
	Zectran®	4-(N,N-dimethylamino)-3,5-dimethylphenyl N-methylcarbamate
	Mesurol®	4-methylmercapto-3,5-dimethylphenyl N-methylcarbamate
	Dimetilan	2-dimethylcarbamoyl-3-methylprazolyl-5-N,N-dimethylcarbamate
	Isolan®	1-isopropyl-3-methylprazolyl-5-N,N-dimethylcarbamate
	Dimetan®	5,5-dimethyl-3-oxo-1-cyclohexenyl-1-N,N-dimethylcarbamate
	Thiram	Tetramethylthiuram disulfide
	Nabam	Disodium ethylene-*bis*-dithio carbamate
	Temik® (aldicarb)	2-methyl-2-(methylthio)propion- aldehyde O-(methylcarbamoyl) oxime
	Lannate® (methomyl)	S-methyl-N-((methylcarbamoyl)oxy) thioacetimidate
	Zineb	([1,2-ethanediylbis(carbamodithioato)][2-]) zinc
	Maneb®	([1,2-ethanediylbis(carbamodithioato)])(2-) manganese

Dimethylated carbamates are exceedingly stable, the half-life of dimetilan being approximately 100 days at pH ranging from 6 to 10.[3] As was indicated, the more alkaline the medium is the more rapid is the hydrolysis of the carbamate ester, this bearing a direct relationship to the hydroxide concentration. In alkaline river water at pH 7.3 to 8.0, the hydrolysis of carbaryl was rapid with only 5% of the original compound being present after 1 week of incubation.[6] In contrast, 10% of the Matacil® was present after 2 weeks of incubation and 5% of the original Baygon® was still present after an interval of 8 weeks.

While carbamates have relatively high melting points and low vapor pressures which enhance the environmental stability of the deposited chemical, decomposition can be markedly enhanced by increased temperatures, a 10°C increase raising the hydrolysis rate two- or three-fold.[7,8] Environmental stability of carbamates is severely affected by photodegradation at short ultraviolet wavelengths (254 nm) and by oxidation upon exposure to air. These aspects of decomposition are covered succinctly by Kuhr and Dorough.[9]

As a consequence of a number of studies involving aliphatic and aromatic esters of unsubstituted, mono- and dimethyl substituted carbamic acid esters, two mechanisms or pathways have been proposed to explain the base-catalyzed decomposition of these

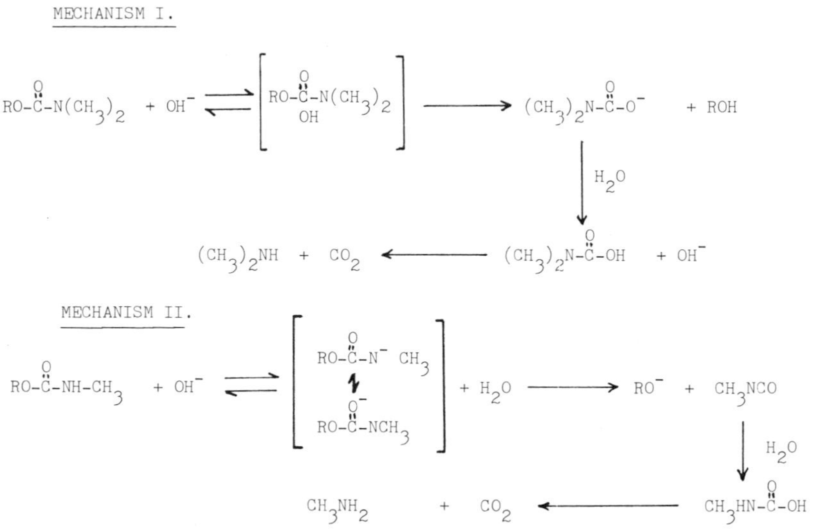

MECHANISM I.

$$RO-\overset{O}{\overset{\|}{C}}-N(CH_3)_2 + OH^- \rightleftharpoons \left[\begin{array}{c} O \\ RO-\overset{\|}{C}-N(CH_3)_2 \\ OH \end{array} \right] \longrightarrow (CH_3)_2N-\overset{O}{\overset{\|}{C}}-O^- + ROH$$

$$\downarrow H_2O$$

$$(CH_3)_2NH + CO_2 \longleftarrow (CH_3)_2N-\overset{O}{\overset{\|}{C}}-OH + OH^-$$

MECHANISM II.

$$RO-\overset{O}{\overset{\|}{C}}-NH-CH_3 + OH^- \rightleftharpoons \left[\begin{array}{c} O \\ RO-\overset{\|}{C}-N^--CH_3 \\ \updownarrow \\ RO-\overset{O^-}{\overset{\|}{C}}-NCH_3 \end{array} \right] + H_2O \longrightarrow RO^- + CH_3NCO$$

$$\downarrow H_2O$$

$$CH_3NH_2 + CO_2 \longleftarrow CH_3HN-\overset{O}{\overset{\|}{C}}-OH$$

FIGURE 2. Mechanisms for the alkaline hydrolysis of carbamate esters. Mechanism I has been suggested as the route of hydrolysis of aliphatic N-methyl- and -dimethylcarbamates and aromatic dimethylcarbamates. Mechanism II has been suggested as the method of hydrolysis of aromatic N-methyl- and unsubstituted carbamates.[10]

chemicals (Figure 2).[10] **Mechanism I,** for aliphatic, methyl- and dimethyl-carbamates as well as aromatic dimethylcarbamates is comparable to the reaction of an OH⁻ with any ester, an unstable tetrahedral intermediate being formed which disintegrates into an alcohol and a methyl- or dimethyl-carbamoyl ion which subsequently picks up a proton and decomposes spontaneously into dimethylamine and carbon dioxide. **Mechanism II,** for aromatic unsubstituted and monomethyl substituted carbamates, appears to involve the removal of an N-proton, resulting in the formation of an unstable methylisocyanate and a phenoxide ion, the former reacting with water to disintegrate into methylamine and carbon dioxide. Some concern has been raised over the existence of a charged form of the carbamate and the formation of the unstable methylisocyanate, Fukuto et al.[7] have developed an alternative mechanism of decomposition which is a slight modification of Mechanism II in which methylisocyanate does not appear as an intermediate product and the carbamate does not become charged. Kinetically, the reaction is indistinguishable from that shown for Mechanism II.

Insecticidal potency is governed by the nature and position of the substituents.[11] The esters of N-alkylcarbamic acid have insecticidal properties, the aryl esters of N-monomethylcarbamic acid being used to control insect pests and showing the maximal insecticidal effect. Replacement of the second hydrogen atom on the carbamoyl nitrogen by an alkyl radical sharply reduces the insecticidal activity. The only exceptions are esters of N,N-dimethylcarbamic acid with complex heterocyclic hydroxy compounds. The alkyl esters of N-arylcarbamic acids are potent herbicides but show no insecticidal activity. The replacement of the carbonyl or ester oxygen by sulfur also decreases the insecticidal potency of carbamates. The derivatives of dithiocarbamic acid are potent fungicides and a few important agricultural fungicidal agents have been developed, i.e., Zineb® and Maneb®, which incorporate divalent metal cations (zinc, manganese) into their structures (Table 1).

C. Biotransformation

There are a number of excellent reviews on the biotransformation of carbamate insecticides including those by Casida,[12] Ryan,[13] Kuhr and Dorough,[14] Matsumura,[15]

Fukuto,[16] Kulkarni, and Hodgson[17] to name only a few. As well, several chapters in the book edited by C. F. Wilkinson give the biochemical background essential for an understanding of the mechanisms involved.[18] While this aspect of the carbamates will not be covered in great detail, it is important to explore some concepts of biotransformation in order to appreciate the toxicity of these insecticides and/or the absence of effects in mammals.

In the short space of 20 years, our knowledge of the biochemistry of carbamate esters in plant, insect, and animal systems has progressed from complete ignorance to a position where the metabolic schemes for degradation are extremely complex and it is questionable whether the number of products can possibly exist in a single species. One should appreciate that many of the studies have been conducted in vitro and, by design, are limited in their metabolic capacity since they were fortified with specific cofactors, essential cations, etc., and conducted at optimal pH values which force or push reactions along selected routes which might prove to be of little importance in the in vivo situation, the particular organism metabolizing the chemical by an entirely different route at such a rapid rate as to nullify the importance of the possibly toxic intermediates identified in vitro. An important feature of metabolic studies is that the same major pathways may be identified in plants, insects, microorganisms, and vertebrates, but the rate of biotransformation may be very species specific. This is important from the standpoint of selective toxicity since it would be desirable to have a low rate of biotransformation in insects and in the plant host and a high rate of metabolism in vertebrate species in order to protect them.

As was observed for the organophosphorus esters in the preceding chapter, the initial response of any species exposed to a carbamate ester is to convert the chemical into a more polar form which then can be excreted or stored away from target organ molecules. To achieve this feat, the organism calls upon Phase I and Phase II detoxification mechanisms which have a surprisingly widespread distribution among various species of plants and animals. Again, as was seen for the organophosphorus esters, the structure of the carbamic acid ester and the nature and position of substituent groups exert a dominant role over the rate and route of degradation. While carbamate esters are susceptible to a wide variety of enzyme-catalyzed detoxification reactions, one finds that the principal mechanisms involved are hydrolysis, oxidation, and conjugation (Figure 3).

1. Hydrolysis

Since carbamate insecticides are esters, they are susceptible to cleavage by esterases, the products formed being identical to many of those achieved via chemical hydrolysis in that an aryl alcohol plus methyl- or dimethylcarbamic acid are formed (Figures 2, 3). The methylated carbamic acids, being unstable, will rapidly decompose into carbon dioxide and methyl- or dimethylamine. Differences in rates of hydrolysis will be dependent upon the structure and, while hydrolysis in vivo should be considered as a potential pathway of degradation, this property plus differences in the properties of tissue enzymes both contribute greatly to the specificity, selective toxicity, and species differences observed. Taking carbaryl as an example, this extensively studied compound is readily hydrolyzed in rat, sheep, dog, and guinea pig in vivo, but is quite resistant to hydrolysis in the monkey and pig.[19,20] That the route of degradation was hydrolytic in nature was equated with the amount of $^{14}CO_2$ formed and exhaled. The human liver as well as other tissues of the body readily hydrolyzes carbaryl both in vivo and in vitro.[21-23] The hydrolysis of carbaryl appears to be a minor pathway of degradation in most insect species, the exception being the German cockroach where hydrolysis is an important route.[24] Krishna and Casida[25] demonstrated the extensive hydrolysis of 10 different carbamates by rats, between 60 and 80% of the carbamoyl

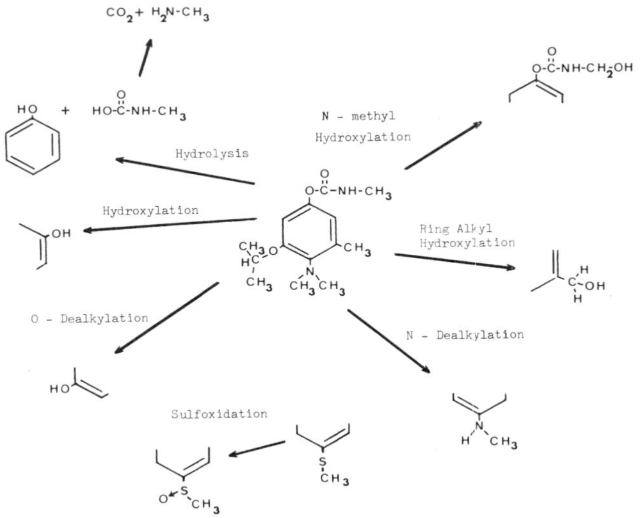

FIGURE 3. Phase I oxidative and hydrolytic detoxification pathways involved in the biotransformation of carbamate ester insecticides by mammalian tissues. See text for details.

^{14}C appearing as $^{14}CO_2$ and signifying that the mechanism shown in Figure 3 was the most probable route. More recent studies of the detoxification of various carbamate esters in the rat revealed a percent hydrolysis of 23% for carbofuran, 31% for propoxur, 39 to 65% for carbaryl, 53 to 62% for aldicarb, 76% for Zectran®, and 99% for Mobam.[26]

Despite extensive study, the nature of the tissue esterases involved in the hydrolysis of carbamate esters remains obscure. The wide distribution of various esterases in body tissues of many species of plants, insects, and animals suggested that hydrolysis would be an important mechanism of detoxification.[27] This assumption would appear to be incorrect and is difficult to test because of the spectrum of results between compounds and animal species. In blood plasma, an esterase associated with the albumin fraction was responsible for the hydrolysis of several carbamate esters.[28-30] This "albumin esterase" has been studied but is not a highly active enzyme as are the plasma pseudocholinesterases, aliesterases, and arylesterases. Hydrolysis appears to play a major role in carbamate metabolism only in the rat, as noted above, but not to the same degree in other mammalian species. Again, carbaryl, the most extensively studied agent, was degraded and eliminated as free or conjugated α-naphthol by rat tissues.[31,32] The variability in quantity of carbamate hydrolyzed by rat tissues has been mentioned.[25,26] The complexity of the hydrolysis problem has been emphasized in recent studies of esterases from American cockroaches, mouse, and human brain which were electrophoretically separated following tissue homogenization and then incubated with carbaryl and Baygon® as substrates.[33,34] A characterized aliesterase and several arylesterases of cockroach origin were found to hydrolyze both carbamates. Mouse and human brain esterases were also active toward carbaryl and Baygon®, the patterns being different for each species.[33-35]

It should be mentioned that carbamate esters can be hydrolyzed by two quite different mechanisms: (1) the hydrolysis of the carbamoylester moiety and (2) hydrolysis of certain side chains if present by an oxidative route. Recent papers have suggested that, in at least insects and mice, direct enzymatic hydrolysis may be a minor route, cleavage of the carbamate ester moiety being achieved by (1) an initial microsomal N-demethylation followed by (2) hydrolysis of the unsubstituted carbamate.[13,36] Any modification of substituent groups on either the carbamoyl nitrogen or on the alkoxy or phenolic or naphtholic portion which leave the ester group intact may increase the susceptibility for hydrolysis to occur. A generalization that carbamates can be hydrolyzed is perhaps somewhat risky in view of the possible complex mechanisms involving oxidative as well as hydrolytic enzymes. Each species to be investigated should be examined rigorously to ascertain whether carbamate-hydrolyzing esterases are present but the specificity toward several carbamates should be assessed before conclusions are drawn.

2. Oxidation

Oxidative reactions are carried out in vivo and in vitro by the cytochrome P-450-related microsomal monooxygenase enzymes which, in mammalian species, have a relatively broad spectrum of substrates ranging from endogenous chemicals (steroids, catecholamines, etc.) to exogenous chemicals (steroids, food additives, environmental contaminants, pesticides, etc.). Using a highly reactive hemoprotein (cytochrome P-450 or P-448) in conjunction with molecular oxygen, electrons donated from flavoproteins and cytochrome b_5 and reduced nicotinamide adenine dinucleotide phosphate (NADPH), a wide variety of reactions can be carried out by this complex enzyme system, depending largely upon the nature of the substituents.[14] Detailed descriptions of the various pathways for different carbamates can be found in Kuhr and Dorough,[14] Matsumura,[15] and Menzie.[38]

The type of reactions observed with carbamate esters can be simpified into two main groups: (1) ring hydroxylation, (2) oxidation of appropriate side chains (Figure 3). As can be seen for this "mythical" methylcarbamate, one can have hydroxylation of the N-methyl group, hydroxylation of methyl groups on an aryl substituent to form a hydroxymethyl group, N-demethylation of secondary and tertiary amines attached to the aryl substituent as well as O-dealkylation of alkoxy side chains and direct ring hydroxylation. Another reaction shown in this scheme involves the S-oxidation (sulfoxidation) of sulfur analogs of methoxy substituents. A reaction, not shown in Figure 3, which pertains particularly to naphthol-containing carbamates such as carbaryl, involves the formation of an epoxide by an oxidative process involving two vicinal or adjacent carbon atoms with the subsequent destruction of this intermediate nonenzymatically or by the enzyme epoxide hydrase to form a trans diol or a hydroxide.

It is important to realize that more than one oxidative route of biotransformation may be operative at any given time in vivo. Considering two structurally similar compounds, Zectran® and Mesurol®, it has been established that both the dog and the rat utilize the same major pathways of metabolism.[39] For Zectran, the major route is the stepwise N-dealkylation of the *para*-N-methyl moiety of the carbamate has been detected for both Zectran® and Mesurol®. Sulfoxidation of the *para*-S-methyl group is the main metabolic pathway for Mesurol®.

Aldicarb is rapidly biotransformed by monooxygenases into the sulfoxide followed by a slow oxidation into the sulfone in plants, whereas in animals the sulfone is not found. Since the carbamoyl ester group is intact, these intermediate metabolites can be hydrolyzed yielding oximes (R—CH=NOH).[40]

These enzymatic mechanisms produce diverse changes in the carbamate molecule resulting, for the most part, in reduced biological activity but, in a few cases, causing increased potency. In the latter reaction, the carbamate ester group is still intact and the oxidative reaction has reduced or modified the size of the insecticide molecule sufficiently to be a better fit for a target site. Matacil® (4-dimethylamino-3-cresyl N-methylcarbamate) is converted by oxidative N-demethylation into 4-amino-3-cresyl N-methylcarbamate which is a considerably more potent anticholinesterase agent than the parent compound.[41] Baygon®, Zectran®, and Banol® all were converted into at least one or more potent anticholinesterase inhibitors.[41]

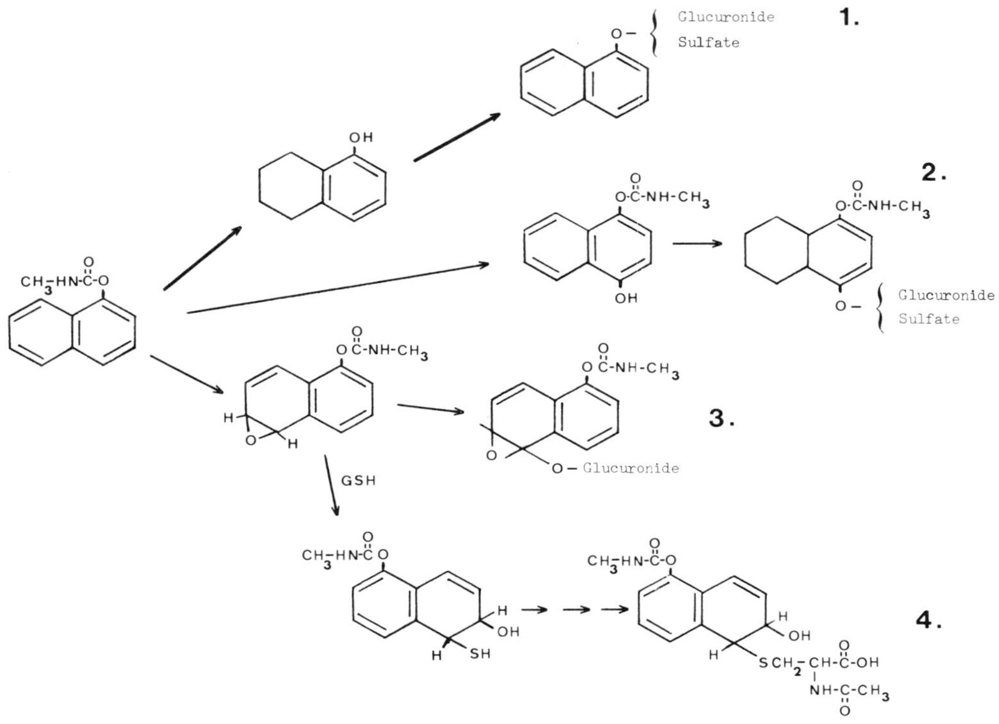

FIGURE 4. Phase II conjugative detoxification reactions involved in the formation of water soluble, readily excreted products of carbaryl. The reactions involve the conjugation of a hydrolytic product (α-naphthol) with glucuronic acid or ethereal sulfate (Reaction 1.); a ring-hydroxylated intermediate with glucuronic acid or ethereal sulfate (Reaction 2, 3); and the reaction of a hydroxylated intermediate with glutathione (GSH) to produce an N-acetylated cysteine derivative (Reaction 4). See text for details.

3. Conjugation

In conjugation reactions, which are Phase II detoxification reactions, a functional group on the molecule (amino, carboxyl, hydroxyl, epoxide, sulfhydryl, etc.), usually introduced as a consequence of hydrolytic or oxidative mechanisms, is enzymatically reacted with an endogenous substance in vivo to form a water-soluble, inactive product.[13] A variety of products can be formed depending upon the species of plant or animal being studied but, generally, the products can be classified as ethereal sulfates, glucuronides, glucosides, amino acid conjugates, acetylated amines, and mercapturic acids. With the exception of plant species in which glucosides are important, the conjugation reactions of major importance in animals are the sulfates, glucuronide and the mercapturic acid derivatives.[42] There have been any number of papers and book chapters written concerning the formation of conjugated products from drugs, etc. but, for the uninitiated, a brief description of the major reactions pertaining to carbamate insecticides will be given. As usual, because of its long usage, carbaryl is the best studied of the carbamate esters (Figure 4).

Glucuronidation of an available hydroxyl group (the acceptor) occurs by the transfer of glucuronic acid, a sugar acid, from the cofactor uridine diphospho-α-D-glucuronic acid (UDPGA), the reaction being catalyzed by the enzyme uridine diphosphoglucuronosyltransferase (Figure 4, reactions 1 to 3).[13,43] Phenols and alcohols form ether glucuronides whereas acids form ester glucuronides and aromatic amines form N-glucuronides. The enzyme glucuronosyltransferase is found localized in the smooth endoplasmic reticulum of the cells of a variety of body tissues, the major source of the enzyme being the liver.[43]

Aryl and alkyl sulfates are formed from phenols, alcohols and aromatic amines by the transfer of sulfate ($-SO_3H$) from the cofactor 3'-phosphoadenosine-5'-phosphosulfate by a sulfokinase (Figure 4, reactions 1 and 2).[44] A number of different sulfokinases have been identified which show distinct specificity for hydroxyl groups attached to various aromatic structures as well as for aromatic amines.

Mercapturic acids may be formed from a number of aromatic and aliphatic compounds. The initial reaction involves the "activated" intermediate product, glutathione (GSH, γ-L-glutamyl-L-cysteinylglycine) and an enzyme glutathione S-alkyl- or aryltransferase, the result being the formation of a glutathione conjugate. This reaction takes place primarily in the liver. This product is then excreted in the bile and subsequently hydrolyzed to a cysteine derivative which is then acetylated to form the final excretory product, an S-substituted N-acetylated cysteine derivative (Figure 4, Reaction 4).[45,46]

Of all of the carbamate insecticides, only carbaryl has received much attention concerning conjugation reactions.[13,42] In most studies, the actual conjugates have not been isolated from urine and feces, only the products following hydrolysis by acid or by the enzymes sulfatase and glucuronidase. Much more needs to be known about conjugation mechanisms in plants, insects, and mammalian species other than the laboratory rat and mouse. For the interested chemist, this would be a fruitful area of research.

In summary, the biotransformation of carbamic acid esters has not received the attention it should. While one can state that these insecticides are modified in vivo by enzymes catalyzing hydrolysis, oxidation and conjugation reactions, species differences in the routes and rates of biotransformation may contribute significantly to the selectivity and the toxicity of the particular agent. These avenues should be explored as an integral part of insecticide development since some studies have revealed unexpected and toxicologically significant findings where biologically potent products have been formed during the above mentioned reactions.[41]

III. TOXICITY

As was stated in the introductory chapter, the origins of carbamate insecticides begin with West African witchcraft and the active principle of the Calabar bean (*Physostigma venenosum*) used in "trial by ordeal". While the extracted active principle of this plant material was used in ophthalmology after 1863, the chemical structure of physostigmine or eserine was not established until 1925 as being the methylcarbamate of a substituted indole derivative.[47] A series of synthetic phenyl carbamates were synthesized and tested as miotic agents in the cat, m-trimethylammonium phenyl methylcarbamate being the most active.[48] This compound proved to be unstable in aqueous solution and subsequently, the dimethylcarbamate neostigmine was synthesized and found to be much more stable.[49] While such compounds were potent anticholinesterase agents in vitro, they and other quaternary derivatives studied proved to be poor contact insecticides because of their inability to penetrate insect cuticle and lipid nerve sheath. Wilson and Cohen[50] found that neostigmine would not penetrate the axon of crab nerve and affect conduction but that the unchanged tertiary amine analog, m-dimethylaminophenyl N,N-dimethylcarbamate, was extremely effective though being only 1/100 as potent a cholinesterase inhibitor.

Gysin[51] initiated a comprehensive study of a series of dimethylcarbamates of alcohols including Dimetan and Isolan (Table 1) and found them to have useful, potent insecticidal properties. In studies to assess why physostigmine and other charged carbamates were not insecticidal, Metcalf and his colleagues modified the structures of simple phenyl methylcarbamates with a view to increasing their lipid solubility and

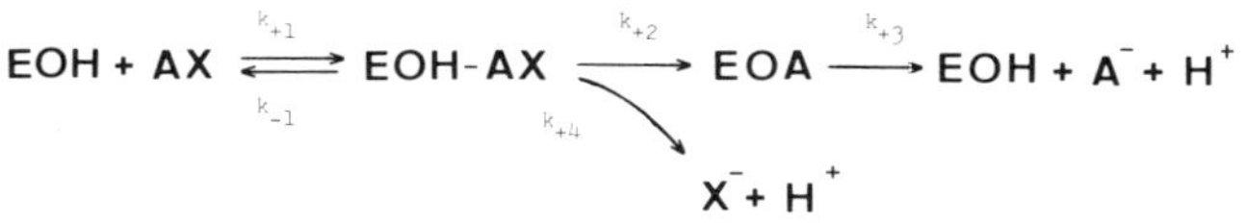

$$\text{EOH} + \text{AX} \underset{k_{-1}}{\overset{k_{+1}}{\rightleftharpoons}} \text{EOH-AX} \overset{k_{+2}}{\underset{k_{+4}}{\longrightarrow}} \text{EOA} \overset{k_{+3}}{\longrightarrow} \text{EOH} + \text{A}^- + \text{H}^+$$
$$\text{X}^- + \text{H}^+$$

FIGURE 5. The mechanism of interaction between a carbamate ester (AX) and acetylcholinesterase (EOH), showing the intermediate complexes and the final products.

found that phenyl methylcarbamate was insecticidal but that the potency and anticholinesterase activity was enhanced by substitution in the *ortho* and *meta* positions with an alkyl group or halogen.[52] The structure-activity relationships of these chemicals as anticholinesterase agents has been reviewed recently.[53] Despite the fact that thousands of carbamate esters have been synthesized only two dozen have found any widespread use as commercial insecticides.

A. Mechanism of Action

Carbamic acid ester insecticides inhibit nervous tissue, erythrocytic, and plasma cholinesterases, this being the mechanism by which toxicity is produced. These esters will inhibit other tissue esterases, primarily nonspecific carboxylesterases or aliesterases but the physiological role of these enzymes remains unknown beyond the fact that they may be involved in the hydrolytic degradation of some of these esters.[27] Several theories have been proposed concerning the mechanism(s) by which carbamates exert their toxicity but most investigators agree that the nervous tissue cholinesterases is the target macromolecule. A schematic diagram of the reaction is shown in Figure 5.

Given the general structure of carbamates as

$$\text{X-CO-N} {<}_R^R$$

the leaving group "X" may be one of a large variety of structures as we have seen earlier but, for the inhibition of cholinesterases, "R" and "R" are usually a methyl and hydrogen (monomethylcarbamate) or methyl and methyl (dimethylcarbamate). In Figure 5, the structure is further simplified so that "A" represents the carbamoyl group while "X" represents the leaving group, the unsubstituted or substituted alcohol, phenol, naphthol, oxime, etc. Looking at acetylcholine and a potent carbamate such as aldicarb (Figure 6) one can see the obvious structural similarity though this is less clear with other members of this class of insecticides. An exhaustive study of the structure-activity relationships of a large number of phenyl N-methylcarbamates revealed that there was a second molecular center that acts as a supplementary fit to cholinesterase at the anionic site, the optimum distance between the carbonyl group and the second center being 5 Å.[54] This fits reasonably well with the known distances between the anionic and esteratic sites on the cholinesterase molecule.[55]

There are two distinct steps in the reaction of a carbamate with the cholinesterase molecule. In step 1, the enzyme (EOH) couples with the carbamate (AX) to form an intermediate which is reversible and can dissociate, releasing free enzyme and insecticide again or proceed forward to step 2. The affinity constant of EOH for AX ($K_A = {}^{k-1}$) is low, indicating that step 1 proceeds rapidly. In step 2, there is a decomposition of the complex into a stable carbamoylated enzyme with the loss of the leaving group "X". In the third and final step, the carbamoylated enzyme hydrolyzes in the presence of water to yield the free and active enzyme again plus a product — mono- or dimethylcarbamic acid which is unstable and degrades to carbon dioxide and a methylated amine.[25,56] In effect, the only distinctive difference between carbamate and organophosphorus esters lies in the rate at which the decarbamoylation and dephosphorylation takes place. This is exceedingly slow for organophosphorus esters, so

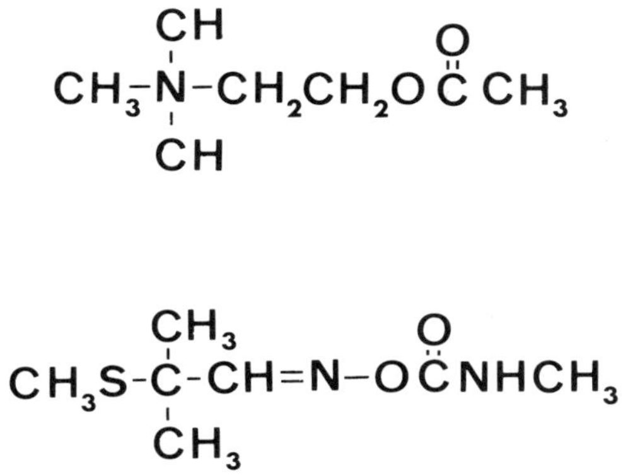

$$\underset{\text{Acetylcholine}}{CH_3-\overset{\overset{CH_3}{|}}{\underset{\underset{CH_3}{|}}{N}}-CH_2CH_2O\overset{O}{\overset{\|}{C}}CH_3}$$

$$\underset{\text{Aldicarb}}{CH_3S-\overset{\overset{CH_3}{|}}{\underset{\underset{CH_3}{|}}{C}}-CH=N-O\overset{O}{\overset{\|}{C}}NHCH_3}$$

FIGURE 6. Comparison of the similar structures of acetylcholine and the carbamoyloxime, aldicarb.

Table 2
KINETICS OF ESTER HYDROLYSIS

$$EH + AB \rightleftharpoons EHAB \rightarrow BH + EA \rightarrow EH + AOH$$

Esters	Complex formation $K_A = k_{-1}/k_{+1}$	Acylation k_2	Deacylation k_3
Substrates	Small	Extremely fast	Extremely fast
Organophosphorus esters	Small	Moderately fast	Slow or extremely slow
Carbamate esters	Small	Slow	Slow

^a Data from Ecobichon.[27]

much so that one often considers them to be irreversible inhibitors. It is sufficiently rapid enough for carbamoylated esterases that they are frequently considered to be reversible inhibitors. Only step 1 is reversible. Looking at the reaction, one can see that carbamate esters could be considered as being poor substrates for cholinesterases with low turnover rates which they are. The characteristics of the rate constants for substrates, organophosphorus, and carbamate esters are shown in Table 2.[27] It is important to appreciate the fact that the rates at which steps 2 and 3 proceed are thousands of times slower with carbamate esters than with the natural substrates of cholinesterases — acetylcholine. This subject has been extensively reviewed by Reiner and Aldridge[57] and the reader is referred to this book for a complete account.

If one avoids administering a large excess of carbamate to an animal, the nervous tissue acetylcholinesterase (AChE) will recover quickly by: (1) reverse dissociation of the complex and (2) decarbamoylation through steps 2 and 3. The half-life for the decarbamoylation of AChE has been variously calculated to be 30 to 40 min depending upon the carbamate being investigated.[58,59] Typically, without the presence of an excess of insecticide, the enzyme will begin to recover within a few minutes and would be completely reactivated after a few hours. Experiments have demonstrated that, unlike the organophosphorus esters, carbamate insecticides are not tightly bound to the AChE active site and can be dislodged easily by washing, dilution, or dialysis, all of which encourage the reverse reaction of step 1. The carbamate-AChE complex (EO-AX) can also be dissociated in vitro by increasing the substrate concentration during enzymatic analysis though this will have no effect on the carbamoylated enzyme

(EOA). The absence of a stable bond to the enzyme protein creates a number of problems for the analysis of inhibition and the differentiation between exposure to an organophosphorus and a carbamate ester. Using the familiar Ellman method[60] with erythrocyte AChE as an indicator enzyme, results with organophosphorus esters are relatively reliable because of the tight binding and "irreversible" nature of the inhibition. However, with carbamate insecticides, dilution of the inhibited enzyme prior to assay, preincubation with substrate, or a prolonged analysis time will all result in the reactivation of the EO-AX complex and considerable confusion for the analyst since the patient may be displaying signs of overt toxicity but his erythrocyte AChE may "appear" normal. These sorts of problems can be eliminated to some extent by: (1) using short assay times of 1 or 2 min; (2) adding the undiluted plasma to the preincubated mixture of buffer and substrate immediately (15 sec) prior to assay; and (3) standardizing the assay on one optimal substrate concentration. The reversal of the EO-AX complex portion of enzyme can be minimized in this way.

Since all N-methylcarbamate insecticides produce the same carbamoylated enzyme, the rate constant (k_3) for step 3 is the same for all compounds tested. The variation in overall rates of inhibition and reactivation then must be related to the properties contributed by the leaving group and associated with K_a and k_2. As was found by Metcalf and Fukuto,[54] the addition of electron-accepting substituents on the phenyl ring of N-methylcarbamates resulted in a reduction of inhibitory potency whereas the addition of electron-donating substituents on the phenyl ring of N-methylcarbamates resulted in a reduction of inhibitory potency whereas the addition of electron-donating substituents enhanced the inhibitory power. The structural complementarity of carbamate esters for the AChE molecule has recently been reviewed and the spatial arrangements of the second site on the enzyme molecule have been explored.[11,53,61]

B. Acute Toxicity — Animal

As a consequence of the reversibility of the cholinesterase inhibition and the reported rapid biotransformation of this group of insecticides, the major difficulties with carbamate esters lie in the acute toxicity following administration or exposure to a large dose. Unfortunately, no good rule of thumb can be applied concerning the toxicity of these agents as a whole. The structural diversity of commercially used compounds has conferred considerable variation in toxicity as is shown in Table 3 where the acute oral LD_{50} values for male rats range from 0.93 mg/kg for aldicarb to 7.500 mg/kg for Maneb.[11,62,63]

An early study in which high doses of carbaryl were administered orally to dogs revealed a typical pattern of cholinergic symptoms (see Table 3) though the dogs survived the highest dose (500 mg/kg body wt).[31] The description of the observed symptoms is succinct and characteristic of poisonings observed in most species. No effect was observed in the first 30 min after treatment but, subsequently, salivation began and the respiratory rate increased. Between 30 and 90 min following administration, lacrimation, muscle fasciculations, urinary incontinence, and defecation occurred. Within 150 min of treatment, the dogs showed constricted pupils, ataxia, diarrhea, loss of bladder control, and a markedly increased respiratory rate. After 3 hr, vomiting, accompanied by severe intestinal spasms and contractions, muscular spasms, and marked muscular weakness were seen. The signs began to decrease at 5 hr after ingestion although lacrimation, salivation, pupillary miosis and fasciculations could still be seen. The dogs all appeared normal some 7 hr after treatment and no signs were visible 24 hr after carbaryl administration. Similar signs of almost equivalent duration and order of appearance have been observed in rats treated with Matacil®,[64] in swine fed carbaryl,[65,66] and in the cat, a particularly susceptible species.[67] In a study of the tox-

Table 3
ACUTE ORAL TOXICITY OF SOME CARBAMATE INSECTICIDES TO MALE RATS[a]

Insecticide	LD_{50} (mg/kg bw)
Banol® (carbanolate)	293
Baygon® (propoxur)	83
Carbofuran	8—14
Dimetan®	150
Dimetilan	47—64
Isolan	11—123
Lannate® (methomyl)	17—24
Matacil® (aminocarb)	30—40
Maneb®	6750—7500
Mesurol®	100
Mobam	150
Nabam	395
Pyrolan®	62
Sevin® (carbaryl)	550—850
Temik® (aldicarb)	0.93
Zectran®	15—60
Zineb®	5200

[a] Values presented were obtained from Melnikov[11] and from Wiswesser[63].

icity of a series of N-monomethylcarbamates in rats via three routes of administration, the intensity of different cholinergic symptoms varied from compound to compound as did the duration of the signs.[68] It was found that the carbamates inducing symptoms of short duration (Baygon®, carbanolate, promecarb) persisted in the bloodstream for 1 to 2 hr whereas those inducing prolonged symptoms (carbaryl, Landrin)® persisted in the blood for more than 6 hr. The comparison of intravenous, intraperitoneal, and oral routes of administration revealed surprising differences in distribution of some agents in the body and in the apparent rate of biotransformation.

The acute toxicity of different carbamate esters correlated well with their anticholinesterase activity, particularly with the inhibition of erythrocyte AChE.[68] Frequently, one sees reports in which obvious cholinergic signs of toxicity are observed though the analysis of anticholinesterase reveals little inhibition. This inconsistency may arise from any one or all of at least 3 problems: (1) the selection of the proper enzyme to assay, the plasma pseudocholinesterase being less sensitive than the erythrocyte AChE;[68,69] (2) the ease with which the carbamoylated cholinesterase spontaneously reactivate following dilution, lysis (in the case of erythrocytes) or the addition of substrate, all of which are related to the analytical method to be used; (3) the rate at which many carbamates can be degraded in vivo while the scientist is attempting to obtain adequate blood samples from animals in some discomfort. These problems and the means by which the artifacts can be minimized have been discussed by several investigators.[68,70-72]

There are reports in the literature which suggest that carbamates possess another mechanism of action in addition to the inhibition of nervous tissue cholinesterases.[68,73] In these experiments, the animals died almost instantly following intravenous injection. Within a very few minutes of treatment, carbaryl and other carbamates produced a marked anesthetic response accompanied by severe respiratory difficulty (dyspnea) and eventual respiratory failure. Artificial respiration applied early in the apneic periods was beneficial in restoring normal respiration though the characteristic cholinergic symptoms subsequently developed. A more moderate effect was produced following the intraperitoneal injection of some carbamates but no such effect was seen after oral administration. The "narcotic" effect was produced only by carbamates of low toxicity and, while the mechanism of action remains unknown, a hypothesis has been suggested that, with the high concentrations of agent which must be injected into the bloodstream to exert an effect, nerve conduction and motor endplates are blocked by an action at the level of sodium ion transport across the axon membrane.[74] This effect was first noted following the intravenous injection of some organophosphorus esters of low toxicity.[74]

C. Subacute and Chronic Toxicity — Animal

In studies involving the subacute or chronic feeding of carbamate insecticides, the most important fact to be recognized is that these agents are remarkably free of toxicity except at very high doses. Carbaryl, the oldest of the commercially available agents, was administered orally at a concentration of 200 mg/kg body wt for 3 days a week for a period of 90 days without producing any overt toxicological signs in male rats.[75] There were no significant histological or biochemical changes in the testes, epididymis, liver, or kidney and no effect on the fertility of male rats was observed at this dose. Slight reductions of blood and brain AChE, of the order of 35 and 12%, respectively, were measured. In contrast, the administration of carbaryl at 7.0, 14.0, and 70.0 mg/kg body wt for a 12-month period caused marked functional and structural changes in the endocrine system of the rats.[76] The oral administration of carbaryl to female rats at 5.0, 10.0, and 20.0 mg/kg body wt for a period of 6 months resulted in a shift in the phases of the estrous cycle.[77] In a three-generation study in rats, carbaryl at 10,000 ppm in the diet impaired fertility, no litters being produced for the second mating of the second generation.[78]

Evidence of teratogenicity was observed in the offspring of rabbits given carbaryl by gavage from days 6 through 18 of gestation since there was a significant increase in omphalocele observed at doses of 150 mg/kg/day.[79] No such effect was observed in mice. A low irregular incidence of malformations was found among the offspring of beagle dogs given 6, 12, 25, or 50 mg carbaryl/kg/day in the diet.[80] Experiments in pregnant guinea pigs were equivocal in that vertebral malformations were observed in one study where the animals were receiving a near-toxic dose (300 mg carbaryl/kg/day) whereas nothing was seen in another study at a carbaryl concentration of 200 mg/kg/day by gavage or 300 mg/kg/day in the diet.[81,82] Monkeys receiving 2 or 20 mg carbaryl/kg/day throughout gestation exhibited a higher rate of abortions but no increased incidence of fetal malformations.[83]

It would appear that female rats are more susceptible to carbaryl-induced toxicity than are male rats and that embryotoxicity and fetal abnormalities might occur. The placental transfer of carbaryl and other carbamates has been measured in rats and accumulation in fetal tissues (eyes, liver, and central nervous system) as well as inhibited fetal tissue AChE have been reported.[84-87] Neither Weil et al.[88] nor Collins et al.[89] found morphological teratogenic changes in fetal rats. No mention was made of any observed neurological deficiencies in any of these studies. The "safety" of carbaryl may be related to the rapid hydrolysis of this ester by esterases of maternal tissues and subsequent distribution of the much less toxic α-naphthol to the fetuses. Such a hypothesis has been tested using ^{14}C-ring- and ^{14}C-carbonyl-labeled carbaryl.[90] The results obtained from nonpregnant and pregnant rats revealed that, with ^{14}C-carbonyl carbaryl, significantly more $^{14}CO_2$ was exhaled by the pregnant than by the nonpregnant rat, suggesting that hydrolysis was a major pathway of biotransformation in the former group. The fetuses, when the dams received ^{14}C-ring-labeled carbaryl, showed much higher tissue concentrations of the label, presumably α-naphthol, than did the dams which suggests that, on the maternal side of the placenta, the α-naphthol was excreted while it was stored in tissues (brain, heart, and lung) on the fetal side.

With other pesticidal carbamate esters, such as Zectran® and Mesurol®, the mammalian placenta presents no physiological barrier, these agents being rapidly distributed to the fetus.[91] Of particular interest in this study was the fact that fetal brain was capable of N-demethylating Zectran® (some 12%) to the more potent anticholinesterase 4-methylamino-3,5-xylyl N-methylcarbamate. In addition, fetal liver showed a marked capability of converting Mesurol® by sulfoxidation (some 23.1%) to the more potent sulfoxide. The authors raised a significant point concerning the formation of more toxic anticholinesterase agents by fetal tissues in vivo subsequent to the placental transfer of the parent compound. The embryotoxicity and teratogenicity of these two carbamate insecticides has not been tested rigorously enough to verify whether fetal biotransformation adds to the risks of exposure to these agents. Other cholinesterase inhibitors such as DFP and eserine alter neonatal cholinesterase activity and development when given during pregnancy.[92] The carbamate pyridostigmine, when administered to pregnant women, may be linked to the appearance of myasthenia gravis in neonatal humans.[93] Such symptoms have not been reported in any of the above-mentioned animal studies which might lead one to conclude that neurological problems are not encountered in the perinatal animal, a hypersensitive individual.

Looking at the subacute and chronic toxicity of one of the more acutely potent carbamate insecticides, Lannate® (methomyl), it was found that this ester exhibited a low order of chronic toxicity.[94] In 90-day feeding studies at 10, 50, 125, and 250 ppm, the weight gain of male rats at the high dose was significantly reduced though no clinical, hematological, biochemical, urinary, or pathological evidence of toxicity was observed in either male or female rats. In a 22-month feeding study in rats at dietary levels of 50, 100, 200, and 400 ppm, decreased hemoglobin values were noted in the higher level female test group and a significantly higher testis/body wt ratio was observed in the highest level males. Histopathologic alterations were observed in the kidneys of male and female rats fed diets containing methomyl at 400 ppm and changes were noted in the spleens of female rats receiving 200 and 400 ppm methomyl in the diet. Beagle dogs of both sexes on dietary feeding studies for 90 days and 2 years at levels of 50, 100, and 400 ppm methomyl showed no nutritional, clinical, urinary, or biochemical evidence of toxicity though histopathologic changes were observed after 2 years treatment in the kidney, spleen, and liver at the highest level of exposure. There was no evidence of methomyl being teratogenic in rabbits or in a three-generation study in rats. There was no evidence of carcinogenicity in the chronic rat and dog studies. No neurological symptoms were observed in the mammalian studies and methomyl did not cause any neurotoxicity in hens treated at a dose of 28 mg/kg body wt.

It is beyond the scope of this chapter to go into details concerning the influence of carbamate insecticides on tissue function and morphology or biochemical and pharmacological effects beyond that of particular interest to the theme of the book. Sporadic reports of neurological problems involving carbamates administered to animals have appeared in the literature and will be discussed under a separate heading. It will suffice, at this point, to say that when enough of any carbamate is administered, altered enzymatic activities reflective of tissue injury can be measured as can changes in hematological parameters, alterations in endocrine functions, or tissue morphology. This aspect of carbamate toxicity was succinctly reviewed by Kuhn and Dorough and the reader is referred to this chapter for a concise overview of the diverse effects elicited following carbamate treatment.[95]

D. Acute Toxicity — Human

Despite the above evidence which indicates that pesticidal carbamate esters are relatively safe to animals at chronic low dose but are highly potent though producing only transient, short-term toxicity following acute administration, carbamate toxicity does occur in humans. Invariably such reports in the literature appear as a consequence of an attempted suicide by ingestion of a large quantity of concentrated insecticide preparation or by accidental exposure to large concentrations of diluted material. At best, there are few cases described in the literature.

In a 1975 report dealing with pesticide poisonings in Mexico, 6.9% of the 847 reported cases involved methomyl while 11% involved carbaryl. An account of a fatal poisoning with carbaryl has been reported in Hungary.[97] Other cases reported include three suicide attempts of which two were successful, one being caused by propoxur (Baygon®).[98] A recent incident of three fatalities following the accidental ingestion of methomyl has been reported in detail.[99] The acute exposure of spraymen to Baygon® resulted in transient symptoms of poisoning and marked decreases in whole-blood cholinesterase.[100] Except for a very few cases of carbaryl poisoning, most of the severe poisonings have involved the more potent carbamate insecticides such as Lannate® or Baygon® (Tables 1 and 2). Other poisonings may have occurred but are difficult to assess since they frequently involve exposure to a number of agents, resulting in toxicological confusion concerning symptoms.

It should be appreciated that carbaryl poisoning can be fatal. In the best documented case, a 39-year-old man purposefully drank approximately 500 mℓ of Sevin-80 solution (80% concentrate of carbaryl).[97] Death occurred some 6 hr after ingestion even though there was immediate hospitalization, prompt gastric lavage, and antidote administration. Quantitative analysis of tissues and fluids revealed the following concentrations of carbaryl (in mg%): stomach lavage fluid (244.6), stomach contents (14.8), intestinal contents (17.6), blood (1.4), liver (2.9), kidney (2.5), and urine (3.1). A problem concerning the antidotal treatment in this case will be discussed in a subsequent section since it may have contributed to the toxicity.

A recent poisoning due to carbaryl has come to the attention of one of us (D.J.E.) and, while incomplete and unmonitored as far as cholinesterase activities, electromyography, etc. bears relating in some detail as it points out the fact that we still do not know everything that we should about the long-term as well as the short-term effects of carbamate insecticides.

A 55-year-old male Caucasian farmer, with no known previous history of pesticide exposure or any illness related to the following symptoms, handsprayed a vegetable garden once with a water-wettable preparation of carbaryl for an infestation of armyworms. He admitted becoming literally soaked with the solution and was not wearing protective clothing, gloves or mask, etc. He was admitted to a regional hospital a few days later suffering what was suspected to be "bacterial meningitis", was treated with antibiotics, appeared to recover and was discharged within a few days.

The farmer was seen by his physician 3 to 4 weeks later with a bizarre and unexplainable set of symptoms, including severe vertigo, blurred vision, photophobia, a peripheral numbness and tingling sensation, loss of recent memory, muscle weakness in the hands as well as in the legs, a feeling of lethargy and continual tiredness. The physician was concerned by the loss of recent memory and forgetfulness but was equally surprised at the behavioral changes manifested in frustrated rage and inability to control temper which were accompanied by severe headaches and short periods of blackout. The individual was unable to control coordination and drive his automobile.

The physician, on questioning the patient closely, discovered that the patient had been spraying with an insecticide but could not ascertain how much agent had been absorbed. The regional hospital had no facilities to measure blood cholinesterase activities and it is unlikely that, 4 weeks after exposure, there would have been any reduction in activity. No neurological examination was conducted, again for lack of facilities and, to my knowledge, no such examination has been carried out to date.

The surprising aspect of this case is that now, over a year after the exposure, the patient still cannot drive his automobile because of the severe photophobia he suffers when going from a shaded portion of the road into bright sunlight. The blurred vision has disappeared but he still has problems remembering points which he wishes to raise with his physician within the time period (60 min or so) of leaving his home and arriving at the physician's office. He has compensated for this deficiency by writing everything down. The behavioral problems are still present but are partially controlled by anticonvulsant and antipsychotic drugs. Mild paresthesia is still present and the patient still suffers from lassitude, lethargy and extreme tiredness, and muscle weakness to the point where he has had to change his farming practices from raising hogs to chickens, the latter industry being less labor intensive. While his condition appears to have stabilized, he has not "recovered" and recently applied for and received a provincial disability pension.*

I have seen no other references in the literature which have remarked on long-term, persistent effects of any of the carbamate esters and am at a loss to adequately explain the neurological manifestations when, as one will see below, the exposure to toxic concentrations of other carbamate esters resulted in only transient symptoms of a few days duration. Hayes has estimated that a single oral dose of 250 mg of carbaryl (2.8 mg/kg body wt) in an adult man caused moderately serious poisoning.[101]

In one of the most detailed and documented studies of the administration of Baygon® to human volunteers, it was observed that the symptoms, while severe initially, began to disappear within 1 hr of treatment and, by 2 hr after ingestion, the patient felt well and his erythrocyte AChE levels were normal.[68] The observed signs and symptoms closely paralleled those described in other mammalian studies as will be seen in the following description.

A 42-year-old male volunteer (90 kg body wt) ingested 1.5 mg Baygon®/kg body weight some 2 hr after a light breakfast. Within 15 min of ingesting the dose and concomitant with the lowest measured erythrocytic AChE level (27% of normal), moderate discomfort described as "pressure in the head" was experienced. Little change was observed in the plasma pseudocholinesterase activity. The subject experienced blurred vision and nausea, became pale accompanied by facial sweating 18 min after dosing. The pulse rate was 140/min (normal = 76/min) and the blood pressure was 175/95 mm Hg (normal systolic/diastolic ratio = 135/90). Within the next 10 min (30 min after treatment), pronounced nausea with repeated vomiting and profuse sweating developed which persisted for the next 15 min. By 60 min, the subject was feeling better, the sweating was less pronounced but he felt nauseated, tired, and complained of muscle weakness. By 70 min after ingestion of the Baygon®, the pulse and blood pressure were normal and, at 120 min after treatment, the volunteer felt normal and had a complete lunch and later a dinner without experiencing any discomfort.**

Table 4 shows the erythrocyte AChE activity of the above individual, the estimated concentrations of Baygon® necessary in the bloodstream to achieve this level of inhibition and the measured levels of insecticide eliminated in the urine.[68] While it is known that Baygon® is more acutely toxic than carbaryl, it is also biotransformed into inactive products at a more rapid rate and should show a shorter duration of action. In a similar experiment, the volunteers took 5 oral doses of Baygon (0.2 mg/kg body wt) at 30 min intervals over a 2-hr period. In these individuals, the erythrocyte AChE activity was depressed to 50 to 60% of normal without overt toxic symptoms being observed. Again, little change in plasma pseudocholinesterase activity was observed. Following the termination of treatment at 2 hr, the AChE activity recovered rapidly and was comparable to normal levels within 3 hr. The results from these two acute studies have been observed in Baygon-exposed spraymen in the field where profuse sweating, nausea, and vomiting were the major symptoms reported.[100,102]

An interesting and useful description was recorded of the more immediate effects of a single exposure to 3-isopropoxyphenyl N-methylcarbamate (UC 10854) spray for approximately 30 min duration.[102] This particular experimental carbamate proved to be quite toxic to the spraymen who had to work at temperatures of 35 to 40°C with protective clothing and masks and who complained of symptoms which could have been associated with the heat and heat-exhaustion. An entomologist on the site tested this by spraying the roof of a large house while wearing only a shirt, shorts, socks, heavy work boots, and a sou'wester but no mask. He reported the following

* I wish to thank Dr. J. Young, Elmvale, Ontario for sharing his observations and case history of his patient with me and for the fruitful discussions which we have had.
** From Vandekar, M., Plestina, R., and Wilhelm, K., *Bull. W.H.O.*, 44, 241, 1971. With permission.

Table 4
PARAMETERS MEASURED IN A HUMAN VOLUNTEER
STUDY FOLLOWING THE ADMINISTRATION OF A
TOXIC DOSE OF BAYGON® (PROPOXUR)[a]

Time after ingestion (min)	Erythrocyte acetylcholinesterase (% of control activity)	Estimated Baygon® concentration ($\times 10^{-7} M$)	Urinary Baygon® ($\mu g/m\ell$)
15	27.0		
30	50.4	5.0	
45	55.5		
60	65.0		
90	80.0		
120	93.0	3.2	177.5
4.75 hr			195.6
7.5 hr	100.0	1.5	

[a] Data from Vandekar, et al.[68]

Heavy sweating, hazy vision on coming into the sunshine, dizziness, a slight twitching of the muscles around the knees which lasted for a minute or two and then disappeared, a few griping stomach pains, vomition about 90 min after terminating the spraying and experiencing a "washed out feeling the next day."*

In the field trials of this new pesticidal carbamate 3-isopropoxyphenyl N-methylcarbamate (UC 10854), spraymen began applying a 5% suspension on house walls but had to terminate the treatment within 2 to 3 hr when severe symptoms occurred.[102] The eight operators and two supervisors were protectively dressed in overalls, sou'westers, gum-boots, and masks. The complaints were recorded and are summarized along with the frequency of occurrence in Table 5. The cholinesterase (whole blood, most of the activity being contributed by erythrocyte AChE) was inhibited by more than 50%. Within 24 hr of exposure, the cholinesterase activities were approximately 80% of normal but weakness, tightness in the chest, and headache continued to persist, the first-mentioned symptom for as long as 4 days after exposure. Some signs of toxicity were observed among the 70 inhabitants of the villages sprayed but the symptoms were mild though 5 people did reach the stage of vomiting. A serious and persistent rash broke out on the skin of 17 of 34 males but this disappeared within 2 to 3 weeks of exposure. Whether or not it is an indication of increased sensitivity of children to carbamate insecticides, severe signs of poisoning were observed in a 14-month-old child which included semicoma, miosis, rapid respiration with excessive rales and crepitations in the lungs, profuse sweating, and rapid pulse rate. Most of these symptoms responded rapidly to a single dose (0.5 mg) of atropine and, within 24 hr, the child appeared normal.

In the recorded acute poisoning in Jamaica involving methomyl, three of the five affected individuals were dead on arrival.[99] It was reported that one of the deceased, having finished his meal, had perspired profusely, jumped up, defecated, and fell down. One of the other deceased had vomited, complained of visual disturbances, trembled, and defecated before dying. One of the survivors was asymptomatic while the other showed generalized twitching and spasms, muscle fasciculations, and respiratory embarrassment. It was ascertained by analysis that methomyl was used in the preparation of the unleavend bread, in place of common salt. The bread contained approximately 11,000 ppm of methomyl and, from an estimate of the amount of bread eaten, each individual dying had consumed 0.82 to 1.10 g of agent (12 to 15 mg/kg body wt).[99]

* From Vandekar, M. *Bull. W.H.O.*, 33, 107, 1965. With permission.

Table 5
SIGNS OF TOXICITY IN
SPRAYMEN FOLLOWING
EXPOSURE TO 3-
ISOPROPOXYPHENYL N-
METHYLCARBAMATE IN
A FIELD TRIAL
EXPERIMENT

Symptom shown	Frequency (n = 10)
Weakness	10
Dizziness	9
Tightness in chest	8
Blurred vision	7
Headache	6
Miosis	6
Nausea	5
Profuse sweating	5
Abdominal cramps	4
Vomiting	2

E. Subacute Toxicity — Human

While definitive evidence of subacute toxicity can be found in the literature associated with organophosphorus insecticides, no references were encountered which were clearly related to carbamate esters. Studies of orchard and forest spraymen, mixers, and handlers, individuals in close contact with carbamate esters, have been unrewarding primarily because of the variety of chemical pesticides and solvents which many of these people handle during a relatively short period of time. Studies have revealed that exposures do occur and that there is a direct relationship between the length of time in the occupation and the severity of neurological defects but it was impossible to identify individual causative agents from the melange of materials used.[103-105]

One study of a group of workers involved in the manufacture of carbaryl has been published though the population was not ideal since it was constantly changing throughout the period of sample collection.[106] The study did reveal that those having the highest exposure (bagging operations, production workers, shippers, and cleaners) occasionally had slightly reduced levels of cholinesterases but experienced no signs or symptoms of poisoning. During the active phases of carbaryl production, the urinary excretion of α-naphthol increased dramatically, signifying that significant exposure had taken place (41% of sample urines contained in excess of 1000 μg α-naphthol/100 mℓ of urine, the normal excretory level being 150 to 400 μg/100 mℓ), but that metabolism was efficient. Hydrolysis appeared to represent the major pathway of biotransformation. Individuals at greatest risk proved to be production operators, those bagging the insecticide, and those cleaning the area in which bagging operations took place, the urinary levels of α-naphthol ranging from 100 μg to 4000 μg/100 mℓ.

In a more recent study, volunteers took daily oral doses of carbaryl (0.06 mg or 0.13 mg/kg body wt) for a period of 6 weeks with concomitant monitoring of plasma and whole-blood cholinesterase, urinary α-naphthol, electroencephalograms, bromosulfophthalein elimination rates, hematology, and urinalysis.[107] The hematological, biochemical, and urinary analyses revealed no significant deleterious effects which could be attributed to carbaryl administration. There were no significant changes in the plasma or whole-blood cholinesterase levels, though this is not surprising when one looks at the methodology used. It has been reported that carbamates have little effect on plasma pseudocholinesterase.[68] The whole-blood cholinesterase, the bulk of the activity being contributed by the erythrocytic AChE, was quantitated by a pH-stat titrimetric technique which is (1) a slow technique, (2) requiring dilution of the blood sample, (3) a lengthy reaction time of 5 to 10 min with the substrate, all factors which have been presented as contributing to the reversal of the enzyme-carbamate complex and leading to incorrect results. Despite this problem, there was no evidence of deleterious effects at the smaller dose and no clearly attributable effect at the larger dose other than a slight, reversible decrease in the ability of the proximal convoluted tubule to reabsorb amine acids. Excretory products, measured by spectrophotofluorometry of urine samples, revealed that approximately 23% of the dose was eliminated as a glucuronide (15%) or a sulfate (8%) of α-naphthol.[108] Approximately 4% was excreted in the urine as 4-(methylcarbamoyloxy)-1-naphthyl glucuronide and the presence of 1-naphthyl methylimidocarbamate O-glucuronide was detected in urine.[108]

IV. NEUROLOGICAL LESIONS

There is little evidence of prolonged neurological effects following the exposure of humans to carbamate insecticides other than the bizarre situation recorded above in which a sprayer drenched himself in an aqueous preparation of carbaryl. To the best of the physician's ability, he could identify no causative organic agent to which the patient might have been exposed other than the spraying incident. In the light of this extremely poor data base, one is forced to examine animal studies more closely in search of neurological effects related to specific carbamate esters.

The neurotoxic effects of dithiocarbamate compounds such as sodium diethyldithiocarbamate (a chelating agent used in Wilson's disease) and tetraethylthiuram-disulfide (disulfiram, used as a vulcanizer and rubber accelerator in industry and as antabuse, a therapeutic measure in treating chronic alcoholism) are well known.[108,109] Detailed descriptions of anatomic lesions including degeneration and vacuolization in the peripheral and central nervous systems of rabbits and hens after chronic treatment with sodium diethyldithiocarbamate have been published.[111-113] Degeneration in the rostal parts of the spinal cord in the long ascending spino-cerebellar tracts, in the dorsal columns, and in the caudal regions of the spinal cord (descending tracts) have been found.[111,112] Vacuolization of nerve cells in the dorsal root ganglia of treated rabbits was also observed.[112] Cavanagh[114] suggested that the overall pattern of lesions caused by this agent was similar to the "dying back" process described for certain organophosphorus esters though this does not suggest that similar mechanisms of action occur. The neurotoxicity of disulfiram, including peripheral neuritis, has been known for many years, associated usually with neuropathies in rubber workers,[115-117] but also observed in fowl.[118] With this information, it is not unexpected that one might see neurological problems following exposure to carbamate esters.

Tecoram® (an oxidation product of disodium ethylenebisdithiocarbamate and sodium dimethyldithiocarbamate with ammonium persulfate) was administered to fertile hen eggs in doses of 0.01, 0.1, 1.0, and 10 mg dissolved in propylene glycol.[119] Profound changes were observed in some of the chick embryos, the effects being most pronounced in the distal peripheral nervous system in the small nerve bundles rather than in the larger ones. Degeneration of small numbers of nerve cells in the spinal dorsal root ganglia and in some cranial nerves was noted, the degenerative changes manifesting themselves as a progressive vacuolization of the cytoplasm, a shrinkage of the cytoplasm, and pyknosis of the nucleus. Hyaline, pale-staining swellings of the distal peripheral nerve fibers were observed. The changes were widespread in nerve fibers associated with muscles of the back, neck, and head though they were more pronounced in the legs. The authors also mentioned studies in rats treated with Tecoram® that showed symptoms of a lower motor paralysis.[119] While the usual comments about the acceptability of the chick embryo as an animal model can be raised, i.e., the inability to biotransform and excrete a toxic chemical, etc., the results do suggest that this chemical may cause a "dying back" neuropathy. Carpenter et al.[31] reported leg weakness in chickens following the administration of high doses of carbaryl.

Thiram (tetramethylthiuram disulfide), a relatively old fungicide, has been reported to cause neural dysfunction in rats following long-term feeding.[120] Diets contained 0.01, 0.04, and 0.1% of the test compound. The characteristic crossing or clasping of the hindlegs when picked up by the tail was seen in approximately 50% of the animals during the 11th through the 27th week of treatment, and several developed an ataxia and paralysis during this period. Electromyography was strikingly different from that of control animals and showed that the nerves innervating the cranial tibial muscle were unable to conduct the stimulus and that most of the motor units had ceased firing. Demyelination with irregular beading and fragmentation was observed in the sciatic nerves of ataxic rats and the axis cylinders showed a similar degree of degeneration. Degenerative changes in the ventral horn of the lumbar cord were also seen. Behavioral changes in open field tests of hyperactivity occurred in otherwise normal-appearing, thiram-treated rats. Conclusions based on a detailed histological examination suggested that the primary lesion appeared to be the peripheral axon. In an extension of the original study, Lee et al.[121] studied the neurotoxic potential of the metal-containing dithiocarbamates ferbam, nabam, maneb, and zineb. They found thiram to be more neurotoxic than ferbam and suggested that ferbam was converted in vivo to the toxic dimethyldithiocarbamate moiety and that this or another metabolite caused toxicity. Female rats fed ferbam (96 mg/kg/day) or thiram (67 mg/kg/day) developed ataxia which led to a paralysis of the hindlimbs. These studies confirmed the earlier reports of Hodge et al. who studied the chronic toxicity of ferbam.[122] Zineb exhibited little toxicity and Maneb® was relatively nontoxic.[121] It should be pointed out that, in these studies, exceedingly high doses of the order of 100 mg/kg body wt/day were administered over an extended period of time in order to elicit these effects and that the results should be interpreted with caution even in the absence of significant changes in hematological and clinical blood studies.

In a subacute feeding study of carbaryl to swine, severe effects on the neuromuscular system were observed.[123] The treated animals received carbaryl at daily doses of 150 mg/kg for 72 or 83 days in one experiment and, in a second experiment, at a dietary level of 150 mg/kg daily for 28 days followed by 300 mg/kg for the balance of the feeding period of 46 or 85 days. The clinical syndrome of intoxication was characterized by a progressive myasthenia, incoordination, ataxia, tremor, and clonic muscular contractions terminating in paraplegia, and prostration. Female swine required a larger total dose of carbaryl to induce paralysis and death than did males. Although the number of daily doses required to initiate signs varied, the total amount of carbaryl consumed when toxicity first appeared was relatively constant as was the order of appearance of signs. Reluctance to stand was observed first followed by a peculiar stance in which the rear legs were carried well forward under the body. The animals appeared to be walking on their dew-claws, the phalanges being nearly horizontal. There was a greatly exaggerated flexion of the rear legs, the hogs experiencing great difficulty in sitting down or backing up. Effects were minimal when resting but, when forced to move, marked incoordination, ataxia, tremors, and clonic contractions were elicited. Prostration followed and the animals were unable to stand even when assisted. While gross lesions were minimal, histopathological examination revealed lesions confined to the CNS and skeletal muscles. The muscular lesions consisted of a myodegeneration of traumatic or ischemic type, an acute hyaline and vacuolar degeneration, and an acute degenerative process associated with dystrophic calcification. The predominant lesions in the CNS involved the myelinated tracts of the brain stem and cerebellar peduncles. Fragmentation of myelin sheaths occurred but no demyelination had taken place. Axons were ruptured or swollen and moderately basophilic. In the spino-cerebellar tracts, there was evidence of necrosis of cellular components with congestion of small arterioles and venules, edema, hypertrophic endothelium, and hyalinized eosinophilic walls. At the higher dose of carbaryl, some pigs showed similar, though milder, white-matter lesions involving the myelinated tracts of the cervical spinal cord and nerve roots with vascular degeneration and hemorrhage in the gray columns. No analysis of blood and tissue cholinesterases was done but the fact that feed containing the largest dose (300 mg/kg) of carbaryl initially induced vomiting suggests that a near-toxic dose was being administered. While this study demonstrated that high doses of carbaryl elicited severe neurological and neuromuscular effects in swine, interpretation of the results in terms of low-dose, subacute, or chronic exposure is difficult since lower doses, which might have provided an indication of dose-response relationships for the lesions, were not administered. In an earlier study, Miller et al.[124] demonstrated that single doses of carbaryl (20 mg/kg body wt) administered to miniature swine caused a 44% inhibition of cerebral cortex cholinesterase and a 75% inhibition of the brain stem enzyme and caused a hindlimb paralysis even though no obvious effects were discerned upon histopathological examination.

Carbaryl and Baygon® were tested for neurotoxic effects in subacute experiments using male rats receiving the former agent at dietary levels of 100 and 200 ppm/day (equivalent to 10 and 20 mg/kg body wt) and the latter at levels of 12.5 and 25 ppm (1.25 and 2.5 mg/kg body wt) for up to 50 days.[125] The learning of a task and the performance of a previously learned task were assessed using a maze system and electroencephalographic patterns were measured under resting conditions and under conditions of rhythmic light stimulation.[125] By the sensitive behavioral methods used, disturbances in the various functions of the CNS could be assessed. In the process of learning a task (finding food in a maze) while being treated with the agents, the rats took a shorter time to find the food initially but performance deteriorated as treatment continued and they took longer to fulfill the task, even forgetting what they had been learning. In experiments involving performance of a previously learned (pretreatment) task, the "learned" treated rats' performance decreased whereas the performance of the control, untreated rats improved. A computerized examination of recorded electroencephalograms demonstrated permanent though minute changes, carbaryl producing an increase in both slow and fast wave components. Slow and fast wave components are dominant at the lower and higher excitation levels of the nervous system, respectively.[125] Baygon® appeared to decrease all wave components, suggesting that the two agents may have slightly different modes of action. Light flashes at 18 Hz induced frequency acceleration in electroencephalogram patterns of treated rats and an increased irritability which was roughly proportional to the dose of pesticide administered. Variable levels of AChE inhibition in different parts of the brain were noted in the treated rats though little effect was observed with the plasma or erythrocyte enzymes. On the basis of these results, it is entirely possible that subacute exposure to low levels of carbamate insecticides might cause behavioral changes in the rats which would not correlate with erythrocytic AChE measurements, the behavioral assessment techniques being much more sensitive. The "forgetfulness" of the rats that had learned to perform a task before exposure is interesting in light of the recent memory loss of the patient described above.

In an attempt to assess whether the CNS was a target organ for carbamate, organochlorine, and organophosphorus pesticides under conditions of prolonged, low-level exposure, Santolucito and Morrison[126] studied the electroencephalograms of Rhesus monkeys following chronic treatment with DDT, dieldrin, parathion, or carbaryl for 18 months. Carbaryl was administered daily at levels of 0.01 and 1.0 mg/kg body wt/day. A significant reduction in the abundance of waves of the 0.5 to 3.0 Hz frequency class was also observed. In Squirrel monkeys chronically treated with 0.007 mg/kg carbaryl for 26 months, interval histograms showed an increase in the abundance of intervals longer than 1.5 sec along with an increased slow-wave amplitude.[127] The interpretation of these observations is complicated by the fact that the recordings were obtained in animals immobilized by phencyclidine and anesthesized with sodium thiamylal, both agents possibly masking changes induced by the carbaryl. The importance of the small changes in electroencephalograms cannot be assessed at this time.

The fact that learning appeared to be affected by small chronic doses of carbaryl led Anger and Setzer[128] to initiate a different type of behavioral assessment in carbaryl-treated monkeys. They reasoned that maze learning was a "one time" learning phenomenon and was not a good baseline for the evaluation of chemicals because it was not repeatable in the same subject. Using a technique called "chain acquisition" task in which monkeys learned one set of equivalent response sequences each day, the animals (*Macaca fascicularis*) were trained to make 4 out of the 12 possible responses in a certain order, the 4 correct responses being changed every day. Carbaryl was administered to the fully trained monkeys either orally (doses as high as 50 mg/kg body wt) or intramuscularly (1.0, 3.0, 5.0, and 10 mg/kg body wt). Intramuscular carbaryl injections resulted in statistically significant decreases in total session time and increases in errors at the two higher doses but not at the 1.0 mg/kg dose. Changes in performance (increased errors) appeared after injections of 3.0 mg/kg. Oral doses as high as 50 mg/kg body wt did not consistently cause changes in performance. The results also suggested that monkeys were more tolerant of carbaryl than were rats. This technique of repeated chain acquisition task has been used in testing behavioral responses to drugs and has been found to be quite sensitive to drug administration.[129] However, in the present experiment, there was considerable variability in the behavioral baseline and it was considered that it would be unwise to choose the present task to determine the lowest effective level of a chemical unless an acquisition decrement is the major expected effect. It also raises the question of the influence of accumulated acetylcholine from acute carbaryl treatment on the ability of an animal to remember.

From the literature, it would appear that behavioral tests, if well designed, are the mechanisms by which subtle changes induced by environmental contaminants such as metals, pesticides, etc. may be detected. The field of behavioral toxicology is still in its infancy but the remarkable success with these techniques in the hands of Russian investigators leads one to believe that, eventually, great strides will be made in this dimension.[130]

In conclusion, while one cannot state that carbamate insecticides are highly dangerous neurotoxic agents, there is sufficient evidence to signify that they can initiate neurological and behavioral changes at dosages producing no obvious clinical signs. This point was highlighted in recent experiments using rats injected with 1 to 5 mg/kg body wt of Mobam (4-benzothienyl-N-methylcarbamate, acute oral LD_{50} in male rats of 150 mg/kg) for the study of changes in spontaneous motor activity and performance of a rapidly acquired conditioned avoidance response.[131] Significant effects were observed in both tests at dosages producing no overt clinical signs. A dosage of 2 mg/kg significantly depressed plasma and erythrocyte cholinesterases and decreased motor activity 15 min after injection. Only higher dosages (3 and 5 mg/kg) significantly depressed brain cholinesterase activity and avoidance performance. A possible explanation for the results was that the depression in motor activity was related to the inhibition of the peripheral nervous system cholinesterase while the depression of conditioned avoidance performance was related to decreased brain enzyme activity. The subtle behavioral and electroencephalographic changes observed following subacute, low-dose administration of carbamate esters to animals have been described but may be extremely difficult to quantitate in humans similarly exposed. The signal danger from carbamate compounds associated with neurological problems appears to involve acute single exposure to massive doses of the agents or repeated exposure to large doses.

It would be fair to state that the mechanism(s) by which carbamates induce neurotoxicity may bear no relationship to that of the organophosphorus esters. While it has been shown that carbamate esters will inhibit "neurotoxic esterase", the inhibition was temporary and certain structural features (two aromatic groups, one attached to the carbonyl group as an ester and one attached to the nitrogen, i.e., phenyl phenylcarbamate or benzyl phenylcarbamate) were essential for activity.[132] The authors did state that it might be possible that repeated ingestion would prolong the "temporary" inhibition to such a degree that neurotoxic effects would follow. This situation might also arise if an extremely high concentration of carbamate, as in a suicide attempt, accidental ingestion, or spray exposure, were absorbed and efficient elimination of the chemical was reduced. This again would likely restrict the problem of covert neurological toxicity to the high-dose acute exposure cases, more than to the low-dose chronic exposure except, perhaps, in situations involving the more potent carbamate esters such as aldicarb, methomyl, etc.

V. TREATMENT

As was observed for the poisonings with organophosphorus insecticides, the symptoms associated with carbamate toxicity are due to the accumulating, unmetabolized, neurotransmitter acetylcholine at the nerve endings of the parasympathetic and sympathetic autonomic ganglia, the postganglionic parasympathetic nerve endings, and at the neuromuscular junctions of the somatic motor nerves (Figure 4, Chapter 5, Table 3, Chapter 5). Atropine is the antidote of choice, antagonizing the action of acetylcholine by blocking the receptor sites for this neurotransmitter. The predominant action of atropine is directed toward the postganglionic, parasympathetic nerve fibers innervating exocrine glands, gastrointestinal tract, respiratory tract, eyes, bladder, and heart. Atropine also exerts a remarkable central effect, appearing to have some direct action on the respiratory center.

Treatment with atropine follows the same course outlined for organophosphorus ester poisoning, frequent small doses (0.5 to 1.0 mg) being administered subcutaneously until there is a dilatation (mydriasis) of the pupils and the face becomes flushed and/or sweating disappears.[133] The patient should be carefully titered using these signs as a physiological basis. Excess atropine can cause severe toxicity. This is of particular importance in carbamate poisoning cases where the enzyme-insecticide complex is unstable and the carbamate can be readily metabolized in a few hours unlike the far more stable situation encountered with organophosphorus poisoning in which the inhibited enzyme may take days or weeks to reactivate.

In contrast to the situation encountered with organophosphorus ester poisoning, the use of specific oxime antidotes such as 2-PAM (2-methylpyridine-2-aldoxime chloride or pralidoxime chloride) is contraindicated in carbamate poisoning. An examination of the literature for the scientific basis for this statement which is seen primarily in conjunction with carbaryl intoxication leads one to three main publications. In the first two reports involving carbaryl toxicity in dogs and rats, the protective effect of atropine was reduced by the concomitant administration of 2-PAM.[31,134] In the second report involving a suicide attempt by drinking carbaryl, it was noted that the patient's condition deteriorated rapidly after the administration of 2-PAM.[97] While experience has dictated that oximes are of little benefit in carbaryl intoxication, the contraindication cannot be extended to include all other carbamate insecticides. In a study of the acute toxicities of eight anticholinesterase carbamates in male rats, atropine reduced the observed toxicity of all of the agents and the oximes obidoxime (Toxogonin, oxy-bis-[4-hydroxyiminomethylpyridinium-1-methyl]dichloride) and P2S (2-hydroxyiminomethyl-N-methylpyridiuium methyl methanesulfonate), when used in combination with atropine, generally enhanced the therapeutic efficacy of atropine.[135] The exception was with carbaryl. In this instance, treatment with the oximes enhanced the toxicity and, when used in combination with atropine, obidoxime markedly reduced the protection afforded by atropine. The authors concluded that carbamate intoxication was best treated with atropine sulfate and that the concomitant use of oximes might be synergistic or ineffective. Only in the case of intoxication by carbaryl was therapy with oximes contraindicated.

REFERENCES

1. Kenaga, E. E. and Allison, W. E., Commercial and experimental organic insecticides, *Bull. Entomol. Soc. Am.*, 15, 85, 1969.
2. Melnikov, N. N., Chemistry of pesticides, *Residue Reviews*, Vol. 36, Gunther, F. A. and Gunther, J. D., Eds., Springer-Verlag, New York, 1971, chap. 16.
3. Kuhr, R. J. and Dorough, H. W., *Carbamate Insecticides: Chemistry, Biochemistry and Toxicology*, CRC Press, Boca Raton, Fla., 1976, 17.
4. Abdel-Wahab, A. M., Kuhr, R. J., and Casida, J. E., Fate of ^{14}C-carbonyl-labeled aryl methylcarbamate insecticide chemicals in and on bean plants, *J. Agric. Food Chem.*, 14, 290, 1966.
5. Casida, J. E., Augustinsson, K.-B., and Jonsson, G., Stability, toxicity, and reaction mechanisms with esterases of certain carbamate insecticides, *J. Econ. Entomol.*, 53, 205, 1960.
6. Eichelberger, J. W. and Lichtenberg, J. J., Persistence of pesticides in river water, *Environ. Sci. Technol.*, 5, 541, 1971.
7. Fukuto, T. R., Fahmy, M. A. H., and Metcalf, R. L., Alkaline hydrolysis, anticholinesterase and insecticidal properties of some nitro-substituted phenyl carbamates, *J. Agric. Food Chem.*, 15, 273, 1967.
8. Aly, O. M. and El-Dib, M. A., Studies on the persistence of some carbamate insecitcdes in the aquatic environment I. Hydrolysis of Sevin, Baygon, Pyrolan and Dimetilan in waters, *Water Res.*, 5, 1191, 1971.
9. Kuhr, R. J. and Dorough, H. W., *Carbamate Insecticides: Chemistry, Biochemistry and Toxicology*, CRC Press, Boca Raton, Fla., 1976, 228.
10. Dittert, W. and Higuchi, T., Rates of hydrolysis of carbamate and carbonate esters in alkaline solution, *J. Pharm. Sci.*, 52, 852, 1963.
11. Fukuto, T. R., Carbamate insecticides, in *The Future for Insecticides. Needs and Prospects*, Metcalf, R. L. and McKelvey, J. J., Jr., Eds., Wiley Interscience, New York, 1976, 313.
12. Casida, J. E., Mixed-function oxidase involvement in the biochemistry of insecticide synergists, *J. Agric. Food Chem.*, 18, 753, 1970.
13. Ryan, A. J., The metabolism of pesticidal carbamates, *CRC Crit. Rev. Toxicol.*, 1, 33, 1971.
14. Kuhr, R. J. and Dorough, H. W., *Carbamate Insecticides: Chemistry, Biochemistry and Toxicology*, CRC Press, Boca Raton, Fla., 1976, chaps. 6 and 7.
15. Matsumura, F., *Toxicology of Insecticides*, Plenum Press, New York, 1975, 228.
16. Fukuto, T. R., Metabolism of carbamate insecticides, *Drug Metab. Rev.*, 1, 117, 1972.
17. Kulkarni, A. P. and Hodgson, E., Metabolism of insecticides by mixed function oxidase systems, *Pharmacol. Ther.*, 8, 379, 1980.
18. Wilkinson, C. F., Ed., *Insecticide Biochemistry and Physiology*, Plenum Press, New York, 1976.
19. Knaak, J. B. and Sullivan, L. J., Metabolism of carbaryl in the dog, *J. Agr. Food Chem.*, 15, 1125, 1967.
20. Knaak, J. B., Tullant, M. J., Kozbelt, S. J., and Sullivan, L. J., The metabolism of carbaryl in man, monkey, pig and sheep, *J. Agr. Food Chem.*, 16, 465, 1968.

21. Strother, A., Comparative metabolism of selected N-methylcarbamates by human and rat liver fractions, *Biochem. Pharmacol.*, 19, 2525, 1970.
22. Sullivan, L. J., Chin, B. H., and Carpenter, C. P., *In vitro* vs. *in vivo* chromatographic profiles of carbaryl anionic metabolites in man and lower animals, *Toxicol. Appl. Pharmacol.*, 22, 161, 1972.
23. Chin, B. H., Eldridge, J. M., and Sullivan, L. J., Metabolism of carbaryl by selected human tissues using an organ-maintenance techinque, *Clin. Toxicol.*, 7, 37, 1974.
24. Kuhr, R. J., The formation and importance of carbamate insecticide metabolites as terminal residues, *Pure Appl. Chem. Suppl.*, 199, 1971.
25. Krishna, J. G. and Casida, J. E., Fate in rats of the radiocarbon from ten variously labeled methyl- and dimethylcarbamate-C^{14} insecticide chemicals and their hydrolysis products, *J. Agr. Food Chem.*, 14, 98, 1966.
26. Schlagbauer, B. G. L. and Schlagbauer, A. W. J., The metabolism of carbamate pesticides — A literature analysis, I and II, *Residue Rev.*, 42(1), 85, 1972.
27. Ecobichon, D. J., Hydrolytic mechanisms of pesticide degradation, in *Advances in Pesticide Science, Part 3*, Geissbuhler, H., Ed., Pergamon Press, New York, 1979, 516.
28. Casida, J. E. and Augustinsson, K.-B., Reaction of plasma albumin with 1-naphthyl N-methylcarbamate and certain other esters, *Biochim. Biophys. Acta*, 36, 411, 1959.
29. Terriere, L. C., Insecticice-cytoplasmic interactions in insects and vertebrates, *Ann. Rev. Entomol.*, Annual Reviews, Palo Alto, Calif., 1968, 75.
30. Lykken, L. and Casida, J. E., Metabolism of organic insecticide chemicals, *Can. Med. Assoc. J.*, 100, 145, 1969.
31. Carpenter, C. P., Weil, C. S., Palm, P. E., Woodside, M. W., Nair, J. H., III, Smyth, H. F., Jr., Mammalian toxicity of 1-naphthyl-N-methyl carbamate (Sevin insecticide), *J. Agric. Food Chem.*, 9, 30, 1961.
32. Dorough, H. W. and Casida, J. E., Nature of certain carbamate metabolites of the insecticide Sevin, *J. Agr. Food Chem.*, 12, 284, 1964.
33. Matsumura, F. and Sakai, K., Degradation of insecticides by esterases of the American cockroach, *J. Econ. Entomol.*, 61, 598, 1968.
34. Sakai, K. and Matsumura, F., Esterases of mouse brain active in hydrolyzing organophosphate and carbamate insecticides, *J. Agric. Food Chem.*, 16, 803, 1968.
35. Sakai, K. and Matsumura, F., Degradation of certain organophosphate and carbamate by human brain esterases, *Toxicol. Appl. Pharmacol.*, 19, 660, 1971.
36. Douch, P. G. and Smith, J. N., Metabolism of m-tert.-butylphenyl N-methyl-carbamate in insects and mice, *Biochem. J.*, 125, 385, 1971.
37. Douch, P. G., Smith, J. N., and Turner, J. C., NADPH-dependent cleavage of carbamates, *Life Sci.*, 10, 1327, 1971.
38. Menzie, C. M., *Metabolism of Pesticides*, Bureau of Sport Fisheries and Wildlife, Special Scientific Report, Wildlife, No. 127, July 1969.
39. Wheeler, L. and Strother, A., *In vitro* metabolism of the N-methylcarbamates Zectran and Mesurol by liver, kidney and blood of dogs and rats, *J. Pharmacol. Exp. Ther.*, 178, 371, 1971.
40. Cremlyn, R., *Pesticides. Preparation and Mode of Action*, John Wiley & Sons, New York, 1978, 100.
41. Oonnithan, E. S. and Casida, J. E., Oxidation of methyl- and dimethylcarbamate insecticide chemicals by microsomal enzymes and anticholinesterase activity of the metabolites, *J. Agr. Food Chem.*, 16, 28, 1968.
42. Knaak, J. B., Biological and nonbiological modifications of carbamates, *Bull. W.H.O.*, 44, 121, 1971.
43. Yang, R. S. H., Enzymatic conjugation and insecticide metabolism, in *Insecticide Biochemistry and Physiology*, Wilkinson, C. F., Ed., Plenum Press, New York, 1976, chap. 5.
44. Robbins, P. W. and Lipman, F., Isolation and identification of active sulphate, *J. Biol. Chem.*, 224, 837, 1957.
45. Boyland, E. and Chasseaud, L. F., The role of glutathione and glutathione-s-transferases in mercapturic acid biosynthesis, *Adv. Enzym.*, 32, 173, 1969.
46. Chen, K.-C. and Dorough, H. W., Glutathione and mercapturic acid conjugations in the metabolism of naphthalene and 1-naphthyl N-methylcarbamate (carbaryl), *Drug Chem. Tox.*, 2, 331, 1979.
47. Stedman, E. and Barger, G., Physostigmine (eserine). III, *J. Chem. Soc.*, 127, 247, 1925.
48. Stedman, E., Studies on the relationship between chemical constitution and physiological action. I. Position isomerism in relation to the miotic activity of some synthetic urethanes, *Biochem. J.*, 20, 719, 1926.
49. Aeschlimann, J. A. and Reinert, M., The pharmacological action of some analogues of physostigmine, *J. Pharmacol. Exp. Ther.*, 43, 413, 1931.
50. Wilson, I. and Cohen, M., The essentiality of acetylcholinesterase in conduction, *Biochim. Biophys. Acta*, 11, 147, 1953.

51. Gysin, H., Some new insecticides, *Chimia*, 8, 205, 221, 1954.
52. **Kolbezen, M. J., Metcalf, R. L., and Fukuto, T. R.**, Insecticidal activity of carbamate cholinesterase inhibitors, *J. Agric. Food Chem.*, 2, 864, 1954.
53. Metcalf, R. L., Structure-activity relationships for insecticidal carbamates, *Bull. W.H.O.*, 44, 43, 1971.
54. **Metcalf, R. L. and Fukuto, T. R.**, Effects of chemical structure on intoxication and detoxication of phenyl N-methylcarbamates in insects, *J. Agric. Food Chem.*, 13, 220, 1965.
55. **Wilson, I. B. and Quan, C.**, Acetylcholinesterase: studies on molecular complementariness, *Arch. Biochem. Biophys.*, 73, 131, 1958.
56. Aldridge, W. N., The nature of the reaction of organophosphorus compounds and carbamates with esterases, *Bull W.H.O.*, 44, 25, 1971.
57. **Reiner, E. and Aldridge, W. N.**, *Enzyme Inhibitors as Substrates. Interaction of Esterases with Esters of Organophosphorus and Carbamic Acids*, North-Holland Publishing, Amsterdam, 1972.
58. O'Brien, R. D., *Insecticides. Action and Metabolism*, Academic Press, New York, 1967, chap. 5.
59. **Reiner, E. and Aldridge, W. N.**, Effect of pH on inhibition and spontaneous reactivation of acetylcholinesterase treated with esters of phosphorus acids and of carbamic acids, *Biochem. J.*, 105, 171, 1967.
60. **Ellman, G. L., Courtney, K. D., Andres, V., and Featherstone, R. M.**, A new and rapid colorimetric determination of acetylcholinesterase activity, *Biochem. Pharmacol.*, 7, 88, 1961.
61. **Reed, W. D. and Fukuto, T. R.**, The reactivation of carbamate-inhibited cholinesterase, kinetic parameters, *Pestic. Biochem. Physiol.*, 3, 120, 1973.
62. Gaines, T. B., Acute toxicity of pesticides, *Toxicol. Appl. Pharmacol.*, 14, 515, 1969.
63. Wiswesser, W. J., *Pesticide Index*, 5th ed., Entomological Society of America, College Park, Md., 1976.
64. **Vassilieff, I. and Ecobichon, D. J.**, Acute Toxicity of Matacil (aminocarb) in the Rat with Measurement of Tissue Esterase Activities, unpublished data.
65. **Smalley, H. E., O'Hara, P. J., Bridges, C. H., and Radeleff, R. D.**, Effects of chronic carbaryl administration on the neuromuscular system of swine, *Toxicol. Appl. Pharmacol.*, 14, 409, 1969.
66. Smalley, H. E., Diagnosis and treatment of carbaryl poisoning in swine, *J. Am. Vet. Med. Assoc.*, 156, 339, 1970.
67. Yakim, V. S., Data for substantiating the maximum permissible concentration of Sevin in the air, *Gig. Sanit.*, 32, 29, 1967.
68. **Vandekar, M., Plestina, R., and Wilhelm, K.**, Toxicity of carbamates for mammals, *Bull. W.H.O.*, 44, 241, 1971.
69. **Wilhelm, K., Vandekar, M., and Reiner, E.**, Comparison of methods for measuring cholinesterase inhibition by carbamates, *Bull. W.H.O.*, 48, 41, 1973.
70. Berry, W. K., Acceleration by free carbamate of the spontaneous reactivation of carbamylated acetylcholinesterase, *Biochem. Pharmacol.*, 20, 3236, 1971.
71. Reiner, E., Spontaneous reactivation of phosphorylated and carbamylated cholinesterases, *Bull. W.H.O.*, 44, 109, 1971.
72. Iverson, F., Affinity and carbamylation rate constants of propoxur in reaction with erythrocyte and serum cholinesterase, *Biochem. Pharmacol.*, 24, 1537, 1975.
73. **Wilhelm, K. and Vandekar, M.**, Studies in the toxicology of N-methylcarbamates: I. Comparative toxicity tests and estimation of persistence of inhibitor in the body, in *XVth Int. Congress on Occupational Health*, Vol. 2, Vienna, 1966, 517.
74. Heath, D. F., *Organophosphorus Poisons. Anticholinesterases and Related Compounds*, Pergamon Press, London, 1961, 338.
75. **Dikshith, T. S. S., Gupta, P. K., Gaur, J. S., Datta, K. K., and Mathur, A. K.**, Ninety day toxicity of carbaryl in male rats, *Environ. Res.*, 12, 161, 1976.
76. **Shtenberg, A. I. and Rybakova, M. N.**, Effect of carbaryl on the neuroendocrine system of rats, *Food Cosmet. Toxicol.*, 6, 461, 1968.
77. Vashakidze, V. I., Mechanisms of action of pesticides (GranoSan, Sevin, Dinoc) on the reproductive cycle of experimental animals, *Soobshch. Aked. Nauk. Gruz. SSR.*, 48, 219, 1967; abstracted in *Chem. Abst.*, 68, 28750 OX, 1968.
78. **Collins, T. H. X., Hansen, W. H., and Keeler, H. V.**, The effects of carbaryl on reproduction of the rat and gerbil, *Toxicol. Appl. Pharmacol.*, 19, 202, 1971.
79. **Murray, F. J., Staples, R. E., and Schwetz, B. A.**, Teratogenic potential of carbaryl given to rabbits and mice by gavage or by dietary inclusion, *Toxicol. Appl. Pharmacol.*, 51, 81, 1979.
80. **Smalley, H. E., Curtis, J. M., and Earl, F. L.**, Teratogenic action of carbaryl in beagle dogs, *Toxicol. Appl. Pharmacol.*, 13, 392, 1968.
81. Robens, J. F., Teratologic studies of carbaryl, diazinon, norea, disulfiram and thiram in small laboratory animals, *Toxicol. Appl. Pharmacol.*, 15, 152, 1969.

82. Weill, C. S., Woodside, M. D., Bernard, J. B., Condra, N. I., King, J. M., and Carpenter, C. P., Comparative effect of carbaryl on rat reproduction and guinea pig teratology when fed either in the diet or by stomach tube, *Toxicol. Appl. Pharmacol.*, 26, 621, 1973.
83. Dougherty, W. J., Goldberg, L., and Coulston, F., The effect of carbaryl on reproduction in the monkey, (Macacca mulatta), *Toxicol. Appl. Pharmacol.*, 19, 365, 1971.
84. Declume, C. and Bernard, P., Foetal accumulation of [^{14}C] carbaryl in rats and mice autoradiographic study, *Toxicology*, 8, 95, 1977.
85. Cambon, C., Declume, C., and Derache, R., Inhibition of acetylcholinesterase from foetal and maternal tissues after oral intake of carbaryl (1-naphthyl-N-methyl-carbamate) by pregnant rats, *Biochem. Pharmacol.*, 27, 2647, 1978.
86. Cambon, C., Declume, C., and Derache, R., Effect of the insecticidal carbamate derivatives (carbofuran, pirimicarb, aldicarb) on the activity of acetylcholinesterase in tissues from pregnant rats and fetuses, *Toxicol. Appl. Pharmacol.*, 49, 203, 1979.
87. Strother, A. and Wheeler, L., Excretion and disposition of [^{14}C] carbaryl in pregnant, non-pregnant and foetal tissues of the rat after acute administration, *Xenobiotica*, 10, 113, 1980.
88. Weil, C., Woodside, M. D., Carpenter, C. P., and Smyth, N. F., Current status of tests of carbaryl for reproductive and teratogenic effect, *Toxicol. Appl. Pharmacol.*, 21, 390, 1972.
89. Collins, T. F. X., Hansen, W. H., and Keller, H. V., The effects of carbaryl (Sevin) on reproduction of the rat and the gerbil, *Toxicol. Appl. Pharmacol.*, 19, 202, 1971.
90. Wheeler, L. and Strother, A., *In vitro* metabolism of ^{14}C-pesticidal carbamates by fetal and maternal brain liver and placenta of the rat, *Drug Metab. Dispos.*, 2, 533, 1974.
91. Wheeler, L. and Strother, A., Placental transfer, excretion and disposition of [^{14}C] Zectan and [^{14}C] Mesurol in maternal and fetal rat tissues, *Toxicol. Appl. Pharmacol.*, 30, 163, 1974.
92. Srinivasan, R., Karczmar, A. G., and Bernsohn, J., Developmental cholinesterase levels and isozyme patterns in offspring after ingestion of some CNS-active drugs into pregnant mice, *Trans. Am. Soc. Neurochem.*, 3, 124, 1972.
93. Blackhall, M. I., Buckley, G. A., Roberts, D. U., Thomas, J. B., and Wilson, B. H., Drug-induced neonatal myasthenia, *J. Obstet. Gyanecol. Br. Commonw.*, 76, 157, 1969.
94. Kaplan, A. M. and Sherman, H., Toxicity studies with methyl N-[(methyla-mino) carbonyl)oxy] ethanimidothioate, *Toxicol. Appl. Pharmacol.*, 40, 1, 1977.
95. Kuhr, R. J. and Dorough, H. W., *Carbamate Insecticides: Chemistry, Biochemistry and Toxicology*, CRC Press, Boca Raton, Fla., 1976, chap. 5.
96. Reyesnauera, R. and Sanchezoelafuenta, E., Poisonings caused by pesticides in the marshy regions during the agricultural season of 1974, *Salud. Publica. Mex.*, 17, 687, 1975.
97. Farago, A., Suicidale, todliche Sevin (1-naphthyl-N-methyl-karbamat) vergiftung, *Arch. Toxikol.*, 24, 309, 1969.
98. Bomirska, T. and Winiarska, A., Toxicology of carbamates as exemplified by intoxication with Baygon insecticide, *Pol. Tyg. Lek.*, 27, 1448, 1972.
99. Liddle, J. A., Kimbrough, R. D., Needham, L. L., Cline, R. E., Smrek, A. L., Yert, L. W., and Bayse, D. D., A fatal episode of accidental methomyl poisoning, *Clin. Toxicol.*, 15, 159, 1979.
100. Vandekar, M., Heyadat, S., Plestina, R., and Ahmady, G., A study of the safety of o-isopropoxyphenyl-methylcarbamate in an operational field-trial in Iran, *Bull. W.H.O.*, 38, 609, 1968.
101. Hayes, W. J., Jr., *Clinical Handbook on Economic Poisons*, U.S. Department of Health, Education and Welfare, Publ. Hlth, Service Toxicol. Section, Atlanta, 1963, 144.
102. Vandekar, M., Observations on the toxicity of carbaryl, Folithion and 3-isopropoxyphenyl N-methylcarbamate in a village-scale trial in Southern Nigeria, *Bull. W.H.O.*, 33, 107, 1965.
103. Jegier, Z., Health hazards in insecticide spraying of crops, *Arch. Environ. Health*, 8, 670, 1964.
104. Davignon, L. F., St-Pierre, J., Charest, G., and Tourangeau, F. J., A study of the chronic effects of insecticides in man, *Can. Med. Assoc. J.*, 92, 597, 1965.
105. Bellin, J. S. and Chow, I., Biochemical effects of chronic low-level exposure to pesticides, *Res. Commun. Chem. Path. Pharmacol.*, 9, 325, 1974.
106. Best, E. M. and Murry, B. L., Observations on workers exposed to Sevin insecticide: a preliminary report, *J. Occup. Med.*, 4, 507, 1962.
107. Wills, J. H., Jameson, E., and Coulston, F., Effects of oral doses of carbaryl on man, *Clin. Toxicol.*, 1, 265, 1968.
108. Knaak, J. B., Sullivan, L. J., and Wills, J. H., Metabolism of carbaryl in man, *Toxicol. Appl. Pharmacol.*, 10, 390, 1967.
109. Gardner-Thorpe, C. and Benjamin, S., Peripheral neuropathy after disulfiram administration, *J. Neurol. Neurosurg. Psychiatry*, 34, 253, 1971.
110. Moddel, G., Bilbao, J. M., Payne, D., and Ashby, P., Disulfiram neuropathy, *Arch. Neurol.*, 35, 658, 1978.
111. Edington, N. and Howell, J. M., Changes in the nervous system of rabbits following the administration of sodium diethyldiethiocarbamate, *Nature (London)*, 110, 1060, 1966.

112. Edington, N. and Howell, J. M., The neurotoxicity of sodium diethyldiethiocarbamate in the rabbit, *Acta. Neuropathol.*, 12, 339, 1969.
113. Howell, J. M. and Edington, N., The neurotoxicity of sodium diethyldithiocarbamate in the hen, *J. Neuropathol. Exp. Neurol.*, 27, 464, 1968.
114. Cavanagh, J. B., Toxic substances in the nervous system, *Br. Med. Bull.*, 25, 268, 1969.
115. Barry, W. K., Peripheral neuritis following tetraethylthiuram-disulfide treatment, *Br. Med. J.*, 2, 937, 1953.
116. Charatan, F. B., Peripheral neuritis following tetraethylthiuram-disulfide treatment, *Br. Med. J.*, 2, 380, 1958.
117. Bradley, W. G. and Hewer, R. L., Peripheral neuropathy due to disulfiram, Br. Med. J., 2, 449, 1966.
118. Waibel, P. E., Johnson, E. L., Pomeroy, B. S., and Howard, L. B., Toxicity of tetraethylthiuram disulfide in chicks, poults and goslings, *Poul. Sci.*, 36, 697, 1957.
119. Van Steenis, G. and Van Logten, M. J., Neurotoxic effect of the dithiocarbamate Tecoram on the chick embryo, *Toxicol. Appl. Pharmacol.*, 19, 675, 1971.
120. Lee, C.-C. and Peters, P. J., Neurotoxicity and behavioral effects of thiram in rats, *Environ. Health Perspect.*, 17, 35, 1976.
121. Lee, C.-C., Russell, J. Q., and Minor, J. L., Oral toxicity of ferric dimethyldithiocarbamate (ferbam) and tetramethylthiuram disulfide (thiram) in rodents, *J. Toxicol. Environ. Health*, 4, 93, 1978.
122. Hodge, H. C., Maynard, E. A., Downs, W., Coye, R. C., Jr., and Steadman, L. T., Chronic oral toxicity of ferric dimethyldithiocarbamate (ferbam) and zinc dimethyldithiocarbamate (ziram), *J. Pharmacol.*, 118, 174, 1956.
123. Smalley, H. E., O'Hara, P. J., Bridges, C. H., and Radeleff, R. D., The effects of chronic carbaryl administration on the neuromuscular system of swine, *Toxicol. Appl. Pharmacol.*, 14, 409, 1969.
124. Miller, E., Reinwall, J., Brouwer, J., Ear, F. L., and Loon, E. J., Effects of acute administration of carbaryl on cholinesterase levels in the CNS of swine, *Toxicol. Appl. Pharmacol.*, 10, 622, 1969.
125. Děsi, I., Gönczi, L., Simon, G., Farkas, I., and Kneffel, Z., Neurotoxicologic studies of two carbamate pesticides in subacute animal experiments, *Toxicol. Appl. Pharmacol.*, 27, 465, 1974.
126. Santalucito, J. A. and Morrison, G., EEG of Rhesus monkeys following prolonged low-level feeding of pesticides, *Toxicol. Appl. Pharmacol.*, 19, 147, 1971.
127. Santalucito, J. A., Comparison of chronic and acute low-level exposure effects of carbaryl on the EEG of Squirrel monkeys, *Ind. Med. Surg.*, 39, 315, 1970.
128. Anger, W. K. and Setzer, J. V., Effect of oral and intramuscular carbaryl administrations on repeated chain acquisition in monkeys, *J. Toxicol. Environ. Health.*, 5, 793, 1979.
129. Thompson, D. M., Repeated acquisition of behavioral chains under chronic drug conditions, *J. Pharmacol. Exp. Ther.*, 188, 700, 1974.
130. Teichner, W. H., Methodology in behavioral toxicology research, in *Behavioral Toxicology*, Xintaras, C., Johnson, B. L., and de Groot, I., Eds., National Institute of Occupational Safety and Health, Ohio, 1974, 441.
131. Kurtz, P. J., Behavioral toxicity of pesticides: physiological and behavioral effects of a carbamate compound, *Toxicol. Appl. Pharmacol.*, 37, 106, 1976.
132. Johnson, M. K. and Lauwerys, R., Protection by some carbamates against the delayed neurotoxic effects of di-isopropyl phosphorofluoridate, *Nature (London)*, 222, 1066, 1969.
133. Namba, T., Nolte, C. T., Jackrel, J., and Grob, D., Poisoning due to organophosphate insecticides, *Am. J. Med.*, 50, 475, 1971.
134. Sanderson, D. M., Treatment of poisoning by anticholinesterase insecticides in the rat, *J. Pharm. Pharmacol.*, 13, 435, 1961.
135. Natoff, I. and Reiff, B., Effect of oximes on the acute toxicity of anticholinesterase carbamates, *Toxicol. Appl. Pharmacol.*, 25, 569, 1973.

Chapter 7

THE MECURIAL FUNGICIDES

Donald J. Ecobichon

I. INTRODUCTION

Mercury is one of the oldest chemicals used in medicine and industry. Its medicinal uses have been known for many centuries and toxicity, "mercurialism", has been reported in the writings of Hippocrates, Pliny, Dioscorides, and Galen.[1] The industrial use of mercurial compounds is equally ancient. However, the first reported incident of industrial mercury poisoning was in 1557.[2] One early industrial use, that of softening animal hair with a mercuric nitrate solution in the process of manufacturing felt for hats, was prominent in the 1600s and toxicity among this occupational group was well known, giving rise to the terms "mad as a hatter" and "hatters' shakes". Some idea of the diversity of modern industrial utilization of mercurials can be obtained from the data in Table 1 which show a subdivision of the 1968 mercury consumption for the U.S., some 5.7×10^6 lb.[4,5] Of particular interest and concern in this chapter is the use of some 260,000 lb as fungicides in agricultural practices, 803,000 lb used as antifouling agents in marine paints, and 32,000 lb used in the pulp and paper industry as "slimicides". Many reviews have been written concerning the pharmacology,[6,7] toxicology,[8-10] and environmental impact[11-13] of mercurial compounds. The reader is also referred to the book *Mercury Contamination: A Human Tragedy* by D'Itri and D'Itri for an excellent overview of the history and scope of the problems with mercury.[14]

Organic mercurial compounds were introduced into agricultural practice in Europe (Germany) around 1915 as liquid preparations for the treatment of seed grain to prevent fungal disease prior to germination and during the growth of plants, fruits, and vegetables. The first product, "Uspulun" marketed by Bayer was described as a chlorophenolmercury.[15,16] Over the next few years, a wide variety of alkyl, alkoxyalkyl, and aryl compounds were introduced, starting with simple salts (iodide, nitrate, phosphate) of methylmercury and switching to more complex agents when the hazards of these agents were recognized. This chapter will deal primarily with fungicidal mercurials and the neurotoxicity produced in humans as a consequence of poisoning. Experiments involving animal models will be discussed in conjunction with mechanism(s) of action of mercurials on nervous tissue. It should be realized that these agents may exert effects on other sensitive target organs in the body and, while these effects will be mentioned, discussion will not be in detail.

II. CHEMISTRY

Mercury and its derivatives can be classified into *inorganic*, as metallic (elemental) mercury or inorganic ionic salts and *organic*, in which the mercury is bound covalently to at least one carbon atom. Table 2 shows the structure, nomenclature, and some examples of the various organomercurial compounds used in world agriculture. Inorganic salts of mercury (mercuric [Hg^{2+}] or mercurous [Hg_2^{2+}]) are rarely used in agriculture. The organomercury compounds may be divided, chemically, into three groups: alkylmercury, alkoxyalkylmercury, and arylmercury agents. The organomercury compounds may all be regarded as salts of the moderately strong bases methylmercury, ethylmercury, and alkoxyalkylmercury hydroxides, or of the weaker base phenylmercury hydroxide with acids such as hydrochloric, hydrobromic, hydroiodic, nitric,

Table 1
MAJOR MERCURY-CONSUMING INDUSTRIES IN THE U.S. BASED ON 1968 CONSUMPTION FIGURES[a]

Industry	Mercury ($\times 10^3$ lbs)	% of National consumption
Electrical apparatus	1500	26.6
Chlor-Alkali industry	1300	23.1
Industrial controls	606	10.6
Paint industry	803	14.4
General laboratory use	151	2.6
Agriculture	260	4.6
Dental	234	4.1
Catalytic use	145	2.5
Pulp and paper	32	0.6
Pharmaceutical	32	0.6
Amalgamation	20	0.35
Diversified uses	628	11.0

[a] Data taken from West, J. M., Minerals Yearbook 1968. U.S. Government Printing Office, Washington, D.C., Vol. L-II, 1969.

Table 2
THE NOMENCLATURE, STRUCTURES AND EXAMPLES OF AGRICULTURAL ORGANOMERCURIALS[a]

Nomenclature	Structure	Examples
Alkylmercurials	R-Hg$^+$ R-Hg-R (R = alkyl group)	Methylmercury sulfate, acetate, propionate, chloride, bromide, iodide, p-chlorobenzoate, dicyandiamide, benzoate Ethylmercury silicate, chloride, bromide, phosphate, urea, pentachlorophenolate, p-toluene sulfonamide
Alkoxyalkylmercurials	R^1-O-R^2-Hg$^+$ (R^1 = methyl, ethyl R^2 = ethyl, propyl, etc.)	Methoxyethylmercury chloride, silicate, dicyandiamide, acetate, lactate, benzoate Ethoxyethylmercury chloride hydroxide, silicate hloromethoxypropylmercury acetate
Arylmercurials	R-Hg$^+$ R = C$_6$H$_5$ (R = aryl group)	Phenylmercury acetate, dimethyldithiocarbamate, chloride, dinaphthyl methane sulfonate, nitrate, iodide, 8-hydroxyguinolate, hydroxide, lactate, propionate, salicylate, naphthenate, formamide N-tolylmercury-p-toluene sulfanilide Tolylmercury chloride Diphenylmercury dodecenyl succinate

[a] Information taken from Smart.[15]

acetic, propionic, lactic, salicyclic, benzoic, and silicic.[15] Alkylmercury, in which a strong carbon-mercury (C—Hg) bond exists does not readily dissociate. Other organic compounds, alkoxyalkyl- and arylmercury tend to degrade more easily into mercuric ion compounds.

Table 3
TRADENAMES, MANUFACTURERS AND ACTIVE INGREDIENTS OF SOME COMMONLY USED ORGANOMERCURIAL FUNGICIDES

Trade name	Manufacturer	Active ingredient
Uspulun®	Bayer A.G.	Sodium sulphate of chlorophenylmercury
Germesan®	Saccharin Fabrik A.-G.	Cresylmercuric cyanide
Ceresan® (Cerosan®)	I.G. Farbenindustrie A.-G.	Phenylmercuric acetate
Agrosan-G®	Imperial Chemical Ind. Ltd.	Tolylmercuric acetate
Panogen®	Morton Chemical Co.	Methylmercuric dicyandiamide
Granosan-M®	Dupont	Ethylmercury p-toluene sulfonanilide
Agrosan-GN®	Imperial Chemical Ind. Ltd.,	Phenylmercuric acetate and ethylmercuric chloride

It is of interest to examine the concentrations of mercury found in dry-dressed and liquid-dressed seed grain. In samples of wheat and barley taken in the U. K., the levels were between 6 and 23 ppm (μg of mercury per gram of grain).[17] Washing of the dressed seed removed only 10% of the fungicide except in the case of one dry-dressing where 50% was removed, presumably the nonadhering dust portion. Another investigator found 34 ppm of mercury in a fungicide-treated grain; washing reduced this level to approximately 20 ppm.[15] In the 1956 mercurial poisoning epidemic in Iraq, the seed wheat had been treated with Granosan M® (Dupont) at a concentration of 50 g of preparation per 100 kg of grain.[18] The commercial product contained 7.7% ethyl mercury p-toluene sulphonanilide with a total mercury content of 3.2%, which yields a concentration of 16 ppm of mercury in the wheat. It would appear that, in most epidemic incidents of ingestion of treated grain, the levels of mercury lie somewhere between 15 and 30 ppm (μg/g grain). Table 3 presents the trade names, suppliers, and active ingredients of some commonly used seed disinfectants.

Three properties, liposolubility, volatility, and rate of biotransformation make mercury unique as a toxicant. The toxic effects of inorganic mercury are attributed to the action of ionic mercury existing either as mercurous, mercuric, or ionic methylmercury. Elemental mercury (Hg°) cannot form chemical bonds but it is oxidized in vivo through the mercurous form which is unstable and readily dissociates into the mercuric ion.[19] The oxidative process is poorly understood but is considered to be enzymatic.[20,21] Until the 1960s, it was generally thought that mercurials released into the environment were simply assimilated and diluted to such a degree that they would no longer pose a problem. It was subsequently found that most mercurial compounds were transformed directly or indirectly into mono- and dimethylmercury, the reaction being catalyzed by anaerobic, methanogenic bacteria which utilized a methylcobalamin (alkylated vitamin B_{12}) as a donor of the alkyl (methyl group).[22-26] The reactions which were postulated to occur are shown in Figure 1. Several bacterial species are capable of methylating inorganic mercury.[27,28] The methylation of mercuric chloride by fish liver homogenates suggests that this same reaction may occur in vivo.[29] It has also been shown that, under mild reducing conditions, the methylcobalamin compounds can serve as intermediates for a rapid, nonenzymatic methylation of ionic mercury. As is shown in Figure 1, both mono- and dimethylmercury are formed, the relative amounts of each produced being dependent upon the microbial species involved, the amount of mercury present, the volatility of the product, and ease of "escape", the temperature, and the pH of the

1. $Hg^{2+} + 2(CH_3-R) \longrightarrow CH_3-Hg-CH_3 \longrightarrow CH_3-Hg^+$

2. $Hg^{2+} + CH_3-R \longrightarrow CH_3-Hg^+ \xrightarrow{CH_3-R} CH_3-Hg-CH_3$

FIGURE 1. The biological transformation of mercuric ions into methyl- and dimethylmercury by methanogenic bacteria using methylcobalamin (CH_3-R). Two reactions have been postulated to occur to account for the varying amounts of the agents formed in the environment.[24-26]

incubation "mixture".[30] At low levels of mercury contamination, dimethylmercury is the major product but this can escape into the atmosphere due to its high volatility. At high levels of contamination, monomethylmercury is formed. Neutral and alkaline environments favor the formation of dimethylmercury which will decompose into monomethylmercury in mildly acidic environments.

In animals, the biotransformation of alkyl and alkoxyalkylmercurials appears straightforward. The metabolism of methoxyethylmercuric chloride has been studied in rats, using a sample labeled with ^{14}C in the ethyl group.[31] Some 60% of the radioactivity was exhaled via the lungs as ethylene with a small amount as carbon dioxide. Twelve percent of the label was excreted in the urine as a mercury-free product. Mercury was eliminated in the urine as an inorganic salt and little bioaccumulation occurred. In contrast, the biotransformation of phenylmercuric acetate with a ^{14}C label in the benzene ring was quite different. Most of the radioactivity (85%) rapidly appeared in the urine as the sulfate and glucuronide conjugated derivatives of *ortho-* or *para*-hydroxyphenol, the hydroxylation of the aryl group occurring in the liver. Most of the mercury (50 to 60%) was excreted readily in the feces, with about 12% in the urine. Most of the phenylmercuric acetate was degraded to inorganic mercury in tissues.[32] It has been suggested that the cleavage of a C—Hg bond is not enzymatic but can occur in an acidic medium in the presence of cysteine.[33] In a study of the toxicity of the seed dressing methylmercury dicyandiamide, the pattern of biotransformation was quite different; there was little breakdown to inorganic mercury and there was a progressive accumulation in body tissues, including the erythrocytes, hair, and brain.[31]

Alkylmercury appears to be especially hazardous because the mercury is firmly bonded to a carbon atom, the molecule not undergoing degradation and, because of its high degree of liposolubility is able to penetrate into nervous tissue.[34] The slow elimination (marked bioaccumulation) of monomethylmercury compounds from various species including man, support the concept of the high stability of the covalent C—Hg bond.[35] The homogenous distribution of monomethylmercury between various body organs also supports this concept as does the fact that the greatest percentage of material isolated from the tissues was organic in nature. A particular characteristic property of alkylmercurials is an affinity for and long-term residence in blood, primarily in the erythrocytes bound to sulfhydryl groups.[36] There is some evidence that a small amount of breakage of the C—Hg bond, of the order of 3%, may occur.[37] Other studies have demonstrated that there was a slow increase in inorganic mercury in renal tissue over 22 days to a level representing approximately 30% of the original methylmercury administered.[38] In contrast, following the intravenous administration of dimethylmercury to mice, there was a rapid exhalation of 80 to 90% of the dose and, after 16 hr, no dimethylmercury could be detected in the body, though there was a small amount of a nonvolatile metabolite in tissues which was identified as a monomethylmercury ion.[37] In studies in various species using ethylmercury salts, the overwhelming evidence was that the material which accumulated in body tissues was almost entirely ethylmercury though small amounts of inorganic mercury were detected which might have resulted from cleavage of the C—Hg bond in the gastrointestinal tract or in the body tissues.[35]

As was noted above in the studies with methoxyethylmercury, the evidence for alkoxyalkylmercurial metabolism showed a rapid breakage of the C—Hg covalent bond with exhalation of one product (ethylene) and a rapid accumulation of inorganic mercury in the kidney.[39,40] Urinary mercury, eliminated during the first day after treatment, consisted of approximately 50% organic mercury but this became inorganic in nature in subsequent days. Other studies, described in more detail by Nordberg and Skerfving[35] have confirmed that mercury from methoxyethylmercury was eliminated in the same kinetic fashion as inorganic mercury.

The administration of arylmercury compounds to animals results in an initial kinetic picture similar to that observed for alkoxyalkylmercury compounds but the tissue levels and distribution observed later resemble the situation found after the administration of inorganic mercurials.[35] Some 48 hr after the intramuscular administration of phenylmercuric acetate to rats, only 20 and 10% of the total mercury present in the liver and kidney, respectively, was organic.[41] These results were confirmed by Gage.[32] It has been suggested that arylmercurials can be converted by bacteria in aquatic detritus and sediments into methyl- and dimethylmercury via pathways through ionic mercury.[42] While the above-mentioned studies have demonstrated that ionic mercury may be formed to a limited extent from phenylmercury compounds, there was a rapid biotransformation, and little bioaccumulation of the mercury, and at least two studies have demonstrated that the phenylmercurials were not as toxic as alkylmercurials, and that the signs and symptoms of poisoning were quite different.[18,43]

Upon ingestion, different inorganic mercurials will be absorbed at different rates related to their solubilities in water or gastrointestinal fluid. Depending upon the compound ingested, some 2.0 to 20% may be retained in the body, the bulk of the dose being excreted in the feces with a small amount appearing in the urine.[35,44,45] Inorganic mercury, on being absorbed into the bloodstream, is associated with the plasma proteins and cellular elements and is more amenable to excretion via the urine.[46] The half-life of inorganic mercury in erythrocytes was 16 days in one study.[47] The body half-life administered protein-bound and ionic inorganic mercury in human volunteers was 42 ± 3 days, no difference being observed between the two dosage forms.[48] Inorganic mercury follows a biexponential elimination equation with a rapid initial component, difficult to measure since it is very short but reflecting the blood levels, and a slow, more prolonged component which reflects the intracellular level in other body tissues.[47]

Organic mercurials have a different pattern of distribution and a considerably longer half-life. Organic mercurials (alkyl and aryl compounds) are more completely absorbed from the intestinal tract, an estimated 90 to 95% and 50 to 85% of alkyl- and arylmercurials being absorbed, respectively.[49,50] In the blood, the alkylmercurials have a much greater affinity for erythrocytes, some 90% of the monomethylmercury being bound to erythrocytes in most species though species differences do exist.[50] As with the inorganic mercurials, the elimination of alkylmercurials is biphasic in nature as was demonstrated by Ostlund[37] who reported the biological half-life of dimethylmercury in mice to be 60 min for 90% of the dose administered, with the remaining 10% being eliminated slowly over the next 60 days, the half-life of this fraction being approximately 30 days. Ostlund demonstrated that this latter fraction of dimethylmercury had been converted to ionic monomethylmercury. The biological half-life of monomethylmercury in the mouse was approximately 25 days.[37] In man, the biological half-life of monomethylmercury has been measured at 50 ± 7 days in erythrocytes,[47] 72 days in hair,[51] 70 to 74 days[52] and 76 ± 3 days in the body.[53]

The pharmacokinetic picture for arylmercurials is not as well defined as that for alkylmercurial compounds. The few studies which have been conducted have been complicated by the fact that agents such as phenylmercuric acetate are rapidly biotransformed into inorganic mercuric ions.[32,54,55] Less than 10% of the dose was eliminated unmetabolized.[41] Subsequent distribution is similar to that observed for inorganic mercury, the ionic form being distributed to the liver and kidneys, and being readily excreted via the feces and urine without significant tissue bioaccumulation.[32,56]

While ingestion is probably the major route of absorption of mercurial fungicides as has been observed in several epidemics, penetration through the skin cannot be entirely ruled out. Metallic mercury and mercuric salts were once used extensively as components of ointments in the treatment of syphilis, the prevention of venereal disease, and for certain dermatological disorders. Systemic effects were reported after prolonged treatment with these preparations.[35,57-60] Alkylmercurials such as methylmercury dicyandiamide or methylmercury thiacetamide can be absorbed through the skin but few other studies have been carried out. The rate of dermal absorption of methylmercury dicyandiamide, some 6% of the mercury in 5 hr, was similar to the rate of uptake of mercuric chloride.[60] Despite the widespread use of arylmercurial fungicides, few studies have examined their dermal absorption. Phenylmercuric dinaphthylmethane disulfonate applied to the skin of rabbits in a buffered aqueous solution was found to penetrate the skin, subdermal connective tissue, and muscle, the concentration in the last tissue being threefold higher than that in the solution applied.[61] The application of phenylmercuric acetate intravaginally in rats resulted in a recovery of 25% of the mercury in the liver and kidneys after 24 hr.[62] Phenylmercuric salts may be absorbed through the intact skin and mucous membranes, but relatively high concentrations must be applied before significant absorption takes place.[16] In summary, it would appear that dermal absorption of most mercurial fungicidal preparations is possible if protective clothing is not worn. In many studies of dermal absorption of mercurial preparations, it was impossible to eliminate the chance of inhalation as a concomitant route of absorption, particularly inhalation of the more volatile compounds such as the alkyl- and alkoxyalkylmercurials.[35,63]

III. TOXICITY

Mercurial intoxication can be separated into acute and chronic poisoning on the basis of the type of compound, the quantity of material ingested, and the duration of exposure. While the syndrome of "mercurialism" can be induced by either inorganic or organic compounds, there are essential differences in the signs and symptoms which can be correlated with the physicochemical and pharmacological properties of the agents, with the body distribution of the chemicals, and the unique sensitivity of certain organs to mercury[56,64,65] (Table 4). Inorganic (ionic) mercurials such as mercurous and mercuric salts tend to concentrate in and affect the kidney and liver before neurological signs appear. Alkylmercury such as methyl- and ethylmercuric salts, with a high degree of liposolubility and low rate of degradation, will cross biological membranes easily and exert their major toxicological effect on the nervous system. The salts of aryl- and alkoxyalkyl compounds are biotransformed into mercuric ions and show a pattern of toxic signs similar to inorganic mercury, the liver and kidney being affected before any effect is observed in the nervous system.[16,66]

A. Acute Mercurialism

The signs and symptoms of acute poisoning arise essentially from the mercuric cation. The acute symptoms generally result from the ingestion of large quantities of agent and include burning (eschar) of the mouth and throat, extreme salivation, thirst, nausea, vomition, severe gastrointestinal irritation with abdominal pain, bloody diar-

Table 4
A COMPARISON OF THE SYMPTOMATOLOGY OF INORGANIC AND ORGANIC MERCURIAL POIONING: ACUTE AND CHRONIC SIGNS[a]

Inorganic mercury	Organic (alkyl) mercury
Acute symptoms	**Early symptoms**
Thirst	Fatigue
Metallic taste	Headache
Inflammation of mouth	Paresthesias in distal
Nausea	extremities, perioral area
Abdominal pains	tongue
Tenesmus	
Bloody diarrhea	
Stomatitis	
Gastritis	
Colitis	
Renal tubule degeneration	
Chronic symptoms	**Later symptoms**
Excessive salivation	Intellectual deterioration
Loosened teeth	Insomnia, depression,
Gingivitis	Anxiety
Nervousness	Spasticity
Irritability	Intention tremors
Tremors	Irritability
Slurred speech	Paralysis
	Cerebellar ataxia
	Deafness
	Constricted visual fields
	Gastrointestinal disorders
	and others associated
	with inorganic mercury
	will also be observed

[a] Data taken from Koos and Longo.[112]

rhea, shock, loss of fluids and electrolytes, rapid and weak pulse, cardiac arrhythmias, clammy cold skin and pallor, peripheral vascular collapse, and slow breathing. Delayed toxicological signs will be observed some 2 to 7 days after the acute exposure and will include swelling of the salivary glands, persistent excessive salivation, metallic taste, stomatitis, soft spongy gums with loosened teeth, and showing the characteristic blue-black gum line caused by mercury-sulfhydryl complexes. Oliguria is often present with anuria, uremia, albuminuria, hematuria, proteinuria, and acidosis. The classical picture of acute poisoning emerges as effects arising from two organ systems, the alimentary tract and the kidneys, since inflammation and corrosion along the intestinal tract as well as severe renal tubular necrosis can be observed in autopsy. The liver will also be affected, a central necrosis being observed.[64,65] It should be appreciated that the above description is related only to the ingestion of large amounts of inorganic mercury salts or possibly aryl- and alkoxyalkylmercurials which can be converted into mercuric ions.

Acute alkylmercury intoxication is rare but, considering the physicochemical properties of these agents, it is not surprising that essential differences in symptoms are seen, the results being manifested in motor and sensory nerve damage, irreversible in the case of severe alkylmercurial poisoning and resulting in permanent brain damage. Even after acute exposure to toxic concentrations of alkylmercurials, several weeks or months may pass before the characteristic clinical signs appear.

Table 5
SYMPTOMATOLOGY OF ORGANOMERCURY POISONING[a]

Paresthesia of the mouth, lips, tongue, hands, feet, fingers, and/or toes
Constriction of visual fields; abnormal "blind" spots
Hearing difficulties, especially in picking out one voice from a group, i.e., sound discrimination
Speech disorders; difficulty in articulating words and swallowing
Neurasthenia; weakness, fatigue, and inability to concentrate
Inability to write, read, or recall basic things such as familiar addresses, telephone numbers
Emotional instability; fits of anger, depression, or agitation
Ataxia: diadochokinesia, stumbling gait, clumsiness in handling familiar objects (fork, shoelaces, buttons, etc.); grotesque, uncoordinated movements
Spasticity: rigidity and partial paralysis
Stupor, coma, death (in extreme cases)

[a] The above symptoms were taken from Zepp et al.[68]

B. Chronic Mercurialism

Hamilton and Hardy[67] described three characteristic features of industrial, chronic mercurialism: inflammation of the mouth, muscular tremors, and psychic irritation. Any or all of these symptoms may be present in different cases. Chronic organic mercury poisoning, by whatever source of contamination, is very slow and insidious in onset and eventually will involve most of the organ systems of the body: gastrointestinal tract, genito-urinary tract, respiratory tract, muscles, eyes, skin, and most of all, the peripheral and central nervous system. Many of the symptoms noted above for acute mercurialism will be found in chronic intoxication. The respiratory tract is affected, with the loss of smell, inflammation of the nose (rhinitis), cough, and fever being observed. Dermatitis can occur with erythematous papular, vesicular lesions, urticaria, and a weeping condition being observed.

Chronic mercurialism usually begins with a progressive numbness of the distal parts of the extremities as well as the lips and tongue. Muscular coordination suffers with spasticity and rigidity being observed, with hyperactive muscle stretch reflexes and extensor plantar responses being elicited in later stages. Generalized muscular weakness, tremors, and pain are observed. Ataxia, dysarthria, dysphagia, and deafness are experienced. Other centrally involved symptoms include insomnia, agitation, hypomania, impairment or slurring of speech, loss of emotional control and, in some patients, stupor and coma are observed. Blurring of vision, accompanied by constriction of the visual field, is frequently reported and, ultimately, tunnel vision and blindness may occur. Voluntary movements are reduced in many patients but most patients have abnormal involuntary movements including myoclonus, choreoathetosis, and coarse tremors. The symptoms, stemming primarily from an involvement of the central nervous system, may include any or all of those shown in Table 5.[68]

A distinct form of psychic disturbance known as "erethism" occurs with chronic mercury intoxication and is characterized by abnormal irritability or responsiveness to stimuli. The symptoms include stammering, irritability, blushing, shyness or timidity, anxiety, restlessness, excessive perspiration, resentment of criticism, loss of ambition and energy, insecurity, melancholia, hallucinations, evidence of mental deterioration and depression. Before the last stages appear, patients may experience severe emotional disturbances including unpredictable outbursts of anger. The mechanisms giving rise to these bizarre and frightening signs and symptoms will be explored in a later section.

A large body of literature dealing with inorganic and organic mercurial poisoning in animals has revealed that most species show the same order of appearance of signs and symptoms as recorded above for the human. The early literature has been succinctly reviewed by D'Itri[64] and Skerfving[56] and showed that inorganic, alkoxyalkyl-, and arylmercurials exerted an influence on a variety of organ systems though primarily affecting the liver and kidneys, whereas the prime target organ of the alkylmercurials was the nervous system. With the methylmercury epidemics at Minamata and Niigata, Japan, many studies have been conducted to explore the mechanisms of action of alkylmercuric salts on the central and peripheral nervous systems of cats,[69,70] monkeys,[71-74] rats,[75,76] and rabbits.[77] Much of this data will be reviewed in a later section.

C. Fungicide-Induced Toxicity in Humans

The cases of organomercurial fungicide-induced toxicity in humans fall into two quite distinct classes. By far the largest group consist of spectacular epidemics which have occurred as a consequence of the consumption of cereal grains treated with mercurial fungicides. More limited in scope is the second class, where workers have become poisoned while manufacturing, handling, or applying organomercurial fungicides to seed grain. Before 1972, it was estimated that, during the last 2 centuries, some 1800 to 2000 individuals suffered from some form of mercury poisoning and some 120 to 150 people died. The number of poisonings attributed to organomercurials has been estimated at 800 with some 125 deaths occurring. Between 1865 and 1954, only 39 cases of organomercurial poisoning had been reported in the literature, most of these involving seed disinfectants.[78] Between 1953 and 1972, some 700 cases of organomercurial fungicide poisoning occurred with approximately 120 fatalities. In 1972, a catastrophic epidemic of organomercurial poisoning occurred in Iraq, with some 6530 individuals being hospitalized throughout the country and some 459 people dying of symptoms. A chronological list of the more important poisonings by organomercurial fungicides and the individual chemical involved are presented in Table 6. It is worthwhile to examine some of these tragic incidents in detail, listing the signs and symptoms and attempting to correlate them with the severity of the exposure.

Kurland[66] described two cases of alkylmercury poisoning which occurred in 1865 and which appear to be the first of many tragic incidents. Two young laboratory assistants inhaled the volatile vapors of diethylmercury which they were synthesizing, one dying 11 days after the onset of symptoms (numbness of hands, deafness, poor vision, sore gums, unsteady gait, and mental slowness) and the second after a year during which he was blind, deaf, ataxic, unable to speak, and almost completely demented.[79]

In 1940, Hunter et al.[80] reported four cases in which the individuals involved worked in a plant for the manufacture of mercury compounds including seed dressings. The causative agents were identified as methylmercury salts (nitrate, iodide, phosphate). The description of one clinical history suffices to describe what has become known as the Hunter-Russell Syndrome.

...A month after starting to work with seed dressings, he developed thirst, polyuria and intermittent glycosuria which lasted 3 weeks. After about 3 months, he complained that his whole body was going numb and tingling. He began to notice weakness of his arms and legs and unsteadiness in his gait. He became clumsy, dropped trays, began to stagger about and collapsed on the floor on several occasions. His speech became difficult and slurred and it was noted that he sometimes could not see objects held in front of his face. He was a thin, worried man of hysterical temperament. There were no abnormalities in respiratory, cardiovascular or gastrointestinal systems. Nervous system: lies in bed, apathetic and dazed; speech indistinct and explosive in character; he hears a watch normally but he cannot quickly comprehend the meaning of spoken speech. The chief symptoms at 5 months after onset of the signs were a need to listen carefully to speech in order to understand its meaning, difficulty in performing coordinated movements with hands, unsteady gait and difficulty in speaking. He could feed himself but only slowly and clumsily. His gait was slow with short mincing steps: it resembles a hysterical gait, but it was definitely ataxic. After 3 years, there was little change in the physical signs and the visual fields were constricted but the fundi were normal.*

---

* From Hunter, D., Bomford, R.R., and Russell, D., *Quart. J. Med.*, N. S., 9, 193, 1940. With permission.

Table 6
CHRONOLOGICAL OCCURRENCE OF MERCURIAL FUNGICIDE POISONINGS

Location	Year	Agent involved	Cases	Fatalities	Ref.
Great Britain	1940	Methylmercuric salts (NO_3, I, PO_4)	4	0	80
Canada	1942	Diethylmercury	2	2	81
Iraq	1956	Ethylmercury	100	14	18
	1960	p-toluene sulfonanilide	370	?	14
West Pakistan	1961	Methylmercury dicyandiamide	100	4—9	84
Guatemala	1962	Methylmercury dicyandiamide	45	20	12
U.S.	1969	Methylmercury dicyandiamide	7	0	86—88
Iraq	1972	Methylmercury dicyandiamide	6530	459	85

Unique to this clinical study was an experimental animal study in rats and monkeys which were treated repeatedly with methylmercury iodide or nitrate and showed a clinical course parallel to that shown by the human subjects and a similar pathological basis such that it was concluded that the mechanism of action was common to all three species.[80] The results indicated a sensory nervous disturbance of wide distribution. A latent, asymptomatic period of 2 to 3 weeks occurred in the animals between the time of initiating treatment and the onset of neurological symptoms. In both the workers and animals, the symptoms of poisoning by inorganic mercury were absent with the exception of tremor. The nervous system alone was affected. Severe generalized ataxia, dysarthria, and gross constriction of the visual fields were present in all cases. Plantar responses were extensor in type in two of the workers studied. Memory and intelligence were unaffected. The neural histological lesions in the animals and in one worker who died after an interval of 15 years will be discussed in a later section.

In a bizarre poisoning in Calgary, Canada, two young stenographers, working in the office of a warehouse containing some 20,000 lb of diethylmercury seed dressing, were fatally poisoned by the highly volatile fumes.[81] One young woman showed symptoms within 3 months of starting her job, the other within 4 months of beginning work in this office. Both died 30 to 40 days following initial illness. It was determined that the air concentrations of this highly volatile alkylmercurial ranged from 2.7 to 5.4 mg/m^3 near the stockpile and was approximately 1.1 mg/m^3 in the poorly ventilated office area. Many of the symptoms of mercurial poisoning were present in degree in both cases at autopsy, the gastrointestinal disorders being most pronounced.

Not all cases of organomercurial poisoning have been associated with inhalation exposure to volatile alkylmercurial seed dressings. In addition to such poisonings, Lundgren and Swensson[82] reported on an incident involving a workman employed for a period of 5 years impregnating timber with an alkyl mercury preparation (Fibrosan®), the task being carried out under extremely primitive hygienic conditions. Their description of this case is interesting.

> The patient was taken ill at the beginning of December 1944 with a disagreeable taste in his mouth and a feeling of numbness in his hands and lower arms and round his mouth. Further symptoms developed rapidly: giddiness, clumsiness, inability to button his clothes and an unsteady gait. In the middle of December he went to the hospital, where his condition rapidly deteriorated. Ataxia increased; his speech became slurred and difficult to understand. His sight and hearing got worse and he became more and more apathetic. Finally he went quite blind, after his visual fields had become concentrically limited. An examination of the

ears showed defective hearing of a central type on both sides though there was no definite change of the vestibular function. At the end of of December, the patient was transferred to another hospital and was unconscious on arrival, the pupils being miotic and not reacting to light. He died on January 3, 1945, without having regained consciousness.*

This particular case demonstrates the latent period before onset of symptoms, and also shows the rapid deterioration of the condition once symptoms appear. A urine sample, analyzed on Dec. 30, 1944, contained 160 µg of mercury/ℓ and a blood sample taken January 2, 1945 contained 400 µg/100 cc. Analyses, carried out post-mortem on various tissues, revealed high levels of mercury in most tissues, the maximum concentrations appearing in the liver (1410 µg/100g) and in certain parts of the brain (340 to 510 µg/100g).[82] One can only assume that the route of uptake in this chronic-toxicity case was a combination of dermal absorption and inhalation.

In the mid-1940s and early 1950s, one begins to see reports of poisonings caused by eating mercurial-treated grain. Lundgren and Swensson[82] reported one such incident. In 1952, Eagleson and Herner[83] reported a case of a child poisoned following the consumption of porridge prepared from flour which had been treated with Panogen® (methylmercury dicyandiamide). It was apparent that the child had consumed this grain almost daily for approximately 8 months. As this case was studied, it became obvious that the entire family had been exposed to the mercurial. In addition to the 1½-year-old child who proved to be mentally retarded and who could not sit up or get up by himself, the mother was delivered of an apparently healthy infant, a girl who showed no clinical signs of mercury poisoning initially but, with continued follow-up examination, was found to be mentally retarded and showed the same signs as her older brother. The father also had been treated for mercurialism.

From 1956 until 1972, the world scientific literature has contained frequent reports on epidemics of fungicide-related poisonings, almost totally restricted to the Third World developing countries and all involving the inadvertent consumption of mercurial-treated seed grain. Invariably, the use of such grain in preparing homemade bread occurred during periods when grain was in short supply and famine was imminent. Spectacular epidemics have occurred in 1956, 1960, and 1972 in Iraq; in West Pakistan in 1961; and in Guatemala in 1965. In all of these cases, there are important lessons to be learned about supplying organomercurial-treated grain to near-illiterate people who do not understand how, why, and with what the grain was treated and who cannot read the label on the bags because it is in a foreign language or bears unfamiliar symbols.

In West Pakistan, the outbreak began in mid-February 1961, and poisoning was first noted a month later.[84] Approximately 100 individuals were affected and between four and nine patients died. Fourteen patients showed mild toxicity. Seven were moderately severe, and four were seriously affected. The reasons why the seed wheat was purchased for food were that wheat was scarce in January and February of that year and it was of higher quality than that available and some of the wealthier families bought it for that reason. It was confirmed that the wheat had been treated with Agrosan GN®. In typical cases, the symptoms started with malaise, lethargy, and tiredness. Within a few days, these symptoms were accompanied by a burning sensation in the mouth and stomach, nausea and vomiting, occasional fever, excessive thirst, and loss of appetite. At the same time, the patients experienced weakness in the limbs, difficulty in walking, slow cerebration, confusion, flight of ideas or thought block, slurred speech, and difficulty in swallowing. At a later stage, there was inability to stand, walk, or even sit; greater disability in speech, mild to severe visual impairment followed by extreme confusion, spontaneous crying, sucking, and chewing movements; spastic-

---

* From Lundgren, K. -O. and Swensson, A., *J. Ind. Hyg.*, 31, 190, 1949. With permission.

ity, coma, and death. This clinical picture developed within 1 to 3 weeks from the onset of illness. Moderate to severe anemia was observed as was wasting and emaciation of the extremities. Visual impairment, when present, was permanent and the fundi, in some cases, showed optic atrophy. Paraesthesia and pain in the extremities or muscle tenderness were infrequently seen.

Iraq has been plagued with three separate epidemics of organomercurial poisoning related to the consumption of fungicide-treated (Granosan M®) wheat in times of privation. In 1956, many cases were observed in the north of Iraq, more than 100 individuals being admitted to the Mosul Hospital with 14 deaths occurring.[18] Again in 1960, families from the central part of Iraq were affected and some 221 patients were admitted to one hospital alone in Baghdad. Many patients went to other hospitals and there were an uncounted number of fatalities, both in hospitals, and in the localities and villages where poisoning occurred. An estimated 1000 patients were affected and some 370 were admitted to hospital. To give some idea of the dosage, post-mortem mercury analyses carried out on liver from 22 patients revealed levels of 6.58 ± 1.85 mg/100 g tissue. Many of the affected people had been warned not to eat this treated wheat. Some individuals washed it, unsuccessfully, to get rid of the fungicide. Others, encouraged by the fact that nothing untoward occurred to chickens that consumed the treated grain for a few days or to their friends who had eaten bread prepared from the grain, consumed considerable quantities of the seed wheat which had been treated with ethylmercury p-toluene sulfonanilide (7.7% preparation containing some 3.2% mercury). One case report is worth relating in detail because of the order of appearance and onset of symptoms.

> A landworker purchased the treated wheat from fellow farmers and mixed it with ordinary wheat. Ten days after eating the bread made from it, the symptoms began to appear. Most of his diet consisted of home-prepared bread. At first he experienced severe pain in the knees; 10 days later he had generalized deep-seated pains; three weeks later he developed pruritis in the soles, palms and genitalia and his vision was affected. Bowels moved once or twice daily and there was some oliguria. He was admitted to hospital 2 months after the onset of symptoms. Examination revealed a thin patient suffering from pain. There was a gray zone on two lower incisors, coarse twitchings in the limb muscles, but no tenderness along the course of nerves; there was no oedema, jaundice or anemia. Tendon reflexes were present and the plantars were flexor. There was no ataxia or nystagmus. Fundi showed no lesion at this stage.*

The clinical picture of the different individual cases fell into distinct groups, depending on the amount of poison consumed and the system or systems mainly involved. Oliguria developed at the beginning of the illness if large amounts of organomercury were consumed, polyuria, and polydypsia with a weight loss and dehydration being observed. Pruritis was not uncommon and, in severe cases, exfoliative dermatitis of the hands and feet occurred. Nervous tissue lesions were always seen, paresthesia, loss of power in the legs, difficulty in walking, cerebellar ataxia, paraplegia, and general spasticity with brisk reflexes being recorded in most cases. Disturbance of speech, restriction of visual fields, and loss of vision with optic atrophy were not infrequent. Tremors or intention tremors were sometimes found. Patients experienced headache, insomnia, confusion, excitation, and hallucinations. Manic states were also seen. In severe cases, cardiac involvement was encountered and included bradycardia, ventricular ectopic beats, prolongation of Q-T interval, depression of the S-T segment, and T inversion. Deep skeletal pain of a generalized nature over the muscles and bones was frequently found and was not alleviated by analgesics. The gastrointestinal symptoms were not usually severe though epigastric or hypogastric pain and colic were frequently observed. An interesting comment was made that the symptoms and signs of

---

* From Jalili, M. A. and Abassi, A. H., *Br. J. Ind. Med.*, 18, 303, 1961. With permission.

poisoning by the ethylmercury p-toluene sulfonanilide were different from those described by Hunter et al.[80] for methylmercury poisoning. In their cases, no cardiovascular abnormalities were noted and the visual fields were constricted and the fundi remained normal. With the Iraq situation, the heart was affected, optic atrophy was seen and skeletal pain, muscular twitching, polyuria, and dermatitis were encountered frequently.

Unfortunately, a catastrophic epidemic of organomercurial poisoning occurred again in Iraq in late 1971 and early 1972, with a total of some 6530 persons hospitalized throughout the country and 459 individuals dying of overt toxicity.[85] This particular outbreak was associated with 100,000 metric tons of wheat and barley treated with methylmercuric dicyandiamide. The methylmercury content of the grain ranged from 3.7 to 14.9 μg/g of grain. Oral ingestion of this seed grain occurred via homemade bread, meat and other animal products obtained from livestock given treated barley, vegetation grown in contaminated soil or stored in sacks that had contained treated grain, game birds that had fed on sown grain in fields, and fish caught in water into which farmers had dumped the treated grain. The magnitude of this poisoning incident and its ramifications has been described extensively and the reader is referred to two main references, that of Bakir et al.[85] and that of D'Itri and D'Itri[14] for a detailed account. An interesting point to be raised here concerns the commercial practice of dyeing such treated grain red as a warning and the natural mistake made by the afflicted people who washed the grain to remove the dye, thinking that they were removing the poisonous substance when, in actuality, they were accomplishing very little. The practice of shipping bags of treated grain stamped with symbols (skull and crossbones) and with warning labels printed in a language not understood by the majority of the people has been criticized as well. However, many illiterate Iraqi peasants would not have been able to read warning labels printed in Arabic either. Lastly, it is important to note the threshold body burdens at which characteristic signs of "mercurialism" occurred. Peripheral numbness occurred at a body burden of 25 mg of methylmercury. Ataxia began at 55 mg, speech impairment at 90 mg, deafness at 170 mg, and death at 200 mg.[85]

One should not complacently think that such poisonings occur only in the remote and underdeveloped countries of the world. This was forcefully brought to the attention of all toxicologists during the autumn and winter of 1969 to 1970 when seven members of a family of nine living in Alamagordo, N.M. were poisoned by pork from animals which had received rations containing grain treated with Panogen.®[86-88] A major widespread epidemic was averted by tracking down and destroying more than 200 fungicide-contaminated hogs which had already been marketed. Three children in the family became ill with classic symptoms of methylmercury poisoning. The least affected was an 18-year-old daughter who partially recovered whereas the younger children, a 14-year-old boy and an 8-year-old girl were severely afflicted and remain blind or partially blind, with speech difficulties, are mentally retarded, and partially paralyzed.[88] The pregnant mother of the family was asymptomatic but delivered an infant boy who soon proved to be suffering from mercury poisoning; intermittent tremors, weak high-pitched cry, irritability, blindness, and mental retardation being observed as the child became older. This poisoning incident has been extensively studied and several reports exist in the literature.[86-89] Other incidents of individuals in the U.S. consuming fungicide-treated, dyed grain have been documented by D'Itri and D'Itri.[89]

Interestingly, there appear to have been no human poisonings published in the literature involving phenylmercurial compounds.[90,91] Studies involving such agents as phenylmercuric acetate, benzoate, and oleate all point to little toxicity associated with this group of fungicides.[91]

D. Perinatal Poisoning

Strong evidence that alkylmercurials can cross the placental barrier and accumulate in the fetus has been presented for several animal species, including mice,[92,93] rats[94-96] guinea pigs,[97] hamsters,[98] cats,[99] and subhuman primates.[100,101] There is little evidence that the placenta forms any sort of effective protective barrier, the fetus generally accumulating concentrations of methylmercury higher than those found in the maternal tissues.[94,97]

Following the Minamata poisonings by methylmercury, several children born between 1955 and 1959 exhibited mental retardation and severe motor disturbances. The clinical symptoms differed markedly from those of adult cases and were similar to symptoms of severe congenital cerebral palsy or cerebral dysfunction syndrome (disturbance of mind, incoordination, problems of speech and hearing, constriction of visual fields, impairment of chewing and swallowing, enhanced tendon reflexes, involuntary movement, primitive reflexes, superficial sensation, excessive salivation, forced laughing). The severity of the disease in children ranged from mild to moderate ataxia and spasticity to individuals showing all of the above-listed symptoms. The frequency of "cerebral palsy" in the Minamata Bay area was 5 to 6% in the total number of births compared to an expected frequency of 0.1 to 0.6%.[102] In one village, 12% of the children were so afflicted. While some 25 cases of fetal Minamata disease have been diagnosed, only one mother showed overt symptoms of the condition, the remainder being asymptomatic.[103-105] The evidence that alkylmercurials can produce fetal damage is quite strong. Tissue analysis of infants who died revealed high levels of mercury in the kidneys, liver, and brain. It would appear that transplacentally acquired alkylmercurials affected the central nervous system during the critical stage of growth causing irreversible brain damage. It was quite characteristic of many cases that the newborn infant appeared normal at birth but deteriorated within the next month or so, developing neurological symptoms including chorea, tremors, seizures, and showing mental retardation.[88,103,106]

The effects observed in perinatal individuals exposed in utero to environmental methylmercury have been observed for fungicidal mercurials, primarily alkylmercurials. Ten cases of prenatal poisoning by ethylmercury have been reported in the U.S.S.R.[107] The mothers had shown symptoms of ethylmercuric chloride (Granosan®) poisoning from having eaten treated grain up to 3 years prior to the pregnancy and delivery and the children showed various degrees of decreased birth weight and muscle tone and severe mental retardation. Six cases of prenatal transplacental exposure to methylmercury have been studied in Iraq, the infants being born blind, deaf, and suffering from severe motor dysfunction.[108] Further evidence of placental transfer and prenatal intoxication comes from the case in Alamagordo, N.M., in which the pregnant mother consumed mercury-contaminated pork (fed Panogen®-treated grain) during the 3rd to 6th months of pregnancy.[87,88] Urinary mercury levels ranged from 0.09 to 0.18 µg/mℓ and serum concentrations of 2.91 and 0.47 µg/mℓ were measured in the last months of pregnancy.[86] The amniotic fluid revealed less than 0.02 µg mercury/mℓ.[86] The male infant appeared normal at birth but showed some persistent tremulous movements of the extremities and had a urinary mercury level of 2.7 µg/mℓ. Initial electroencephalograms were normal in appearance but were abnormal by 3 months of age and the infant suffered severe and bizarre neurological symptoms.[87] Postnatal exposure of this child to mercurials was excluded as the infant was not breast fed. All of the evidence to date has pointed toward the sensitivity of the fetal nervous system to alkylmercurials and the importance of the intrauterine exposure in causing severe and permanent damage to the central nervous system.

In addition to the obvious transplacental exposure of the fetus to the toxic agent, postnatal exposure via breast milk also occurred. In the 1972 Panogen®-associated demic in rural Iraq, a number of infants were exposed to the methylmercury in utero. The level of alkylmercury in the blood of each infant was found to be persistently higher than that of their respective mothers during the first 4 months of life.[108,109] In addition to the prenatal exposure, all infants except three were exposed to methylmercury postnatally via nursing, possibly explaining why the mothers were asymptomatic while their infants showed overt toxicity which could be related to milk levels and maternal and fetal blood levels.[85] Some of the mercury present in the milk was inorganic in nature and was less of a risk for the infant since gastrointestinal absorption of inorganic mercury is low as a consequence of complex formation with free sulfhydryl groups on milk proteins. It was demonstrated that, in babies not exposed in utero, the blood content could reach toxic levels following suckling.[109] In breast-fed infants, the decline of blood mercury was slower than that of the mother who was no longer exposed to the toxic agent, due to the maintenance of the infant's body burden by additional chemical received in the milk.[109] Many of the Minamata Bay children were breast fed, possibly explaining why the mothers, who were continually eating contaminated fish, showed few symptoms since the alkylmercury was being eliminated via the breast milk.[106]

The fetal central nervous system appears to be much more vulnerable to the effects of methylmercury than that of the mother or that of older children. After the cessation of exposure, there was a gradual improvement in the clinical picture for the adults and older children, the prenatally exposed individuals showed evidence of permanent damage due, probably, to the fact that the developing fetal brain is more susceptible to the toxic effects. In a more extensive study of 32 prenatally exposed infants over a 5-year period, the investigators found that some infants recovered from the ataxia, 5 of 17 blind children recovered partial sight, but 7 of 18 children who suffered very severe poisoning remained physically and mentally handicapped.[110,111]

IV. NEUROLOGICAL LESIONS

When one compares the lists of signs and symptoms of poisoning by aryl- and alkoxyalkylmercurials and inorganic mercury with that for alkylmercurials, one cannot help but be impressed with the degree of neurological involvement observed with alkylmercurials (Table 4).[112] It is obvious that the central nervous system does accumulate all forms of mercury but the levels of inorganic mercury (from inorganic mercuric salts or biotransformed aryl- and alkylmercurials) are always much lower than those of the alkylmercuric compounds. Methylmercury accumulates in the brain and persists for a long time whereas "inorganic" mercuric ions penetrate less readily and do not persist for as long a time.[113,114] In a recent study, it was shown that there was a tendency for a strong residual deposit of methylmercury in metabolic organs (liver, kidney) and in the reticuloendothelial system while mercury tended to accumulate slowly in the nervous tissues and to remain in them over much longer periods of time.[115] While it can be shown that mercuric chloride (oral LD_{50} = 37 mg/kg in the rat) is equitoxic to methylmercury (oral LD_{50} = 30 to 40 mg/kg in the rat), the latter compound is considerably more neurotoxic, the essential quality of methylmercury neurotoxicity residing in the high degree of lipid solubility rather than in any other property.[92,93,114,116-118] According to Cavanagh,[114] it may be irrelevant for toxicity which form enters the cell provided that a minimal quantity is present and though divalent mercuric cations form more stable complexes than do monovalent alkylmercury ions, the more rapid penetration and accumulation of the latter in lipid membranes probably is closely linked to its toxicity.

In the post-mortem examination of the central nervous tissue of an individual who died some 15 years after industrial exposure to methylmercury compounds, Hunter and Russell found that the generalized ataxia (evident in this victim for years) was associated with cerebellar cortical atrophy, selectively involving the granule-cell layer of the neocerebellum while the concentric constriction of the victim's visual fields was correlated with a bilateral cortical atrophy in the area striata.[119] Since the outbreak of the methylmercury-associated Minamata disease, extensive study has been carried out on the pathology of this condition and a number of excellent reviews have been published to which the interested reader is referred, particularly those by Takeuchi,[106] Shiraki and Takeuchi,[120] Shiraki and Nagashima,[121] and Chang.[122] Clinically afflicted individuals manifested various signs and symptoms, including the Hunter-Russell syndrome, while morphological examination revealed severe involvement of the cerebral cortices, particularly the calcarine cortex and pre- and postcentral cortices, the superior temporal gyrus, cerebellum, and peripheral sensory nerves.[123] A brief summary of the main clinicopathological findings in human cases is worthwhile.[96,122]

In chronic cases, atrophy of the brain was most evident in the medial aspect of the occipital lobes, particularly in the calcarine regions and in the cerebellar folia in the lateral lobes and in the vermices. Extensive thinning of the cerebellar gray matter was a consistent finding. Pathological changes were observed in the cerebellar cortex and disintegration of the granule cells began adjacent to or below the Purkinje cell layer, with disintegration of Purkinje cells being observed in chronic cases. Basket, climbing, and parallel fibers were usually severely involved and proliferation of Bergmann's glia and astrocytes was observed. No remarkable pathology was found in the white matter with the exception of some thinning. Lesions in the calcarine cortex were found in all cases of Minamata disease along with degeneration of nerve cells and myelin sheaths in the second and upper-third layers. The number of surviving neurons in less severe lesions was reduced and many had a shrunken and distorted appearance.

Studies in a number of species of animals have revealed morphological effects of methylmercury poisoning comparable to those observed in humans. Cerebellar changes were observed in young chicks following exposure to diethylmercury or methylmercuric nitrate with the earliest changes being dilatation and disintegration of the endoplasmic reticulum and Golgi complex in both Purkinje and granule cells.[124] Extrusion of nuclear material and destruction of the synaptic complex were also observed. Brown and Yoshida also suggested that the altered cell membrane structures were a consequence of mercury interfering with protein production in these nerve cells. Similar findings were obtained in mice exposed in utero to methylmercury.[125] Ultrastructural changes were observed in the cerebellum of rats exposed to dimethylmercuric sulfide, particularly disintegration of the granule cells.[126] In subacute studies (2 weeks of treatment) in rats with methylmercury chloride, no conspicuous histochemical changes were seen in the cerebrum but the pronounced changes in the cerebellum consisted of a severe degeneration of the granular cells in which the more central parts seemed to be affected first (urula, nodulus, and lobulus medius) and all stages of degeneration (karyopcycnosis, karyolysis, and cytolysis) were observed.[75] In the brain stem, ballooning, swollen axons, and axon fragmentation occurred with myelophages being seen. Similar changes were observed in the spinal cord, in the *Fasicular gracilis*, and *F. cuneatus*. In the peripheral nervous system, changes consisting of ballooning and degeneration of the myelin sheath, swelling and fragmentation of axons, proliferation of Schwann's cells, and fibrosis were observed but appeared to involve sensory tracts rather than motor tract groups of nerves. In cats treated subacutely with methylmercuric chloride, lesions were found in the cerebellar vermis and the cerebral cortex, the changes consisting of loss of nerve cells and their replacement by reactive and fibrillary gliosis. In a follow-up chronic study in cats receiving 176 μg Hg/kg/day, lesions were again observed in the cerebral cortex and cerebellum, the former consisting of patchy and also diffuse neuronal degeneration, microglial proliferation, and occasional perivascular cuffing, while the latter consisted primarily of granular cell degeneration.[70] At lower doses (74 µg Hg/kg/day) administered chronically to cats, the lesions were less severe in degree, but a distinctive difference was seen between the cerebrum and cerebellum, the latter being more severely affected than at the higher dose level and consisted of a marked loss of granular cells in the cerebellar vermis at the bottom of the folia, in the centralis and culmen of the anterior lobe, and part of the lobulus simplex. The most severely affected parts of the folia exhibited almost complete loss of granular and Purkinje cells with Bergmann glial cell proliferation. There was also moderate to severe involvement of some dorsal root ganglia consisting of degenerative changes of ganglia cells and proliferation of satellite cells. In studies using alkylmercury in monkeys, marked atrophy of the calcarine cortex was observed with severe neuronal changes and astroglial proliferation predominantly in the occipital, parietal, and temporal lobes and in the lateral geniculate nucleus similar to that seen in Minamata disease cases.[127] The most severe damage, spongy degeneration, was seen in the calcarine and insular cortices as well as pyknosis, karyolysis with ultimate cytolysis. Occasional mild atrophy was observed in Purkinje cells in the cerebellum. Focal myelin destruction and axonal changes were observed in the optic nerves. Similar observations have been made in infant monkeys treated with methylmercury.[101]

In the Minamata cases, human fetal lesions differed significantly from those of the adults in that fetal cerebral cortical involvement appeared more diffuse and fetal brain hypoplasia and malformations occurred while never present in adults.[103,112,128] The cortical lesions were distributed more widely and were more severe in the fetal form of Minamata disease than in nonfetal infantile cases. It appeared that the earlier the involvement, the more widespread the pathology is, and the later the involvement, the more localized the lesions were.[106,122] The differences could be observed also in the signs and symptoms in fetal and infantile cases, such parameters as involuntary movement, hypersalivation, and primitive reflex being more significant in fetal poisoning. While severe cerebral cortical and cerebellar granule cell lesions were observed in both fetal and infantile poisonings, atrophy, cytoarchitectural changes, and malformed neurons were much more prominent in fetal cases and almost absent in infantile and adult poisonings.[122] It would appear that minute amounts of methylmercury, which by itself may produce no significant injury to maternal tissues nor any observable gross teratology in the fetus, may still be hazardous to the developing nervous system.[122]

Why are the lesions associated with methylmercury poisoning localized, for the most part, in the largest nerve cells of the nervous system (primary sensory cells) and in the smaller cells of the central nervous system? In discussing this, Cavanagh suggested that the capillary bed of sensory ganglia is permeable to plasma constituents owing to the presence of fenestrations between endothelial cells.[114] If Evans Blue-albumin complex and horseradish peroxidase can readily enter the extracellular spaces and be found between satellite cells and lying against neuronal surfaces, it would not be difficult to envisage penetration of a highly liposoluble small molecule such as ionic inorganic or organic mercury and achieving relatively high concentrations in primary sensory neurons. The reasons for the selective damage of the small cells in the central nervous system is not obvious and no clear-cut results demonstrating increased vascular permeability have been generated to date. The liposolubility of methylmercury over that of other forms of the toxic agent may also play an important role in achieving high concentrations in cells which, as Cavanagh states, cannot tolerate the disorganization of subcellular organelles by mercury.[114] The larger surface-to-volume ratio of small neurons might result in these cells absorbing more mercuric ion from the extracellular space than larger neurons.[122] Berlin and Ullberg[92] showed that, following a single intravenous injection of mercuric chloride in mice, large portions of the radiolabel were detectable in the cerebellar gray matter, area postrema, hypothalamus, and areas near the lateral ventricles. When methylmercury was administered, the initial uptake was slow and accumulation occurred gradually over a few days and distribution was quite different.[93] In contrast to the heterogeneous distribution of the inorganic mercury, the distribution of monomethylmercury salt was uniform, the areas showing the maximum concentrations being the hypocampus and the cerebellar gray matter. Such results have been confirmed in other studies using rats, pigs, dogs, etc.[122] Somjen et al. found that, following the administration of methylmercuric hydroxide to rats, the highest concentration was detected in the spinal dorsal root ganglia, followed closely by the cerebral cortex and the cerebellum and then the subcortical part of the forebrain.[129] Spinal cord and peripheral nerves contained much less mercury than the sensory ganglia. The essential point to be made here is that the distributional pattern of mercury correlated well with the pathological findings in certain areas of the brain.

As has been mentioned previously, the signs and symptoms caused by one form of mercury will be seen following exposure to another form of the toxic agent, only the order and rate of appearance and the severity of the symptoms being associated with the form of the agent and whether acute or chronic exposure has occurred. In other words, there is a broad overlapping of signs and symptoms for inorganic and organic mercurialism. At the level of mercury analysis and light microscopic examination of nervous tissue, distinct differences between inorganic and organic mercury salts can be observed. First, as indicated above, the distribution was somewhat different. The distribution of methylmercuric chloride among nerve cell types was found to be dorsal root ganglion neurons > neurons of the calcarine cortex < Purkinje cells of the cerebellum < anterior horn motoneurons < granule cells of the cerebellum. After mercuric chloride administration, the distribution of mercury in the nerve cells was dorsal root ganglion neurons < Purkinje cells of the cerebellum < anterior horn motoneurons < calcarine cortical neurons < granule cells of the cerebellum. The lesions and patterns of degeneration in the dorsal root spinal ganglia were found to differ in inorganic and organic mercury intoxication, vacuolization, and fragmentation of the neurons being induced by mercuric chloride and focal cytoplasmic degradation being the most characteristic lesion observed after methylmercury chloride treatment.[117,118] The lesions produced in dorsal root fibers were different in the two types of mercurial intoxication as well. After mercuric chloride treatment, large axonal spaces were created in many axons as a result of detachment of the axolemma from the myelin sheath and axonal shrinkage. Axonal vacuolation, degeneration, and collapse were observed. After methylmercuric chloride poisoning, the myelin sheaths appeared to lose their laminar appearance, acquiring a smudged or solid-looking composition accompanied by axoplasmic degeneration, axonal collapse, and degeneration of the myelin, and marked pathological changes at the nodes of Ranvier.[122] It will be some time before the reasons for these differences can be established, investigations at the subcellular level being essential to unravel the complexities.

Essential to such studies are investigations concerning the subcellular distribution of mercury. Biochemical analysis has indicated that the order of mercury concentration was mitochondrial fraction < microsomal fraction < supernatant fraction < nuclear fraction with a gradual increase in the supernatant or cytosolic fraction as intoxication progressed.[130,131] These studies, done in two different laboratories, used phenylmercuric acetate and mercuric acetate in one and methoxyethylmercuric chloride in the other, both organomercurials which are actively biotransformed into inorganic mercuric ions. The selective appearance of the mercury in membranes can probably be associated with the high content of sulfhydryl group to which the mercuric ions could bind. In nervous tissue, mercury was found bound to most membranous structures, i.e., the mitochondria, the endoplasmic reticulum, the Golgi apparatus, the nuclear envelope, the myelin sheath, and to a lesser extent bound to neurofilaments and neurotubules. Again, as intoxication progressed, increasing levels of mercury were detected in the cytoplasm. Only very limited amounts of mercury were found inside the nucleus or nucleolus.[122]

The possible mechanisms by which ionic mercury (inorganic or organic) might exert cellular toxicity are numerous.[122] Treatment with mercurials can result in a number of tissue enzyme changes, decreases in renal lysosomal enzymes (acid phosphohydrolase, β-glucuronidase, N-acetylglucosaminidase), and mitochondrial cytochrome c-oxidase.[132] Elevated plasma glutamic oxalacetic transaminase and lactic dehydrogenase have been reported as have increases in plasma catalase and galactosidase and decreases in erythrocytic acetylcholinesterase, the latter effect possibly being due to the binding of ionic mercury to membrane and enzyme protein sulfhydryl groups.[133,134] Methylmercury and inorganic mercury cause an increase in nerve fiber ribonucleic acid content of the anterior horn motor neurons, this being thought to be a repair mechanism to damage caused by mercurials, since both inorganic and organic mercury have been shown to reduce the ribonucleic acid content of spinal ganglion neurons.[117,118,122,135,136] Mercury can react with phosphate and base groups of nucleic acids, denaturing both animal and bacterial DNA and thereby interfering with transcription of genetic information, ribosome, and protein synthesis by the formation of stable complexes.[112,122] A hypothesis for a cellular mechanism of action of mercurials has been proposed on the basis of the inhibition of protein synthesis by methylmercury and ultrastructural damage to ribosomes, both of which are early features of the toxicity in neurons and occur independently of the axonal degeneration which appears to be a secondary event.[137-139] The hypothesis, presented by Cavanagh, suggests that small neurons, which have a small total quantity of ribosomes in comparison with that of larger neurons, are more susceptible to mercury when a certain proportion of the ribosome content is damaged and cannot be readily replaced.[114] Neurons appear to tolerate a certain amount of loss of ribosomes but the critical level may be reached more quickly in small neurons, whereas the larger neurons are less likely to be overwhelmed. Thus, the intracellular lesion may begin in the ribosome and correlate with the ability or inability to initiate repair mechanisms to produce new ribosomal material and proteins. Alternatively, Chang[122] has postulated that the pathogenic mechanism of mercury as a neurotoxin may be two-fold; (1) disruption of cellular metabolism and uptake of nutrient materials which will eventually lead to cellular death, and (2) mercury, being a strong protein denaturing agent may penetrate into the neuronal cytoplasm and cause a direct degradation of cellular constituents.

V. TREATMENT

The protocol of treatment of mercury poisoning is governed by the acute or chronic nature of the exposure. Alkylmercury poisoning is extremely difficult to treat successfully since there is often a latent period of several weeks between exposure and the appearance of initial signs of toxicity. While the reason for this latency is not known, it may be linked to cellular metabolism and the fact that, by the time treatment is initiated, the cellular functions have been irreversibly damaged and treatment would be useless.

In acute poisoning with inorganic, aryl-, or alkoxyalkylmercurials, treatment is directed at the precipitation and removal of the chemical from the gastrointestinal tract, the inactivation of the absorbed mercuric ions, and general supportive measures to maintain fluid and electrolyte balance. Prompt administration of milk and/or egg whites may delay mercuric ion absorption by providing sufficient sulfhydryl groups for the formation of stable mercury complexes. The use of 5% sodium formaldehyde sulfoxylate (rongalite) with a 3% sodium biocarbonate (200 mℓ) is recommended to

7. **Chang, L. W.**, Mercury, in *Experimental and Clinical Neurotoxicology,* Spencer, P. S. and Schaumberg, H. H., Eds., Williams & Wilkins, Baltimore, 1980, chap. 35.
8. **Miller, M. W. and Clarkson, T. W.**, Eds., *Mercury, Mercurials and Mercaptans,* Charles C Thomas, Springfield, Ill., 1973.
9. **Chang, L. W.**, Neurotoxic effects of mercury — a review, *Environ. Res.,* 14, 329, 1977.
10. **Tsubaki, T., Hirota, K., Shirakawa, K., Kondo, K., and Sato, T.**, Clinical, epidemiological and toxicological studies of methylmercury poisoning, in *Proceedings of the First International Congress on Toxicology. Toxicology as a Predictive Science,* Plaa, G. L. and Duncan, W. A. M., Eds., Academic Press, New York, 1978, 339.
11. **Friberg, L. and Vostal J.**, Eds., *Mercury in the Environment,* CRC Press, Boca Raton, Fla., 1972.
12. **D'Itri, F. M.**, *The Environmental Mercury Problem,* CRC Press, Boca Raton, Fla., 1972.
13. **Goldwater, L. M. and Clarkson, T. W.**, Mercury, in *Metabolic Contaminants and Human Health,* Lee, D. H. K., Ed., Academic Press, New York, 1972, chap. 2.
14. **D'Itri, P. A. and D'Itri, F. M.**, *Mercury Contamination: A Human Tragedy,* John Wiley & Sons, New York, 1977.
15. **Smart, N. A.**, Use and residues of mercury compounds in agriculture, *Residue Rev.,* Vol. 23, Gunther, F. A., Ed., Springer-Verlag, New York, 1968, 2.
16. **Goldwater, L. J.**, Aryl- and alkoxyalkylmercurials, in *Mercury, Mercurials and Mercaptans,* Miller, M. W. and Clarkson, T. W., Eds., Charles C Thomas, Springfield, Ill., 1973, chap. 3.
17. **Smart, N. A.**, Retention of organomercury compounds on dressed grain after washing, *Nature (London),* 199, 1206, 1963.
18. **Jalili, M. A. and Abassi, A. H.**, Poisoning by ethyl mercury toluene sulphonalide, *Br. J. Ind. Med.,* 18, 303, 1961.
19. **Clarkson, T. W.**, Biochemical aspects of mercury poisoning, *J. Occup. Med.,* 10, 351, 1968.
20. **Nielsen Kudsk, F.**, Uptake of mercury vapour in blood *in vivo* and *in vitro* from Hg-containing air, *Acta Pharmacol. Toxicol.,* 27, 149, 1969.
21. **Nielsen Kudsk, F.**, Factors influencing the *in vitro* uptake of mercury vapour in blood, *Acta Pharmacol. Toxicol.,* 27, 161, 1969.
22. **Westöö, G.**, Determination of methylmercury compounds in foodstuffs I, *Acta Chem. Scand.,* 20, 2131, 1966.
23. **Westöö, G.**, Determination of methylmercury compounds in foodstuffs II, *Acta Chem. Scand.,* 21, 1790, 1967.
24. **Wood, J. M., Rosen, C. G., and Kennedy, S. F.**, Synthesis of methyl-mercury compounds by extracts of a methanogenic bacterium, *Nature (London),* 220, 173, 1968.
25. **Jensen, S. and Jerneloy, A.**, Biological methylation of mercury in aquatic organisms, *Nature (London),* 223, 753, 1969.
26. **Ridley, W. P., Dizikes, L. J., and Wood, J. M.**, Biomethylation of toxic elements in the environment, *Science,* 197, 329, 1977.
27. **Laner, L.**, Biochemical model for the biological methylation of mercury suggested from methylation studies *in vivo* with Neurospora crassa, *Nature (London),* 230, 452, 1971.
28. **Wollast, R., Billen, G., and MacKenzie, F. T.**, Microbiological transformations of mercury in aquatic environments, in *Ecological Toxicology Research. Effects of Heavy Metal and Organohalogen Compounds,* McIntyre, A. D. and Mills, C. F., Eds., Plenum Press, New York, 1975, 151.
29. **Ukita, T.**, Research on the distribution and accumulation of organomercurials in animal bodies, in *Environmental Toxicology of Pesticides,* Matsumura, F., Boush, G. M., and Misato, T., Eds., Academic Press, New York, 1972, 132.
30. **D'Itri, F. M.**, *The Environmental Mercury Problem,* CRC Press, Boca Raton, Fla., 1972, 63.
31. **Gage, J. C.**, The metabolism of methoxyethylmercury and phenylmercury in the rat, in *Mercury, Mercurials and Mercaptors,* Miller, M. W. and Clarkson, T. W., Eds., Charles C Thomas, Springfield, Ill., 1973, chap. 20.
32. **Gage, J. C.**, Distribution and excretion of methyl and phenyl mercury salts, *Br. J. Ind. Med.,* 21, 197, 1964.
33. **Weiner, I. M., Levy, R. I., and Mudge, G. H.**, Studies on mercurial diuresis: Renal excretion, acid stability and structure-activity relationships of organic mercurials, *J. Pharmacol. Exp. Ther.,* 138, 96, 1962.
34. **Goldwater, L. M.**, Mercury in the environment, *Sci. Am.,* 224, (May) 15, 1971.
35. **Nordberg, G. F. and Skerfving, S.**, Metabolism, in *Mercury in the Environment,* Friberg, L. and Vostal, J., Eds., CRC Press, Boca Raton, Fla., 1972, chap. 4.
36. **Rothstein, A.**, Mercaptors, the biological targets for mercurials, in *Mercury, Mercurials and Mercaptors,* Miller, M. W. and Carrkson, T. W., Eds., Charles C Thomas, Springfield, Ill., 1973, chap. 4.
37. **Ostlund, K.**, Studies on the metabolism of methyl mercury and dimethyl mercury in mice, *Acta Pharmacol. Toxicol.,* 27, (Suppl. 1), 1, 1969.

38. Norseth, T., Biotransformation of methyl mercuric salts in the mouse studied by specific determination of inorganic mercury, *Acta Pharmacol. Toxicol.*, 29, 375, 1971.
39. Daniel, J. W. and Gage, J. C., The metabolism of 2-^{14}C methoxyethylmercuric chloride, *Biochem. J.*, 111, 20, 1969.
40. Daniel, J. W., Gage, J. C., and Lefevre, P. A., The metabolism of methoxyethylmercury salts, *Biochem. J.*, 121, 411, 1971.
41. Miller, V. L., Klavano, P. A., and Csonka, E., Absorption, distribution and excretion of phenylmercuric acetate, *Toxicol. Appl. Pharmacol.*, 2, 344, 1960.
42. Jernelöv, A., Conversion of mercury compounds, in *Chemical Fallout, Current Research on Persistent Pesticides,* Miller, M. and Berg, G., Eds., Charles C Thomas, Springfield, Ill., 1969, 68.
43. Gage, J. C. and Swan, A. A. B., The toxicity of alkyl and aryl mercury salts, *Biochem. Pharmacol.*, 8, 77, 1961.
44. Ellis, R. W. and Fang, S. C., Elimination, tissue accumulation and cellular incorporation of mercury in rats receiving an oral dose of ^{203}Hg-labeled phenylmercuric acetate and mercuric acetate, *Toxicol. Appl. Pharmacol.*, 11, 104, 1967.
45. Clarkson, T. W., Epidemiological and experimental aspects of lead and mercury contamination of food, *Food Cosmet. Toxicol.*, 9, 229, 1971.
46. Cember, H., Callagher, P., and Faulkner, A., Distribution of mercury among blood fractions and serum proteins, *Am. Ind. Hyg. Assoc. J.*, 29, 233, 1968.
47. Miettinen, J. K., The accumulation and excretion of heavy metals in organisms, in *Ecological Toxicology Research. Effects of Heavy Metal and Organohalogen Compounds,* McIntyre, A. D. and Mills, C. F., Eds., Plenum Press, New York, 1975, 215.
48. Rahola, T., Hattula, T., Korolainen, A., and Miettinen, J. K., Elimination of free and protein-bound ionic mercury (^{203}Hg^{2+}) in man, *Ann. Clin. Res.*, 5, 214, 1973.
49. Fitzhugh, O. G., Lang, E. P., Nelson, A. A., and Kunze, F. M., Chronic oral toxicities of mercuriphenyl and mercuric salts, *Arch. Ind. Hyg. Occup. Med.*, 2, 433, 1950.
50. Berglund, R. and Berlin, M., in *Chemical Fallout, Current Research on Persistent Pesticides,* Miller, M. and Berg, G., Eds., Charles C Thomas, Springfield, Ill., 1969, 258.
51. Al-Shahristani, H. and Shihab, K. M., Variation of biological half-life of methylmercury in man, *Arch. Environ. Health,* 28, 342, 1974.
52. Aberg, B., Ekman, L., Falk, R., Greitz, U., Persson, G., and Snihs, J. O., Metabolism of methyl mercury (^{203}Hg) compounds in man. Excretion and distribution, *Arch. Environ. Health,* 19, 478, 1969.
53. Miettinen, J. K., Kahola, T., Hattula, T., Rissanen, K., and Tillander, M., Elimination of ^{203}Hg-methyl mercury in man, *Ann. Clin. Res.*, 3, 116, 1971.
54. Prickett, C. S., Laug, E. P., and Kunze, F. M., Distribution of mercury in rats following oral and intravenous administration of mercuric acetate and phenylmercuric acetate, *Proc. Soc. Exp. Biol. Med.*, 73, 585, 1950.
55. Swensson, A., Lundgren, K. D., and Lindstrom, O., Retention of various mercury compounds after subacute administration, *A. M. A. Arch. Ind. Health,* 20, 467, 1959.
56. Skerfving, S., Organic mercury compounds — relation between exposure and effects, in *Mercury in the Environment,* Friberg, L. and Vostal, J., Eds., CRC Press, Boca Raton, Fla., 1972, chap. 8.
57. Almkvist, J., Quecksilberchadigungen, in *Handbuch der Haut- und Geschlechtskrankhecten,* Jadassohn, J., Ed., Springer-Verlag, Berlin, 1928, 18 and 178.
58. Laug, E. P., Vox, E. A., Kunze, F. M., and Umberger, E. J., A study of certain factors governing the penetration of mercury through the skin of the rat and the rabbit, *J. Pharmacol. Exp. Ther.*, 89, 52, 1947.
59. Turk, J. L. and Baker, H., Nephrotic syndrome due to ammoniated mercury, *Br. J. Dermatol.*, 80, 623, 1968.
60. Friberg, L., Skog, E., and Wahlberg, J. E., Resorption of mercuric chloride and methyl mercury dicyandiamide in guinea pigs through normal skin and through skin pre-tested with acetone, alkylarylsulphonate and soap, *Acta Dermatovener.*, 41, 40, 1961.
61. Goldberg, A. A., Shapero, M., and Wilder, E., The penetration of phenylmercuric dinaphthylmethane disulphonate into skin and muscle tissue, *J. Pharm. Pharmacol.*, 2, 89, 1950.
62. Laug, E. P. and Kunze, F. M., The absorption of phenylmercuric acetate from the vaginal tract of the rat, *J. Pharmacol. Exp. Ther.*, 95, 460, 1961.
63. Fang, S. C., Comparative study of the uptake and distribution of methylmercury in female rats by inhalation and oral routes of administration, *Bull. Environ. Contam. Toxicol.*, 24, 65, 1980.
64. D'Itri, F. M., in *The Environmental Mercury Problem,* CRC Press, Boca Raton, Fla., 1972, chap. 8.
65. Gerstner, H. B. and Huff, J. E., Clinical toxicology of mercury, *J. Toxicol. Environ. Health,* 2, 491, 1977.

66. Kurland, L. T., An appraisal of the epidemiology and toxicology of alkylmercury compounds, in *Mercury, Mercurials and Mercaptors,* Miller, M. W. and Clarkson, T. W., Eds., Charles C Thomas, Springfield, Ill., 1973, chap. 2.
67. Hamilton, A. and Hardy, H. L., *Industrial Toxicology,* 3rd ed., Publishing Sciences Group, Acton, Maine, 1974.
68. Zepp, E. A., Thomas, J. A., and Knotts, G. R., The toxic effects of mercury. A survey of the newer clinical insights, *Clin. Pediatr.,* 13, 783, 1974.
69. Charbonneau, S. M., Munro, I. C., Nera, E. A., Willes, R. F., Kuiper-Goodman, T., Iverson, F., Moodie, C. A., Stoltz, D. R., Armstrong, F. A. J., Uthe, J. F., and Grice, H. C., Subacute toxicity of methylmercury in the adult cat, *Toxicol. Appl. Pharmacol.,* 27, 569, 1974.
70. Charbonneau, S. M., Munro, I. C., Nera, E. A., Armstrong, F. A. J., Willes, R. F., Bryce, F., and Nelson, R. F., Chronic toxicity of methylmercury in the adult cat. Interim report, *Toxicology,* 5, 337, 1976.
71. Ikeda, Y., Tobe, M., Kobayashi, K., Suzuki, S., Kawasaki, Y., and Yonemaru, H., Long-term toxicity study of methylmercuric chloride in monkeys, *Toxicology,* 1, 361, 1973.
72. Garman, R. H., Weiss, B., and Evans, H. L., Alkylmercurial encephalopathy in the monkey (*Saimiri sciurens* and *Mamaca arctoides*). A histopathologic and autoradiographic study, *Acta Neuropathol. (Berlin),* 32, 61, 1975.
73. Luschei, E., Mottet, N. K., and Shaw, C. -M., Chronic methylmercury exposure in the monkey (*Macaca mulatta*), *Arch. Environ. Health,* 32, 126, 1977.
74. Evans, H. L., Garman, R. H., and Weiss, B., Methylmercury: exposure duration and regional distribution as determinants of neurotoxicity in nonhuman primates, *Toxicol. Appl. Pharmacol.,* 41, 15, 1977.
75. Verschuuren, H. G., Kroes, R., Den Tonkelar, E. M., Berkvens, J. M., Helleman, P. W., Rauws, A. G., Schuller, P. L., and Van Esch, G. J., Toxicity of methylmercury chloride in rats. I, Short-term study, *Toxicology,* 6, 85, 1976.
76. Verschuuren, H. C., Kroes, R., Den Tonkelar, E. M., Berkvens, J. M., Helleman, P. W., Rauws, A. G., Schuller, P. L., and Van Esch, G. J., Toxicity of methylmercury chloride in rats. III, Long-term study. *Toxicology,* 6, 107, 1976.
77. Carmichael, N., Cavanagh, J. B., and Rodda, R. A., Some effects of methyl mercury salts on the rabbit nervous system, *Acta Neuropathol. (Berlin),* 32, 115, 1975.
78. Kurland, L. T., Faro, S. N., and Siedler, H., Minamata disease. The outbreak of a neurological disorder in Minamara, Japan, and its relationship to the ingestion of seafood contaminated by mercuric compounds, *World Neurol.,* 1, 370, 1960.
79. Edwards, G. N., Two cases of poisoning by mercuric methide, *St. Bart. Hosp. Rep.,* 1, 141, 1865.
80. Hunter, D., Bomford, R. R., and Russell, D., Poisoning by methyl mercury compounds, *Quart. J. Med. N.S.,* 9, 193, 1940.
81. Hill, W. H., A report on two deaths from exposure to the fumes of diethyl mercury, *Can. J. Public Health,* 34, 158, 1943.
82. Lundgren, K. -D. and Swensson, A., Occupational poisoning by alkyl mercury compounds, *J. Ind. Hyg.,* 31, 190, 1949.
83. Engleson, G. and Herner, T., Alkyl mercury poisoning, *Acta Paediatr.,* 41, 289, 1952.
84. Haq, I. U., Agrosan poisoning in man, *Br. Med. J.,* 1, 1579, 1963.
85. Bakir, F., Damluji, S. F., Amin-Zaki, L., Murtadha, M., Khalidi, A., Al-Rawi, N. Y., Tikriti, S., Dhahir, H. I., Clarkson, T. W., Smith, J. C., and Doherty, R. A., Methylmercury poisoning in Iraq, *Science,* 181, 230, 1973.
86. Curley, A., Sedlak, V. A., Girling, E. F., Hawk, R. E., Bartnel, W. F., Pierce, P. E., and Likosky, W. H., Organic mercury identified as the cause of poisoning in humans and hogs, *Science,* 172, 65, 1971.
87. Snyder, R., Congenital mercury poisoning, *New Engl. J. Med.,* 284, 1014, 1971.
88. Pierce, P. E., Thompson, J. F., Likosky, W. H., Nickey, L. N., Bartnel, W. F., and Hinman, A. R., Alkyl mercury poisoning in humans. Report of an outbreak, *J.A.M.A.,* 220, 1439, 1972.
89. D'Itri, P. A. and D'Itri, F. M., in *Mercury Contamination. A Human Tragedy,* J. Wiley & Sons, New York, 1977, chap. 2.
90. Hunter, D., *Diseases of Occupations,* English Universities Press, London, 1955, 280.
91. Ladd, A. C., Goldwater, L. J., and Jacobs, M. B., Absorption and excretion of mercury in man, *Arch. Environ. Health,* 9, 43, 1964.
92. Berlin, M. and Ullberg, S., Accumulation and retention of mercury in the mouse. II. An autoradiographic comparison of phenylmercuric acetate with inorganic mercury, *Arch. Environ. Health,* 6, 602, 1963.
93. Berlin, M. and Ullberg, S., Accumulation and retention of mercury in the mouse. III. An autoradiographic comparison of methylmercuric dicyandiamide with inorganic mercury, *Arch. Environ. Health,* 6, 610, 1963.

94. Null, D. H., Gartside, P. S., and Wei, E., Methylmercury accumulation in brains of pregnant, non-pregnant and fetal rats, *Life Sci.*, 12, 65, 1973.
95. Mansour, M. M., Dyer, N. C., Hoffman, L. H., Schulert, A. R., and Brill, A. B., Maternal-fetal transfer of organic and inorganic mercury via placental and milk, *Environ. Res.*, 6, 479, 1973.
96. Chang, L. W., Reuhl, K. R., and Lee, G. W., Degenerative changes in the developing nervous system as a result of *in utero* exposure to methylmercury, *Environ. Res.*, 14, 414, 1977.
97. Kelman, B. J. and Sasser, L. B., Methylmercury movements across the perfused guinea pig placenta in late gestation, *Toxicol. Appl. Pharmacol.*, 39, 119, 1977.
98. Harris, S. B., Wilson, J. B., and Printz, R. H., Embryotoxicity of methyl mercuric chloride in golden hamsters, *Teratology*, 6, 139, 1972.
99. Khera, K. S., Teratogenic effects of methylmercury in the cat: note on the use of this species as a model for teratogenic studies, *Teratology*, 8, 293, 1973.
100. Reynolds, W. A. and Pitkin, R. M., Transplacental passage of methylmercury and its uptake by primate fetal tissues, *Proc. Soc. Exp. Biol. Med.*, 148, 523, 1975.
101. Willes, R. F., Truelove, J. F., and Nera, E. A., Neurotoxic response of infant monkeys to methylmercury, *Toxicology*, 9, 125, 1978.
102. Narada, Y., in *Minamata Disease, Study Group of Kumamoto University,* Kutsuna, S., Ed., Shuhan Press, Kumamoto, Japan, 1968, 93.
103. Matsumoto, H., Koya, G., and Takeuchi, T., Fetal Minemata disease. A study of two cases of intrauterine intoxication by a methyl mercury compound, *J. Neuropathol. Exp. Neurol.*, 24, 563, 1965.
104. Moriyama, H., A study on the congenital Minamata disease, *J. Kumamoto Med. Soc.*, 41, 506, 1967.
105. Tatetsu, S. and Harada, M., Mental deficiency resulting from mercury intoxication in the prenatal period, *Adv. Neurol. Sci. (Tokyo),* 12, 181, 1968.
106. Takeuchi, T., in *Minamata Disease, Study Group of Mumamoto University,* Katsuna, S., Ed., Shuhan Press, Kumamoto, Japan, 1968, 141.
107. Bakulina, A. V., The effect of a subacute ethylmercury coated grain poisoning on the progeny, *Soviet Med.*, 31, 60, 1968.
108. Amin-Zaki, L., Elhassani, S., Majeed, M. A., Clarkson, T. W., Doherty, R. A., and Greenwood, M., Intra-uterine methylmercury poisoning in Iraq, *Pediatrics*, 54, 587, 1974.
109. Amin-Zaki, L., Elhassani, S., Majeed, M. A., Clarkson, T. W., Doherty, R. A., and Greenwood, M. R., Studies of infants postnatally exposed to methylmercury, *J. Pediatr.*, 85, 81, 1974.
110. Amin-Zaki, L., Majeed, M. A., Clarkson, T. W., and Greenwood, M. R., Methylmercury poisoning in Iraqi children: clinical observations over two years, *Br. Med. J.*, 1, 613, 1978.
111. Amin-Zaki, L., Majeed, M. A., Elhassani, S. B., Clarkson, T. W., Greenwood, M. R., and Doherty, R. A., Prenatal methylmercury poisoning, *Am. J. Dis. Child.*, 133, 172, 1979.
112. Koos, B. J. and Longo, L. D., Mercury toxicity in the pregnant woman, fetus and newborn infant, *Am. J. Obstet. Gynecol.*, 126, 390, 1976.
113. Magos, L. and Butler, W. H., Cumulative effect of methyl mercury dicyandiamide given orally to rats, *Food Cosmet. Toxicol.*, 10, 513, 1972.
114. Cavanagh, J. B., Metabolic mechanisms of neurotoxicity caused by mercury, in *Neurotoxicology,* Vol. 1, Roizin, L., Shiraki, H., and Grčević, N., Eds., Raven Press, New York, 1977, 283.
115. Okabe, M. and Takeuchi, T., Distribution and fate of mercury in tissues of human organs in Minamata disease, *Neurotoxicology,* 1, 607, 1980.
116. Berlin, M., Jerksell, L. G., and Von Ubisch, H., Uptake and retention of mercury in the mouse brain, *Arch. Environ. Health,* 12, 33, 1966.
117. Chang, L. W. and Hartmann, H. A., Ultrastructural studies of the nervous system after mercury intoxication. I. Pathological changes in the nerve cell bodies, *Acta Neuropathol.*, 20, 122, 1972.
118. Chang, L. W. and Hartmann, H. A., Ultrastructural studies of the nervous system after mercury intoxication. II. Pathological changes in the nerve fibers, *Acta Neuropathol. (Berlin),* 20, 316, 1972.
119. Hunter, D. and Russell, D. S., Focal cerebral and cerebellar atrophy in a human subject due to organic mercury compounds, *J. Neurol. Neurosurg. Psychiatr.*, 17, 235, 1954.
120. Shiraki, H. and Takeuchi, T., Minamata disease, in *Pathology of the Nervous System,* Vol. II, Minckler, J., Ed., McGraw-Hill, New York, 1971, 1651.
121. Shiraki, H. and Nagashima, K., Essential neuropathology of alkylmercury intoxications in humans from the acute to the chronic stage with special reference to experimental whole body autoradiographic study using labeled mercury compounds, in *Neurotoxicology,* Vol. I, Roizin, L., Shiraki, H., and Grčević, N., Eds., Raven Press, New York, 1977, 247.
122. Chang, L. W., Neurotoxic effects of mercury — a review, *Environ. Res.*, 14, 329, 1977.
123. Takeuchi, T., Neuropathology of Minamata disease in Kumamoto: especially at the chronic stage, in *Neurotoxicology,* Vol. I, Roizin, L., Shiraki, H., and Grčević, N., Eds., Raven Press, New York, 1977, 235.

124. **Brown, W. J. and Yoshida, N.**, Organic mercury encephalopathy: an experimental electron microscopy study, *Adv. Neurol. Sci. (Tokyo)*, 9, 34, 1965.
125. **Spyker, J. M. and Chang, L. W.**, Delayed effects of prenatal exposure to methyl mercury — brain ultrastructure and behavior, *Teratology*, 9, A-37, 1974.
126. **Miyakawa, T. and Deshimaru, M.**, Electron microscopic study of experimentally induced poisoning due to organic mercury compound, *Acta Neuropathol. (Berlin)*, 14, 126, 1969.
127. **Sato, T. and Ikuta, F.**, Neuropathology of methylmercury intoxication in Niigata and chronic effects in monkeys, in *Neurotoxicology*, Vol. I, Roizin, L., Shiraki, H., and Grčević, N., Eds., Raven Press, New York, 1977, 261.
128. **Matsumoto, H., Koya, G., and Takeuchi, T.**, Fetal Minamata disease. A neuropathological study of two cases of intrauterine intoxication by a methyl mercury compound, *J. Neuropathol. Exp. Neurol.*, 24, 563, 1964.
129. **Somjen, G. G., Herman, S. P., Klein, R., Brubaker, P. E., Briner, W. H., Goodrich, J. K., Krigman, M. R., and Maseman, J. K.**, The uptake of methylmercury (^{203}Hg) in different tissues related to its neurotoxic effects, *J. Pharmacol. Exp. Ther.*, 187, 602, 1973.
130. **Norseth, T.**, The intracellular distribution of mercury in rat liver after methoxyethyl mercury intoxication, *Biochem. Pharmacol.*, 16, 1645, 1967.
131. **Massey, T. H. and Fang, S. C.**, A comparative study of the subcellular binding of phenylmercuric acetate and mercuric acetate in rat liver and kidney slices, *Toxicol. Appl. Pharmacol.*, 12, 7, 1968.
132. **Verity, M. A. and Brown, W. J.**, Hg^{++}-induced kidney necrosis. Subcellular localization and structure-linked lysosomal enzyme changes, *Am. J. Pathol.*, 61, 57, 1970.
133. **Ringoir, S.**, LDH isoenzyme pattern of rat kidney in mercurial intoxication, *Nephron*, 7, 538, 1970.
134. **Lauwerys, R. and Buchet, J.**, Occupational exposure to mercury vapors and biological action, *Arch. Environ. Health*, 27, 65, 1973.
135. **Chang, L. W., Desnoyers, P., and Hartmann, H.**, Quantitative cytochemical studies of RNA in experimental mercury poisoning I. Changes in RNA content, *J. Neuropathol. Exp. Neurol.*, 31, 489, 1972.
136. **Chang, L. W., Martin, A., and Hartmann, H.**, Quantitative autoradiographic study of the RNA synthesis in the neurons after mercury intoxication, *Exp. Neurol.*, 37, 62, 1972.
137. **Cavanagh, J. B. and Chen, F. C. -K.**, Amino acid incorporation into protein during the "silent phase" before organo-mercury and p-bromophenylacetylurea neuropathy in the rat, *Acta Neuropath., (Berlin)*, 19, 216, 1971.
138. **Yoshino, Y., Mozai, T., and Nakao, K.**, Biochemical changes in the brain in rats poisoned with an alkylmercury compound, with special reference to the inhibition of protein synthesis in brain cortical slices, *J. Neurochem.*, 13, 1223, 1966.
139. **Jacobs, J. M., Carmichael, N., and Cavanagh, J. B.**, Ultrastructural changes in the dorsal root and trigeminal ganglia of rats poisoned with methyl mercury, *Neuropathol. Appl. Neurobiol.*, 1, 1, 1975.
140. **McCord, C. P.**, Mercury poisoning in dentists, *Ind. Med. Surg.*, 30, 554, 1961.
141. **Arena, J. M.**, Treatment of mercury poisoning, *Mod. Treat.*, 8, 619, 1971.
142. **Arena, J. M.**, in *Poisoning, Toxicology, Symptoms, Treatments*, 3rd Ed., Charles C Thomas, Springfield, Ill., 1974, 126.
143. **Skerfving, S. B. and Copplestone, J. F.**, Poisoning caused by the consumption of organomercury-dressed seed in Iraq, *Bull. W.H.O.*, 54, 101, 1976.
144. **Berlin, M. and Ullberg, S.**, Increased uptake of mercury in mouse brain caused by 2,3-dimercaptopropanol, *Nature (London)*, 197, 84, 1963.
145. **Longcope, W. T. and Luetscher, J. A., Jr.**, The use of BAL (British Anti-Lewisite) in the treatment of injurious effects of arsenic, mercury and other metallic poisons, *Ann. Intern. Med.*, 31, 545, 1959.
146. **Hirschman, S. Z., Feingold, M., and Boylen, G.**, Mercury in house paint as a cause of acrodynia, *N. Engl. J. Med.*, 269, 889, 1963.
147. **Clarkson, T. W., Small, H., and Norseth, T.**, Excretion and absorption of methyl mercury after polythiol resin treatment, *Arcb. Environ. Health*, 26, 173, 1973.
148. **Matthes, F. T., Kirchner, R., Yow, M., and Brennan, J. C.**, Acute poisoning associated with inhalation of mercury vapor, *Pediatrics*, 22, 675, 1958.
149. **Locket, S.**, Haemodialysis in the treatment of acute poisoning, *Proc. Roy. Soc. Med.*, 63, 427, 1970.
150. **Kostyniak, P. J., Clarkson, T. W., Cestero, R. V., Freeman, R. B., and Abassi, A. H.**, An extracorporeal complexing hemodialysis system for the treatment of methylmercury poisoning I. In vitro studies of the effects of four complexing agents on the distribution and dialyzability of methylmercury in human blood, *J. Pharmacol. Exp. Ther.*, 192, 260, 1975.
151. **Takeuchi, T., Eto, N., and Eto, K.**, Neuropathology of childhood cases of methylmercury poisoning (Minamata Disease) with prolonged symptoms, with particular reference to the decortication syndrome, *Neurotoxicology*, 1, 1, 1979.
152. **Reuhl, K. R. and Chang, L. W.**, Effects of methylmercury on the development of the nervous system: A review, *Neurotoxicology*, 1, 21, 1979.

153. **Shirabe, T., Eto, K., and Takeuchi, T.**, Identification of mercury in the brain of Minamata disease victims by electron microscopic x-ray microanalysis, *Neurotoxicology*, 1, 349, 1979.
154. **Okabe, M. and Takeuchi, T.**, Distribution and fate of mercury in tissues of human organs in Minamata disease, *Neurotoxicology*, 1, 607, 1980.
155. **Shaw, C. -M., Mottet, N. K., Luschei, E. S., and Finocchio, D. V.**, Cerebrovascular lesions in experimental methylmercury encephalopathy, *Neurotoxicology*, 1, 57, 1979.

INDEX

A

Abnormalities, 141
  electroencephalographic, 118
  in gait, 138
  motor, 119
  sensory, 119
Absorption
  dermal, 240
  from intestinal tract, 239
Accumulation, see also Bioaccumulation, 23
  of CHIs, 91
Acethion, 162
Acetylated amines, 212
Acetylcholine, 73, 74—76, 114, 131, 164, 214
Acetylcholinesterase (AChE), 53, 74, 179, 215, 218, 221, 222, 224
  carbamate, 215
  erythrocyte, 189, 217
  half-life for decarbamoylation of, 215
  inhibition of, 171, 226
N-Acetyl-D,L-penicillamine, 254
AChE, see Acetylcholinesterase
Acquisition phase of conditioned avoidance task, 129
ACTH, 73
Action
  cellular mechanism of, 253
  CNS, 103
  DDT analogues, 101
  ionic mercury, 253
  lindane, 131
  mechanisms of, see Mechanisms of action
  organophosphorus esters, 176—182
  potential for, 67—70
  presynaptic site of, 130
  sites of, 101—106, 123—125, 130
Activation of receptors, 73
Acute alkylmercury intoxication, 241
Acute exposure, 121—122, 132—134
  to DDT, 99—101, 116—118
  to DDT analogues, 99—100, 116—118
Acute hepatic porphyria, 123
Acute lethal affects of CHIs, 98
Acute mercurialism, 240—241
Acute oral LD_{50} values, 216
Acute poisoning, 165
  agents of, 132
  alkoxyalkylmercurials, 253
  arylmercurials, 253
  endrin, 134
  hearing loss and, 136
  inorganic mercurials, 253
  mercurial intoxication, 240
  symptoms of, 132
  therapy for, 192
Acute toxicity, 164—167, 219—222
  defined, 10
  in animals, 216—218
  of CHIs, 92
  of DDT, 101
Acylamide amidohydrolase, see Carboxyamidases
Adipose tissue, 39
Adrenal glands, 138, 139
Adulterated fat substitutes, 173
Adulterated oils, 173
Aechmophorus occidentalis, see Western grebes
Aerial forest spraying, 19
A-esterases, see Arylesterases
Afferent, see Sensory
Afterdischarge of synapse, 124
Afterpotential, 102, 110
Aging
  mechanisms of, 187
  neurotoxic esterase and, 190, 191
Agrosan GN, 245
Aircraft pilots, 170
Akinesia, 82
Aldicarb, 8, 205, 210, 211, 214
Aldrin, 4, 91, 96, 119, 122, 132—137
  acute toxicity of, 92
  as nonepoxide compound, 126
  cholinergic action of, 123
  exposure to in utero, 129
  hyperkinesia in children exposed to, 136
  mental retardation in children exposed to, 136
Aldrin-6,7-transdiol (ATD), 131
  peripheral nerve and, 124
Aliphatic acid herbicides, 27
Aliphatic esters, 205
Aliphatic oximes, 205
Alkoxyalkylmercurials, 235, 242—243, 249
  acute poisoning with, 253
  biotransformation of, 238
Alkyl esters, 206
Alkylmercurials, 235, 240, 244
  biotransformation of, 238
  fetal exposure to, 248
  intoxication from, 241
  milk and, 248
  poisoning from, 248
  target organ of, 243
Alkyl phosphate esters, 177
Alkyl sulfates, 213
Alkyltransferases, 160
Allergy, 119
Alpha chymotrypsin, 187
Alpha naphthol, see Aryl
Alpha-receptors, 76
Altered motor function, 115
Amide, 27
Amino acids, 73
  conjugates of, 212
Aminocarb, 8, 15
Ammonia, 114, 131
Ampere, 64
Amphibian myelinated nerve, 110
Analogues of DDT, see DDT analogues
Anatomy, 59—63

Anesthetic response, 217
Angiotensin II, 73
Animal models, 37
Animals, see also specific animals
   acute hepatic porphyria in, 123
   acute hepatic porphyria induced by lindane in, 123
   acute lethal affects of CHIs in, 98
   acute toxicity in, 216—218
   acute toxicity of organophorus in, 164
   alkymercurial poisoning in, 248
   behavioral studies of exposure to organophosphorus esters in, 179
   central nervous system toxicity in, 101
   CHIs and, 98
   chlordecone and, 137
   chronic toxicity in, 218—219
   hepatic porphyria in, 123
   lindane and, 123
   organophosphorus and, 179
   poisoning of, 99—101, 243, 248
   signs and symptoms of mercurial poisoning in, 243
   subacute toxicity in, 218—219
   toxicity in, 101, 137—139, 216—219
Annulospiral formation of muscle spindle, 181
Anterior horn cells, 180
Anticholinesterase, 217
   organophosphorus, 178
Antidotes for oxime, 192
Antifouling agents, 235
Antimony pentachloride catalyst, 137
Aphasia, 89
Apiol, 172
Appearance rate, 27
Application of pesticides, 18
Arc, 126
Area striata, 250
Ariallmercury, 235
Aromatic esterases, see Arylesterases
Arthropod giant axon, 110
Aryl, 8
Arylesterases, 161
Aryl esters, 205, 206
Arylmercurials, 239, 240, 243, 249
   acute poisoning with, 253
Aryl phosphate esters, 174
Aryl sulfates, 213
Ascending tracts, 181
Aspartic acid, 131
Asphyxiation, 164
Astrocytes, 54
Astroglia, 54
Asynergy, 82
ATD, see Aldrin-6,7-transdiol
Atmospheric model, 20
ATPase, 115, 131
   Na^+, K^+-dependent, 114
Atrophy
   cerebellar cortical, 250
   cortical, 250
   distal, 181
   testicular, 138
Atropine, 228
Auditory discrimination task, 129
Auditory sensory systems, 78
Autonomic fibers, 62
   efferent, 82
Autonomic nerves, 84
Autonomic nervous system, 53, 61, 82—87
Aviation lubricant, 173
Avoidance task, 129
Axoaxonal synapse, 58, 72
Axodendritic synapse, 58
Axon, 54—56, 58, 182
   destruction of, 180
   direct affect on, 183
   giant, 110, 113
   unmyelinated, 56
Axonal process, 56
Axon hillock, 56, 71
Axoplasmic transport, 56, 182
Axosomatic synapse, 58
Azinophosmethyl, 160
Azodrin, 162

B

BAL (British Anti-Lewisite), see Dimercaprol
Balance of mass, 20
Banol, 206, 211
Barrier
   blood-brain, 54
   blood-nerve, 181
Basal ganglia, 79, 81—82
Base-catalyzed decomposition, 207
Basic cellular elements, 54
Basic neural unit, 106
Basic structural elements, 54
Baygon, 206, 207, 210, 211, 217, 220, 221, 226
Behavior, 108—110, 129—130
   changes in, 178
   DDT analogues and, 106
   mating, 129
   organophosphorus esters and, 179
   progeny-caring, 129
   progeny from DDT-treated mothers, 109
Behavioral sequelae, 171
Behavioral toxicology, 178, 227
Benzene hexachloride (BHD), 2, 91, 95
   acute toxicity of, 92
Benzoate, 27
Beta-endorphin, 73
Beta-receptors, 76
BHC, see Benzene hexachloride
γ-BHC, see Lindane
Bidrin, 162, 173
Bioaccumulation, see also Accumulation, 21, 25, 42
   CHIs, 91
   DDT, 35, 37
   defined, 26
   marked, 238

Biochemical analysis, 252
Bioconcentration, 26
Biodegradability, 8
Biological activity, 157
Biological half-life, 140, 189, 206
    of dimethylmercury, 239
    of monomethylmercury, 239
Biological magnification, 26
Biological potency, 154
Biological system pharmacokinetics, 98
Biopsies of sural nerve, 141
Biotransformation, 208—213
    alkoxyalkylmercurials, 238
    alkylmercurials, 238
    carbamate insecticides, 208
    methoxyethylmercuric chloride, 238
    methylmercury dicyandiamide, 238
    organophosphorus esters, 157—163
    oxidative, 158—160
    phenylmercuric acetate, 238
Blood-brain barrier, 54
Blood-nerve barrier, 181
Blood vessels, 54
Body burden, 24
Bombesin, 73
Bonds
    phosphoric anhydride, 152
    P-O-aryl, 183
    thiolophosphoric anhydride, 152
    thionophosphoric anhydride, 152
Bordeaux Mixture, 2
Bounding membrane, 54
Boutons terminaux in spinal gray matter, 181
Brain, 250
    Broca's area of, 89
    butyrylcholinesterase in, 182
    damage to, 90, 136
    electrical activity in, 107—108, 125—126
    higher functions of, 87—90
    mammalian, 61
    pathological changes in, 176
    protein in, 189
    white matter in, 182
Brainstem, 79, 81
    DDT and, 105
Breast milk, 248
British Anti-Lewisite (BAL), see Dimercaprol
Broca's area of brain, 89
BuChE, see Butyrylcholinesterases
Buffers, 206
Butyrylcholinesterase (BuChE), 177
    in brain, 182

C

Cage-structured insecticide, 46
Calabar plant (*Physostigma venenosum*), 7, 213
Calcium ions, 74
Carbamate AChE complex, 125, 216
Carbamates, 7—10, 15, 27, 76, 205—233
    biotransformation of, 208
    hydrolysis of, 209—211
    photodegradation of, 207
    poisoning with, 228—229
    treatment of poisoning with, 228—229
Carbanolate, 217
Carbaryl, 8, 205, 206, 209, 210, 217
    manufacture of, 223
    toxicity of, 218
Carbofuran, 8, 205, 210
Carbon 14 labeling, 218
Carboxyamidases, 161, 162
Carboxylesterases, 162
    nonspecific, 161
Carnosine, 73
Caroxylic acid ester hydrolases, see Carboxylesterases
Catabolizing enzymes, 74
Catechol-O-methyltransferase, 76
Cats, 100
    delayed neurotoxicity in, 183
    DFP-treated, 182
    dieldrin and, 126
Cattle "X" disease, 120
Caudal regions of descending tracts, 181
CBOP, 183
Cell body, 54—56
Cells
    anterior horn, 180
    fusiform, 57
    interstitial, 54
    nerve, 224
    neurolemma, 54, 56
    oligodendrocyte, 56
    Purkinje, 57
    pyramidal, 57
    Schwann, see Neurolemma cells
Cellular elements, 54
Cellular mechanism of action, 253
Central compartment, 31
Central excitatory processes, 109
Central nerve cord of cockroach, 124
Central nervous system (CNS), 54, 57, 58, 61, 101—105, 107, 115, 125—130, 132, 171, 181, 249
    cyclodiene derivatives and, 121
    DDT analogues and, 106
    excitability of, 108
    HCH-CYCs and, 123, 124
    hexachlorocylohexane derivatives and, 121
    hyperexcitability of, 129
    increased excitability of, 108
    persistence of effects on, 179
    poisons in, 101
Central nervous tissue, 249
Central sensory-motor reflex status, 126—128
Cerebellar cortical atrophy, 250
Cerebellum, 61, 79, 82, 106—107, 250
    dysfunction of, 107
Cerebral cortex, 87
Cerebral palsy, 248
Cerebrum, 86
Cervical regions of spinal cord, 180

Changes
  behavior, 178
  EEG, 135, 170, 178
  morphological, 180
  neurological, 178
  pathological, 101, 176, 250
  personality, 140
  tissue enzyme, 253
Chemical structure, 184
Chemical synapses, 58
Chemical transection, 182
Chemical warfare agents, 5, 168
  organophosphorus, 165
Chemistry, 93—98
  of mercurial fungicides, 235—240
  of organophosphorus ester insecticides, 151—163
Children, 136
CHIs, see Chlorinated hydrocarbons
Chlordan, 96
Chlordane, 4, 91, 96, 124
  acute toxicity of, 92
Chlordecone, 1, 138
  neurotoxicity of, 137
Chlorfenvinphos, 163
Chlorinated hydrocarbons (CHIs), 1, 8, 91, 93, 116, 120, 136, 137, 173
  accumulation of, 91
  acute lethal effects of, 98
  acute toxicity of, 92
  comparative toxicity of, 98—99
  cyclodiene-type, 4, 103
  environmental persistence of, 92
  exposure to, 92
  hazards in large-scale use of, 91
  hexachlorocyclohexane groups, 103
  large-scale use of, 91
  lethal effects of, 98
  neurological consequences of exposure to, 92
  persistence of, 91, 92
  production of, 15
  toxicity of, 92, 98—99
  translocation of, 91
Chlorobenzylate, 94
  acute toxicity of, 92
Chlorophenolmercury, 235
Cholecystokinin-like peptide, 73
Cholestyramine, 141
Choline acetylase, 74
Cholinergic action of aldrin, 123
Cholinergic neurons, 53
Cholinesterase, 177, 178, 214
Chromatin, 55
Chronic effects of organophosphorus insecticides on humans, 169
Chronic exposure, 123, 134—135
  to DDT, 118, 119
  to DDT analogues, 101, 118
Chronic mercurialism, 242—243
Chronic poisoning, 132
  from mercurial intoxication, 240
  from organochlorine or organophosphate agents, 136
Chronic toxicity, 167—171
  defined, 10
  in animals, 218—219
  of DDT, 118
Chrysanthemum cinerarinefolium, 2
Classification, 93—98
Clear Lake, California, 21
CNS, see Central nervous system
Cockerels
  DDT-treated, 48
  distribution of DDT in tissues of, 35
  White Leghorn, 35
Cockroach central nerve cord, 124
Coho salmon, 35
Collateral inhibition, 130
Colliculi, 61
Communications between cells, 63
Comparative anatomy of nervous system, 59—63
Comparative roles of sympathetic and parasympathetic divisions, 86—87
Comparative toxicity of CHIs, 98—99
Compartmental models, 31
Compound reflex arc, 126
Concentration gradients, 65, 67
Concentrations of DDT in membranes, 113
Conditioned avoidance task, 129
  electrotonic, 70
  ionic, 112
  saltatory, 71
Cones, 78
Congeners, 42
Congenital cerebral palsy, 248
Conjugates of amino acid, 212
Conjugation, 212—213
  Phase II, 163
  Type I, 163
Connectives, 60
Connective tissue elements, 54
Consciousness, 53, 88—89
Contact insecticides, 213
Continued exposure, 92
Control
  of mosquito-borne diseases, 91
  of posture, 81
  supraspinal, 84
Convulsants, 125
Convulsions, 53, 100, 108, 115, 132, 135
  as outstanding feature of toxicity, 122
  causitive role in genesis of, 104
  children and, 136
  DDT analogues and, 106
Cooking oils, 173
Copper arsenite (Paris Green), 2
Copper sulfate, 2
Corpus callosum, 61
Cortex, 107
  atrophy of, 250
  cerebral, 87
  limbic, 86
  motor, 79, 82
Cottonseed oil, 173

Coulomb, 64
Crayfish, 113
Crossed extensor reflex, 80
Current, 64
    peak transient, 110
    steady state, 110
Cutaneous porphyria, 1
Cutaneous sensation, 77—78
Cyclic phosphate structure, 183
Cyclic saligenin phosphates, 185
Cyclization, 183
Cyclodienes, 4, 122—124, 126, 130
    acute toxicity of, 92
    CHIs of, 103
    CNS as locus of action of, 121
    derivatives of, 96—98
    neurotoxicity of, 121—137
    skeletal muscle and, 123
    toxicity of, 92
Cyclodiene-type chlorinated hydrocarbons, 4
CYCs, see Cyclodienes

D

DCPC, 94
DDA, 113
DDD, 21, 25, 48, 94, 103, 113, 119, 137
    acute toxicity of, 92
DDE, 35, 37, 93, 113
DDT, 1, 3, 4, 17, 19—21, 24, 25, 35, 37, 48, 91, 93, 101—103, 106—109, 111, 112, 116, 120, 122, 124, 131, 227
    accumulation of, 35, 37
    acute exposure to, 99, 116—118
    acute toxicity of, 92, 101
    adipose tissue and, 39
    behavior of progeny from mothers treated with, 109
    brainstem and, 105
    chronic exposure to, 118, 119
    chronic toxicity of, 118
    cockerels treated with, 48
    development of, 4
    dietary, 39
    dogs and, 35
    dosage-response to, 116
    exposure to, 109, 116—118, 119
    Green Bay, 25
    ingestion of by volunteers, 117
    membrane concentrations of, 113
    molecular basis for action of, 114
    motor function and, 104, 109
    neurotoxocity of, 116—121
    peripheral actions of, 104
    primary target for action of, 113
    seizure thresholds and, 109
    spinal cord and, 105
    symptoms of poisoning by, 106
    tissues and, 34, 35
    toxicity of, 92, 101, 118
DDR analogues, 102, 104, 132
    acute exposure to, 99—101, 116—118
    basic neural unit and, 106
    behavior and, 106
    central effects of, 106
    chronic exposure to, 101, 118
    convulsive activity of, 106
    development of symptoms of, 115—116
    insects and, 99—102
    mechanisms of action of, 110—115
    neurotoxicity of, 99, 116—121
    sites of action of, 101, 102
    structure-activity studies of, 113
DDT-lindane vaporizor, 136
Dealkylation, 159
Dearylation, 160
Decarbamoylation of AChE, 215
Decerebrate rigidity, 81, 136
Decerebration, 105
Decomposition, 207
Deep sensation, 77—78
Degeneration, 181
    in buffers, 206
    in water, 206
    myelin, 176, 180
    spinal cord and, 176
    Wallerian, 181
Delayed lesion, 11
Delayed neurotoxicity, 183
Demeton, see Systox
Demyelination, 177, 180—182
Dendrites, 54, 56, 58
    morphology of, 57
Dendritic spines, 56
Dendrodendritic synapse, 58
Deoxyribonucleic acid, see DNA
Depressive reactions, 169
Dermal absorption, 240
Dermal penetration, 240
Derris, 2
Derris ellipticus, 2
Desbromoleptophos, 189
Descending tracts, 181
Destruction of axons, 180
Desulfuration, 158—159
Desynchronization, 178
Detoxification, 157
    Phase I, see Phase I detoxification
    Phase II, see Phase II detoxification
Development
    of DDT, 4
    of hematopoietic pathologies, 13
    of hepatic pathologies, 13
    of nervous system pathologies, 13
    of symptoms for DDT analogue poisoning, 115—116
    phylogenetic, 59
    temporal, 122
$DF^{32}P$, 189
DFP, 5, 161, 165, 168, 177, 179, 182, 185, 191, 193
    cats treated with, 182
Diaphragm, 192

muscle in, 191
Diazinon, 160
1,1-Dichloro-2,2-*bis*(p-chlorophenyl)ethane, 94
Dichlorodiphenyl acetic acid, see DDA
Dichlorodiphenyldichloroethane, 94
Dichlorodiphenylethane, 93—95
  acute toxicity of, 92
  neurotoxicity of, 99—121
Dichlorodiphenyltrichloroethane, see DDT
4,4-Dichloro-α-methylbenzhydrol, 94
4,4-Dichloro-α-(trichloromethyl)-benzhydrol, 94
2,2-Dichlorovinyl dimethyl phosphate, see Dichlorvos
Dichlorvos, 7, 160
  neuropathies and, 191
Dicofol, 94
  acute toxicity of, 92
Dieldrin, 4, 38, 48, 53, 91, 96, 122—124, 129, 130, 132, 133, 135, 136, 227
  acute toxicity of, 92
  as epoxide compound, 126
  cats and, 126
  dogs and, 35
  dosage-response to, 133
  nonlinear model for distribution of, 38
  technical, 35
Diels-Alder reaction, 4
Diencephalon, 61
Dietary DDT, 39
O,O-Diethyl-2-ethylmercaptoethyl phosphorothioate, 7
Diethylmercury, 243, 244, 250
O,O-Diethyl-O-4-nitrophenyl phospate, see Paraoxon
O,O-Diethyl O-4-nitrophenyl phosphorothioate, see Parathion
Differences in species, see Species differences
Diisopropylfluorophosphate, see DFP
Dimercaprol, 254
2,3-Dimercapto-1-propanol, see Dimercaprol
Dimetan, 8, 213
Dimethoate, 162
4-Dimethylamino-3-cresyl N-methylcarbamate, see Matacil
Dimethylcarbamates, 8, 214
N-Dimethylcarbamates, 205
O,O-Dimethyl-S-carbethoxyethyl phosphorothioate, see Acethion
Dimethyl GSH esters, 160
O,O-Dimethyl 1-hydroxy-2,2,2-trichloroethylphosphonate, see Trichlorfon
Dimethylmercuric sulfide, 250
Dimethylmercury, 237
  biological half-life of, 239
  mice and, 238, 239
O,O-Dimethyl-S-(N-methylcarbamoylmethyl) phosphorodithioate, see Dimethoate
O,O-Dimethyl O-(4-nitrophenyl) phosphorothioate, see Methyl parathion
Dimethyl phosphate esters, 173
Dimethyl-substituted phosphate, 160
O,O-Dimethyl(2,2,2-trichloro-1-hydroxyethyl)phosphate, see Trichlorfon
Dimethylyhsphoro-thioic acid, 161
Dimetilan, 8, 205, 207
Dimite, see DMC
Dinitro, 113
Diphenylhydantoin, 45
Disappearance, 26
  from soils, 27
  rate of, 27
Discharge, repetitive, 102—104, 110
Discrimination tasks, 129
Diseases, see also specific diseases
  Minamata, 248, 251
  mosquito-borne, 91
  Parkinson's, 75
  "virus-X", 120
  "X" in cattle, 120
Disinfectants, 237
Dispersion, 20
Distal atrophy, 181, 224
Distribution
  chlordecone in tissues, 138
  DDT in tissues, 34, 35
  dieldrin, 38
  environmental, 17—27
  lipophilic chemicals in tissues, 33
  methylmercuric chloride, 252
  nonlinear model of dieldrin, 38
  subcellular, 252
Disulfoton, 160, 179
Disulfoton sulfone, 160
Disulfoton sulfoxide, 160
Dithiocarbamates, 8, 205, 224
Division, sympathetic, 83
DMC, 94
  acute toxicity of, 92
DMDT, 95
DNA, 55
Dogs, 35, 100
  DDT in, 35
  dieldrin in, 35
Dopamine, 73, 75, 114, 131
Dosage-response
  to DDT, 116
  to dieldrin, 133
Dose-dependence, 192
Dose-response curve, 23
Drugs, responses to, 53
Dyfonate, see Fonophos
Dying back, see Distal atrophy
Dysfunction of cerebellum, 107

E

EEG, 108, 125—126, 133, 136, 248
  abnormalities in, 118
  changes in, 135, 170, 178
  excitatory epileptiform changes in, 178
  sleep, 179
  waking, 179
Efferent, see Motor

Electrical activity
  brain, 107—108, 125—126
  evoked, 107, 108
  spontaneous, 107, 108
Electrical potential, 66
Electrical synapses, 58
Electric current, 64
  peak transient, 110
  steady state, 110
Electrochemical properties of neurons, 63—72
Electroencephalograph, see EEG
Electronic factors, 155
Electronic properties, 157
Electrotonic conduction, 70
Electrotonic potentials, 69—71
Elemental mercury (Hg_o), 237
Elimination
  environmental, 28
  through excretion, 141
Ellman method, 216
Endogenous compounds, 157
Endoplasmic reticulum, 56
Endosulfan, 97, 122
  acute toxicity of, 92
End-plate, 191
Endrin, 1, 4, 53, 91, 96, 119, 122, 123, 132, 134—137
  acute poisonings by, 134
  acute toxicity of, 92
  as epoxide compound, 126
Environment
  dispersion in, 20
  distribution in, 17—27
  elimination through, 28
  kinetics of, 27—49
  persistence of CHIs in, 92
  pollution of, 23
  stability in, 206
Environmental Protection Agency (EPA), 137
Enzymatic oxidation, 237
Enzymes
  catabolizing, 74
  changes in, 253
  hydrolytic, 161
  oligomycin-sensitive mitochondrial, 139
EO-AX, see Carbamate AChE complex
EPA, see Environmental Protection Agency
Epidemics, 245, 246
  fungicide-related poisonings, 245
  Hopewell, 139
  mercurial poisoning, 243
Epidemiological studies, 169
Epileptiform activity, 129
Epileptiform alterations, 178
Epinephrine, 73
EPN, 173
Epoxide compounds, 126
EPSP, see Excitatory postsynaptic potential
Equilibrium potential, 110
Erethism, 242
Erythrocyte AChE, 189, 217
Eserine, 8

Esterase, 209
  neurotoxic, 189—191, 228
  phosphorylation of, 190
  tissue, 210
Esthetics, 90
Ethereal sulfates, 212
Ethics, 90
Ethyl 4,4-dichlorobenzylate, 94
O-Ethyl O-2,4-dichlorophenyl phenylphosphonothioate, see S-Seven
Ethyl-N,N-dimethyl phosphoramidocyanidate, see Tabun
Ethylmercuric chloride, 1, 237, 246, 248
Ethylmercury salts, 238
Ethylmercury p-toluene sulfonanilide, 237, 246, 247
O-Ethylphenyl phosphates, 185
Evoked electrical activity, 107, 108
Evoked responses, 108
Examination
  of biopsies of sural nerve, 141
  pathological, 176
Excitability
  CNS, 108
  neuronal, 115, 125
Excitatory epileptiform alterations in EEGs, 178
Excitatory postsynaptic potential (EPSP), 71
Excitatory processes, 109
Excretion, 141
Execution of movement, 82
Exocytosis, 74
Exogenous compounds, 157
Exposure
  acute, see Acute exposure
  chronic, see Chronic exposure
  continued, 92
  industrial, 140, 179, 243
  postnatal, 248
  prenatal placental, 248
  to aldrin, 129, 136
  to alkylmercurials, 248
  to CHIs, 92
  to DDT, 99, 109, 116—118
  to DDT analogues, 99—101, 116—118
  to dieldrin, 126
  to lindane, 136
  to mercurials, 243
  to organophosphorus esters, 179
Extensor reflex, 80
Extinction phase of conditioned avoidance task, 129

F

Facilitation, presynaptic, 72
Fast, transient phenomena, 72
Fat substitutes, 173
Fenitrothion, 15, 19, 151, 160, 173, 179
Ferbam, 225
Fetus
  exposure of to alkylmercurials, 248

lesions in, 251
Fibers
  autonomic, 62, 82
  motor, 62, 82
  parasympathetic, 62, 164
  peripheral nerve, 102
  postganglionic parasympathetic, 164
  sensory, 61
  sympathetic, 62
  teased, 182
Fibroblasts, 54
First order kinetics, 27
First order motor neurons, 53
First order sensory neurons, 53
First ventral ganglion, 60
Fluorine-containing alkyl phosphate esters, 177
Fonophos, 160
Food chains and CHIs, 91
Forest spraying, 19
Forgetfulness, 226
Four-compartment model, 46
Freely moving subjects, 107
Frog, 110
Frontal lobe damage, 90
Functional organization of nervous system, 77
Functions of brain, 87—90
Fungicides, see also specific fungicides, 8, 205, 235
  chlorinated hydrocarbon, 1
  mercurial, see Mercurials
  organomercurial, 1
  poisonings from, 245
  production of, 16
  toxicity of, 243—247
Fusiform cells, 57

G

GABA, 73, 75, 114, 131
GABA-mediated inhibition, 130
Gait, 138
Ganglia, 59—62
  basal, 79, 81—82
  first ventral, 60
  subesophageal, 60
Gardona, 173
Gastrocnemius muscles, 192
Gastrointestinal symptoms, 166
Gemmules, 56
Generation
  of membrane potential, 65—67
  repetitive discharge, 103
Gesarol, 4
Gesaron, 4
GHK equation, see Goldman, Hodgkin, Katz equation
Giant axons
  in arthropod species, 110
  in crayfish, 113
Ginger Jake Paralysis, 7, 172, 180
Glomerulosclerosis in kidneys, 138

Glucosides, 212
Glucuronidation, 212
Glucuronides, 212
Glutamate, 77
Glutamic acid, 73, 131
Glutamine, 114, 131
γ-L-Glutamyl-L-cysteinylglycine, see Glutathione
Glutathione, 160
Glutathione transferases, 160—161
Glycine, 73, 77, 114, 131
Goldman, Hodgkin, Katz (GHK) equation, 66
Golgi apparatus, 56
Golgi neurons, 57
Golgi tendon organs, 79
Gradients of concentration, 65, 67
Grain, 245
Granosan, see Ethylmercuric chloride
Grebes, 21
Green Bay, 25
GS-CH_3, see S-Methylglutathione
GSH, see Glutathione
Gulls, 25
Gustatory sensory systems, 79
Guthion, 160

H

Half-life
  biological, 140, 189, 206
  of decarbamoylation of AChE, 215
  of dimethylmercury, 239
  of inorganic mercury, 23-
  of mercury, 239
  of monomethylmercury, 239
  of protein-bound mercury, 239
Halogenated biphenyls, 37
Hammett's constants, 155
Hansch model, 156
Hazards in large-scale use of CHIs, 91
HCB, 45
HCH, 2
  acute toxicity of, 92
  CHIs of, 103
  CNS as locus of action of, 121
  derivatives of, 95, 121—137
  neurotoxicity of derivatives of, 121—137
  toxicity of, 92
γ-HCH, 2
HCH-CYCs, 122, 125, 126, 130—132, 135
  changes produced by, 123
  CNS as primary locus action of, 123
  derivatives of, 132
  peripheral actions of, 123
  species differences in susceptibility to, 121
  spontaneous activity of within insect CNS, 124
Hearing loss from organochlorines or organophosphates, 136
Hematopoietic pathologies, 136
Hemicholinium, 191
Hemodialysis, 255
Hens

delayed neurotoxicity in, 183
neuropathies from trichlorfon in, 191
HEOD, 96, 124
Hepatomas, 138
Hepatomegaly, 138
Heptachlor, 4, 35, 97
    acute toxicity of, 92
    as nonepoxide compound, 126
Heptachlor epoxide, 4
1,4,5,6,7,8,8-Heptachloro-3a,4,7,7a-tetrahydro-4,7-methanoindene, 97
Herbicides, see also specific types
    aliphatic acid, 27
    disappearance of from soils, 27
    persistence of, 27
Herring gulls, 25
Heterocyclic enols, 8
Hexachlorobenzene, 1
    in coho salmon and rainbow trout, 35
2,4,4,2′,4′,5′-Hexachlorobiphenyl, see HCB
Hexachlorocyclohexane, see HCH
1,2,3,4,5,6-Hexachlorocyclohexane, 95
γ-1,2,3,4,5,6-Hexachlorocyclohexane, 95
Hexachlorocyclohexane-cyclodienes, see HCH-CYCs
Hexachlorocyclopentadiene, 137
Hg_o, 237
Hillock, 56, 71
Histamine, 73
Histological evidence of skeletal muscle lesions, 191
Histopathological findings, 138
History, 172—174
Hopewell epidemic, 139
Hormones, 73
Horn cells, 180
Human volunteers, 39, 117
Hunter-Russell Syndrome, 243
Hydrocarbon insecticides, see Chlorinated hydrocarbons
Hydrolysis, 161—163, 210
    of carbamates, 209—211
    oxidative, 210
    rates of, 209
Hydrolytic enzymes, 161
Hydrophobic properties, 157
p-Hydroxybenzaldehyde, 185
3-Hydroxy-N,N-dimethylcroton-amide dimethyl phosphate, see Bidrin
Hydroxylation, 211
3-Hydroxy-N-methyl-crotonamide dimethyl phosphate, see Azodrin
Hyperexcitability, 108
    of CNS, 129
Hyperkinesis, 136
Hyperpolarization, 72
Hyperreflexia, 100, 115
Hyperresponsiveness, 126
Hypersensitivity reactions, 119
Hypothalamus, 61, 84, 86
Hypotonus, 82

I

Idiosyncratic responses, 118
Impaired mating performance, 123
Impaired progeny survival, 123
Impaired survival, 123
Inactivation of transmitters, 73
Inclusions, 181
Indirect involvement, 53
Induced myopathy and organophosphorus esters, 191
Industrial exposure, 140, 179
    to mercurials, 243
Ingestion of DDT by volunteers, 117
Inhibition
    acetylcholinesterase, 171
    AChE, 226
    GABA-mediated, 130
    irreversible, 216
    Mg^{2+}-ATPase, 139
    Na^+,K^+-ATPase, 139
    neurotoxic esterase, 190
    postsynaptic, 130
    presynaptic, 72
    protein synthesis, 253
    recurrent collateral, 130
Inhibitory postsynaptic potential (IPSP), 72
Initiation, 68
Inorganic mercurials, 235, 239, 240, 242, 249
    acute poisoning with, 253
    half-life of, 239
    mechanisms of action of, 253
Inorganic mercurial salts vs. organic mercury salts, 252
Insecticides, see also specific types, 53
    cage-structured, 46
    carbamate, see Carbamates
    chlorinated hydrocarbon, see Chlorinated hydrocarbons
    contact, 213
    cyclodiene, see Cyclodienes
    organophosphate, see Organophosphates
    organophosphorus, see Organophosphorus agents
    potency of, 208
Insects, 99—100, 102—103, 121—122
    DDT analogues and, 99—100, 102
    HCH-CYCs and, 124
    nervous system in, 59—61, 124
Integrated central systems, 125—130
Integrative components, 83—86
Intentional paralysis, 108
Intention tremor, 82
Intersegmental reflexes, 80
Interstitial cells, 54
Intestinal polypeptides, 73
Intestinal tract, 239
Intoxication
    acute alkylmercury, 241
    mercurial, 240
    prenatal, 248

Intracellular communications, 63
In-vitro and in-vivo DDT membrane
    concentrations, 113
Ionic conductances, 112
Ionic mercurials, see Inorganic mercurials
Ionic monomethylmercury, 239
IPSP, see Inhibitory postsynaptic potential
Iraq, 246, 247
Irreversible inhibition, 216
Isodrin, 4, 96
    acute toxicity of, 92
Isogam, 2
Isolan, 8, 205, 213
Isolated nerve-muscle preparation, 104
Isolated sciatic nerve, 104
Isopestox, see Mipafox
3-Isopropoxyphenyl N-methylcarbamate, 221, 222

J

Jack-knife reflex, 80
Jake Leg, 172
Japanese quail, 138

K

Kelevan, 137
Kelthane, see Dicofol
Kepone, see Chlordecone
Kepone shakes, 140
Kidneys, 138
Kinetics
    environmental, 27—49
    first order, 27
    of phosphorylation, 186
Knee-jerk reflex, 80

L

Labeling by carbon 14, 218
Lamellated inclusions, 181
Landrin, 217
Lannate, see Methomyl
Laryngeal nerve of DFP-treated cats, 182
LD_{50} values, 216, 249
    oral, see Oral LD_{50} values
Lead arsenate, 2
Learning, 53, 59, 227
    memory and, 89
    of visual discrimination task, 129
Length constant, 70, 71
Leptophos, 174, 188
    metabolites of, 189
Leptophos oxon, 189
Lesions, 251
    delayed, 11
    histological evidence of, 191
    in human fetus, 251

localization of, 251
neurological, 172—182, 224—228, 249—253
psychopathological, 167
skeletal muscle, 191
Lethal effects of CHIs in animals and humans, 98
Leucine-enkephalin, 73
Limbic cortex, 86
Limbic system, 86, 107
Lime sulfur, 2
Lindane, 2, 95, 122, 126, 129, 133—136
    acute hepatic porphyria induced by, 123
    acute toxicity of, 92
    hematopoietic pathologies and, 136
    hepatic pathologies and, 136
    mechanism of action for, 131
    mental retardation and hyperkinesia in children
        exposed to, 136
    nervous system pathologies and, 136
    polyneuritis in workers exposed to, 136
    toxicity of, 92
Lindane-DDT, 136
Lipophilic chemical distribution in tissues, 33
Liver, 138
    chlordecone and, 137
    increased weight of, 123
    lindane and, 136
    pathologies of, 136
    toxicity to, 137
Liver prophyria, 123
Localization of lesions, 251
Lonchocarpus, 2
Long-term toxicity, 11
Losses
    off-target, 19
    of hearing, 136
    of memory, 136
    of weight, 123
Lubricant, 173
Lumbar regions of spinal cord, 180
Luteinizing-hormone releasing hormone, 73
Lysosomes, 56

M

Macaca fascicularis, 227
Macromolecule transport, 182
Macrophages, 54
Magnification, 26
Major targets, 138
Malathion, 1, 7, 25, 160, 162, 173, 179
Malignant tumors, 138
Mammals
    brain of, 61
    nervous system in, 61—63
    toxicity in, 8
Maneb, 208, 225
Manufacture of carbaryl, 223
Marked bioaccumulation, 238
Markets for pesticides, 16
Mass balance, 20
Matacil, 206, 207, 211, 216

Mating, 129
  impaired performance of, 123
Maze test, 129
Mechanisms
  aging, 187
  biotransformation of organophosphorus esters, 157
  inhibition of neurotoxic esterase, 190
  oxidative, 211
  pathogenic, 253
  pharmacological, 180
  Phase I detoxification, 209
  Phase II detoxification, 209
Mechanisms of action, 130—131
  cellular, 253
  of DDT analogues, 110—115
  of inorganic mercurials, 253
  of ionic mercury, 253
  of lindane, 131
  of organophosphorus esters, 176—182
Medulla oblongata, 61
Membrane
  bounding, 54
  DDT concentrations in, 113
  plasma, 54
  specialized, 54
  unit, 54
Membrane potential, generation of, 65—67
Memory, 53, 59
  learning and, 89
  loss of, 136
Mental retardation in children exposed to aldrin and lindane, 136
Mercaptophos, see Systox
Mercapturic acids, 212, 213
Mercurial poisoning, 235, 244
  acute, 240—241
  chronic, 242—243
  epidemics of, 243
  signs and symptoms of in animals and humans, 243
  treatment of, 253—255
Mercurials, 235—261
  chemistry of, 235—240
  grain treated with, 245
  industrial exposure to, 243
  inorganic, see Inorganic mercurials
  intoxication from, 240
  neurological lesions and, 249—253
  organic, 239
  target organs for, 242
  toxicity of, 240—249
Mercuric acetate, 252
Mercuric chloride, 251
  oral LD_{50} of, 249
Mercury
  elemental, 237
  inorganic, 235, 249
  ionic, 253
  organic, 235
  protein-bound, 239
  urinary, 248

Mercury Contamination: A Human Tragedy, 235
Mesencephalon, 61
Mesocephalon, 61
Mesurol, 206, 211, 219
Metabolism, 157
Metabolites, 48
  of leptophos, 189
  TOCP-active, 183
Metencephalon, 61
Methane sulfonyl fluoride, 5
Methionine-enkephalin, 73
Methomyl, 8, 205, 219, 220, 222
Methoxychlor, 25, 95, 103, 113, 122
  acute toxicity of, 92
Methoxy-DDT, 95
Methoxyethylmercuric chloride, 252
  biotransformation of, 238
Methoxyethylmercury, 239
4-Methylamino-3,5-xylyl N-methylcarbamate, 219
Methylation, 237
Methylcarbamates, 205
1-Methylethyl methylphosphonofluoridate, see Sarin
N,N'-bis(1-Methylethyl) phosphorodiamidic fluoride, see Mipafox
bis(1-Methylethyl) phosphorofluoridate, see DFP
S-Methylglutathione, 160
Methylmercuric chloride, 1, 250
  distribution of, 252
Methylmercuric dicyandiamide, 247
Methylmercuric hydroxide, 252
Methylmercuric nitrate, 250
Methylmercuric thiacetamide, 240
Methylmercury, 248, 252
  oral LD_{50} of, 249
  prenatal transplacental exposure to, 248
Methylmercury chloride, 250
Methylmercury dicyandiamide, see Panogen
Methyl paraoxon, 160
Methyl parathion, 40, 151, 160, 170
Mg^{++}-ATPase, 131
  inhibition of, 139
Mice, 138
  dimethylmercury and, 238, 239
Microglia, 54
Microtubules, 56, 182
Mild poisonings, 165
Milk, 248
Minamata disease, 248, 251
Minor oxidative reactions, 160
Mipafox, 168, 173, 175, 177, 181—183, 185
  sciatic nerves and, 181
Mirex, 46, 48
  animal models for, 37
Mitochondrial enzymes, 139
Mobam, 210, 227
Mobilization of PCB, 48
Models
  atmospheric, 20
  compartmental, 31
  dieldrin distribution, 38
  four-compartment, 46

halogenated biphenyls, 37
Hansch, 156
mirex, 37
molecular, 131
nonlinear for dieldrin distribution, 38
one-compartment, 31
three-compartment, 40, 46
toxicokinetic, 31
two-compartment, 31, 40
Moderate poisonings, 166
Molecular basis for DDT action, 114
Molecular modeling, 131
Monkeys, 100
   Rhesus, 179
Monoamines, 73
Monodesmethyl, 160
bis-(Monoisopropylamino)-fluoro-phosphine oxide, see Mipafox
Monomethylcarbamates, 214
N-Monomethylcarbamates, 217
Monomethylmercury, 237
   biological half-life of, 239
   ionic, 239
Monosynaptic stretch reflex, 80
Morphology, 180, 250
   dendritic, 57
Mosquito-borne diseases, 91
Motor autonomic fibers, 62, 83
Motor autonomic nerves, 84
Motor centers, 79
Motor cortex, 79, 82
Motor end-plate, 191
Motor function
   abnormalities in, 119
   altered, 115
   aphasia of, 89
   DDT and, 104, 109
   peripheral actions of DDT and, 104
Motor neurons, 53
Motor reflexes, 80
Motor-sensory reflex, 126—128
Motor systems, 79—83, 104
Movements, 81, 82
Moving subjects, 107
Muscarinic effects, 164, 165
Muscarinic receptor, 74
Muscles
   diaphragmatic, 191
   gastrocnemius, 192
   peripheral, 191
   skeletal, see Skeletal muscles
   soleus, 192
   tonus of, 82
   weakness of, 138
Muscle spindles, 79
   annulospiral formation of, 181
Myasthenia gravis, 219
Myelencephalon, 61
Myelin, 56, 181
   degeneration of, 176, 180
Myelinated nerve, 110
Myelinated neurons, 56

Myopathy, 192
   organophosphorus esters and, 191
   skeletal muscle, 193
Myotonic response in isolated nerve-muscle preparation, 104

N

Nabam, 8, 225
NADPH, 158, 159
NADPH-regenerating system, 158
Na^+,K^+-ATPase, 114, 131
   inhibition of, 139
Neck reflexes, 81
Necrosis of skeletal muscle, 191—192
Negative afterpotential, 102, 110
Neocerebellum, 250
Neocid, 4
Neostigmine, 192, 213
Nernst equation, 66
Nerve axons, 183
Nerve cell vacuolization, 224
Nerve cord, 124
Nerve fibers, 102
Nerve gas, 178
Nerve-muscle preparation, 104
Nerves
   amphibian myelinated, 110
   efferent autonomic, 84
   isolated sciatic, 104
   laryngeal, 182
   peripheral, 53, 61, 104, 176
   sciatic, 110, 181
   sural, 141
   vertebrate peripheral, 104
Nervous system, 54, 138
   autonomic, 53, 61, 82—87
   central, see Central nervous system
   comparative anatomy of, 59—63
   functional organization of, 77—87
   insect, 59—61
   lindane and, 136
   mammal, 61—63
Neural unit, 106
Neurofilaments, 56
Neurolemma cells (Schwann cells), 54, 56
Neurological function
   changes in, 178
   CHI exposure and, 92
   persistent effects on, 170
Neurological lesions, 172—182, 224—228
   mercurials and, 249—253
   organophosphorus esters and, 172
Neurological sequelae persistence, 118—121, 135—137
Neurological symptoms, 106
   severity of, 140
Neuromodulators, 72
Neuromuscular system, 225
Neuronal excitability, 115
   nonspecific increase in, 125

Neurons, 54—57
 cholinergic, 53
 electrochemical properties of, 63—72
 Golgi, 57
 motor, 53
 myelinated, 56
 sensory, 53
Neuropathies, 255
 from dichlorvos, 191
 from trichlorfon, 191
 peripheral, 119
Neuropeptides, 73
Neuropil, 113
Neurotensin, 73
Neurotoxic compounds, 190
Neurotoxic esterase, 189—191, 228
Neurotoxicity, 132, 141, 180, 184
 chlordecone, 137
 cyclodiene derivatives, 121—137
 DDT and its analogues, 116—121
 delayed, 183
 dichlorodiphenylethane derivatives, 99—121
 hexachlorocyclohexane derivatives, 121—137
 organophosphorus esters and, 172
 potential for, 185
 structure and, 190
 structure-activity relationships and, 182—191
Neurotoxic protein, 189
Neurotransmission, 71—77
 phases of, 73
 reuptake, 72
Nicotinamide adenine dinucleotide phosphate, see NADPH
Nicotine, 2
 actions of, 164
Nicotinic receptor, 74
Nitrile, 27
Nodes
 of Ranvier, 56
 single, 110
Nomenclature of organophosphorus ester insecticides, 151—153
Nonenzymatic methylation, 237
Nonlinear model for dieldrin distribution, 38
Nonneurological sequelae, 132
Nonspatial successive discrimination task, 129
Nonspecific carboxylesterases, 161
Nonsynthetic processes, see Phase I detoxification
Norepinephrine, see also Reuptake, 73, 75—76, 114, 131
 synthesis of, 75
Nucleolus, 55
Nucleus, 55

O

1,2,4,5,6,7,8,8a-Octachloro-2,3-3a,4,7,7a-hexahydro-4,7-methano-1H-indene, 96
1,3,4,5,6,7,8,8-Octachloro-3a,4,7-7a-tetrahydro-4,7-methanophthalan, 97
N,N,N,N-Octamethyl phosphorodiamidic anhydride, 152
Octamethylpyrophosphortetramide, see OMPA
Off-target losses, 19
Ohm, defined, 64
Oils, 173
Oil-water (O/W) partition coefficient, 155
Olfactory sensory systems, 79
Oligodendrocyte cell, 56
Oligodendrocytes, 54
Oligodendroglia, 54
Oligomycin-sensitive mitochondrial enzyme, 139
Oligospermia, 140
Olive oil, 173
OMPA, 5
One-compartment model, 31
Operant procedures, 129
Oral LD_{50} values
 acute, 216
 of mercuric chloride, 249
 of methylmercury, 249
Organic mercurials, 239
Organic mercury, 235
Organic mercury salts vs. inorganic mercury salts, 252
Organization, functional, 77—87
Organochlorine agents, 136
Organomercurial fungicides, 1
Organophosphates, 53, 76, 103
 hearing loss and, 136
Organophosphorus agents, 5, 8, 151—203
 behavioral studies in animals exposed to, 179
 biotransformation of, 157
 chemistry of, 151—163
 chronic effects on humans of, 169
 induced myopathy and, 191
 mechanisms of action of, 176—182
 neurological effects of, 170
 neurological lesions and, 172
 neurological lesions and, 172
 neurotoxicity and, 172
 nomenclature of, 151—153
 physiochemical properties of, 154—157
 poisoning and, 192—193
 production of, 16
 psychiatric sequelae, 170
 thioether-containing, 159
 treatment and, 192—193
Oxidation, 211—212
 enzymatic, 237
 side chain, 211
 thioether, 159—160
S-Oxidation, 211
Oxidative biotransformations, 158—160
Oxidative dealkylation, 159
Oxidative dearylation, 160
Oxidative desulfuration, 158—159
 of phosphorothioate, 159
Oxidative hydrolysis, 210
Oxidative reactions, 211
 minor, 160
Oxime antidotes, 192
Oxytocin, 73

P

P2S, 193
    as oxime antidote, 192
PAD, 193
2-PAM, 193
    as oxime antidote, 192
Panogen, 240, 245, 247, 248
    biotransformation of, 238
Paralysis
    Ginger Jake, 172, 180
    intentional, 108
    spastic, 82
Paraoxon, 5, 191, 192
Parasympathetic division, 83
    compared to sympathetic division, 86—87
Parasympathetic fibers, 62
    postganglionic, 164
Parasympathetic system, 86
Parathion, 1, 5, 25, 158—161, 168, 170, 173, 179, 191, 227
Paraxon, 5, 159
Paris Green, see Copper arsenite
Parkinson's disease, 75
Passive tremor, 82
Pathogenic mechanism, 253
Pathological changes, 101, 250
    in brain, 176
Pathological examination of tissues, 176
Pathological findings, 106
PBB, 40
PCB, 40
    congeners of, 42
    mobilization of, 48
    redistribution of, 48
Peak transient current, 110
Penetration
    dermal, 240
    superior, 185
D-Penicillamine, 254
Pentobarbital, 133
Pentylenetetrazol, 130
Pentylenetetrazol-type convulsants, 125
Peptides, 73
Perinatal poisoning, 248—249
Peripheral actions
    of DDT and motor activity, 104
    of HCH-CYCs, 123
Peripheral compartment, 31
Peripheral muscle weakness, 191
Peripheral nerve fibers, 102
Peripheral nerves, 53, 61, 176
    ATD and, 124
    vertebrate, 104
Peripheral neuropathies, 119
Peripheral systems, 104
Perkow reaction, 7
Permanent brain damage, 136
Permeabilities, 64
    potassium, 68
    relative, 67
    selective, 65
    sodium, 68
Persistence
    CHIs, 91, 92
    CNS effects, 179
    environmental, 92
    neurological sequelae, 118—121, 135—137, 170
    soils, 27
Personality changes, 140
Pesticide, defined, 1
Pharmacokinetics
    in biological systems, 98
    parameters of, 27
Pharmacological mechanisms, 180
Phase I detoxification, 157, 158—163
    mechanisms of, 209
Phase II conjugations, 163
Phase II detoxifications, 157, 163
    mechanisms of, 209
    reactions to, 212
Phenoxy, 27
Phenylmercuric acetate, 1, 240, 252
    biotransformation of, 238
Phenylmercuric dinaphthylmethane disulfonate, 240
Phenylmercuric salts, 240
Phenyl methylcarbamates, 213
Phenyl phenylacetate (PPA), 189
Phorate, 160
Phosphate
    dimethyl-substituted, 160
    structure of, 183
Phosphinic acid, 154
Phospho-creosote, 7, 172
Phospholipids, 131
Phosphonic acid, 154
Phosphonothiolothioate, 160
Phosphoric acid, 154
    triaryl esters of, 7
Phosphoric anhydride bonds, 152
Phosphorofluoridates, 185
Phosphorothioate
    esters of, 160
    oxidative desulfuration of, 159
Phosphorylation, 189
    esterase, 190
    kinetics of, 186
Phosphorylphosphatases, 161
Phosphotriesterases, 161
Phosvel, see Leptophos
Photodegradation of carbamates, 207
Phylogenetic development, 59
Physiochemical properties, 185, 206—208
    of organophosphorus ester insecticides, 154—157
Physostigma venenosum, see Calabar plant
Physostigmine, 8, 192
Picloram, 27
Pigmented inclusion bodies, 56
Pilots, 170
Piperonal butoxide, 120
Pirimicarb, 8
Pituitary gland, 61

Placental barrier, 248
Placental transfer, 248
Plasma membrane, 54
Plasma pseudocholinesterase, 189
Plasticizing agent, 173
Poisoning, see also Toxicity
  acute, see Acute poisoning
  alkylmercurial, 248
  by aryl phosphate esters, 174
  carbamate, 228—229
  chronic, see Chronic poisoning
  fungicide-related, 245
  mercurial, see Mercurial poisoning
  mild, 165
  Minamata, 248
  moderate, 166
  organophosphorus esters and, 192—193
  perinatal, 248—249
  severe, 167
  signs and symptoms of, 99—101, 121—123
  treatment of for carbamates, 228—229
Poliesneuropathy, 119
Pollution, 23
Polybrominated biphenyls, 40
Polychlorinated biphenyls, 40
Polyhalogenated biphenyls, 40
Polyneuritis, 136
Polyneuropathies, 118, 119, 136, 141
Polypeptides, 73
Polyvinyl chloride, 173
Pons, 61
Population of world, 91
Porphyria
  acute hepatic, 123
  cutaneous, 1
Postganglionic parasympathetic fibers, 164
Postnatal exposure, 248
Postsynaptic element, 58
Postsynaptic inhibition, 130
Posture, 81
Potassium permeability, 68
Potency
  biological, 154
  insecticidal, 208
Potential
  action, 67—70
  electrical, 66
  electrotonic, 69—71
  equilibrium, 110
  excitatory postsynaptic, 71
  inhibitory postsynaptic, 72
  membrane, 65—67
  neurotoxic, 185
  postsynaptic, 71, 72
  resting, 67
  synaptic, 71—72
PPA, see Phenyl phenylacetate
Pralidoxime chloride as oxime antidote, 192
Predictive agreement, 191
Prenatal intoxication, 248
Prenatal transplacental exposure to
    methylmercury, 248

Presynaptic element, 58
Presynaptic facilitation, 72
Presynaptic inhibition, 72
Presynaptic receptors, 75
Presynaptic site of action, 130
Primary target for DDT action, 113
Primates, 179
Production
  of chlorinated hydrocarbons, 15
  of fungicides, 16
  of organophosphorus esters, 16
Progeny
  behavior of from mothers treated with DDT, 109
  impaired survival of, 123
Progeny-caring behavior, 129
Programming of movement, 81, 82
Prolonged recovery, 135
Promecarb, 217
Propoxur, 8, 210
Prosencephalon, 61
Protein
  brain, 189
  inhibition of synthesis of, 253
  neurotoxic, 189
Protein-bound mercury, 239
Pseudocholinesterase, 182, 217, 224
  plasma, 189
Psychiatric sequelae of organophosphorus esters, 170
Psychiatric symptoms in pilots, 170
Psychological symptoms, 178
Psychopathological effects, 167—171
Psychopathological lesions, 167
Psychopharmacology, 178—180
Purkinje cell, 57
Pyramat, 205
Pyramidal cells, 57
Pyrethrins, 2, 103
Pyrethrum, 2, 120
Pyridine-2-aldoxine methochloride, see 2-PAM
Pyridostigmine, 219
Pyrophosphate esters, 5

Q

Quail, 138

R

Rabbits, 100
  distribution of technical DDT in tissues of, 34
Rainbow trout, 35
Ranvier nodes, 56, 110
Rapid transport system, 56
Rate constants, 215
Rats, 100, 138
Reactions, 237
  conjugation, 212

depressive, 169
detoxification, 212
Diels-Alder, 4
hypersensitivity, 119
minor oxidative, 160
oxidative, 160, 211
Perkow, 7
Phase II detoxification, 212
schizophrenic, 169
toxic, 53
Type II, 163
Reactivation, 47
Receptors, 74, 76, 77
   activation of, 73
   alpha, 76
   beta, 76
   muscarinic, 74
   nicotinic, 74
   presynaptic, 75
   sensory, 102
Recovery, 135
Recurrent collateral inhibition, 130
Redistribution, 47
   of PCB, 48
Reflexes
   arc of, 126
   central sensory-motor, 126—128
   crossed extensor, 80
   intersegmental, 80
   jack-knife, 80
   knee-jerk, 80
   monosynaptic stretch, 80
   motor, 80
   sensory-motor, 126—128
   tonic neck, 81
Relative permeabilities, 67
Relearning of visual discrimination task, 129
Release of transmitter, 73
Repetitive discharge, 102, 104, 110
   generation of, 103
Reproductive system, 138
   toxicity of chlordecone to, 137
Resin, 254
Responses
   anesthetic, 217
   evoked, 108
   idiosyncratic, 118
   myotonic, 104
   species differences in, 100
   startle, 140
   to chemicals, 53
   to drugs, 53
Resting potential, 67
Retardation, 136
Reticulum
   endoplasmic, 56
   formation of, 61
Retina, 78
Reuptake, 72, 76—75
Rhesus monkeys, 179
Rhombencephalon, 61
Rhothane, see DDD
Ribonucleic acid, see RNA

Ribosomes, 253
Rigidity, 81, 136
Rigor, 82
Ring hydroxylation, 211
RNA, 55, 253
Rods, 78
Ronnel, 160
Rotenone, 2

S

Salad oils, 173
Sales of pesticides worldwide, 16
Saligenin cyclic phosphorus esters, 187
Salmon, 35
Saltatory conduction, 71
Sarin, 5, 161, 165, 170, 179
Schizophrenic reactions, 169
Schrader's acyl rule, 154
Schwann cells, see Neurolemma cells
Sciatic nerves
   frog, 110
   isolated, 104
   mipafox and, 181
Second order motor neurons, 53
Second order sensory neurons, 53
Seed disinfectants, 237
Seed dressing, 244
Seizures, 126
   convulsive, 132
Seizure thresholds, 109
Selective permeability, 65
Selective toxicity, 209
Sensation, 77—78
Sensory components of autonomic system, 82—83
Sensory fibers, 61
Sensory-motor reflex, 126—128
Sensory neurons, 53
Sensory receptors, 102
Sensory systems, 77—79, 104
   abnormalities in, 119
   aphasia of, 89
   auditory, 78
   gustatory, 79
   olfactory, 79
   stimulation of, 108
   vestibular, 78
   visual, 78
Serotonin, 73, 75, 114, 131
S-Seven, 185
Severe poisonings, 167
Severity of symptoms, 122, 140
Sevin, 8
Shakes, 140
Side chain oxidation, 211
Signs, see also Symptoms, 164, 174—176
   of acute mercurialism, 240
   of alkylmercurial poisoning, 248
   of mercurial poisoning, 243
   of poisoning, 99, 121—123, 243, 248
   of toxicity, 164

Simple movements, 81
Single nodes of Ranvier, 110
Sites of action, 101—106, 123—125
　of DDT analogues, 101, 102
　presynaptic, 130
Skeletal muscle
　cyclodiene compounds and, 123
　histological evidence of lesions in, 191
　lesions of, 191
　myopathy of, 193
　necrosis of, 191—192
Sleep EEGs, 179
Slimicides, 235
Slow transport, 56
Sodium permeability, 68
Soils
　accumulation of CHIs in, 91
　disappearance of herbicide from, 27
　persistence of herbicides in, 27
Soleus muscles, 192
Soma, see Cell body
Soman, 6, 161, 165, 170
Somatoaxonic synapse, 58
Somatodendritic synapse, 58
Somatosomatic synapse, 58
Somatostatin, 73
Spastic paralysis, 82
Spatial summation, 72
Specialized membrane, 54
Species differences, 209, 239
　in responses, 100
　in susceptibilities to HCH-CYCs, 121
Speech, 89
Spinal cord, 61, 79—81
　cervical regions of, 180
　DDT and, 105
　degeneration of myelin in, 176
　gray matter in, 181
　lumbar regions of, 180
　white matter in, 182
Spinal segmental systems, 115
Spontaneous activity of HCH-CYCs, 124
Spontaneous cyclization, 183
Spontaneous electrical activity, 107, 108
Spraying over forests, 19
Stability in environment, 206
Startle response, 140
Status epilepticus, 100
Steady state current, 110
Steers, 35
Steric constant, 155
Steric limitations, 185
Steric properties, 157
Stimulation
　sensory, 108
　stroboscopic light, 129
Storage, 73
Stretch reflex, 80
Strobane, 98
　acute toxicity of, 92
Stroboscopic light stimulation, 129

Structural elements, 54
Structural properties, 183
Structure-activity relationships, 182—191
Structure-activity studies of DDT analogues, 113
Structures, 151
　chemical, 184
　cyclic phosphate, 183
　neurotoxicity and, 190
Subacute toxicity, 223—224
　in animals, 218—219
Subcellular distribution, 252
Subcortical loci, 107
Substance P, 73
Substituent groups, 154
Subthalamus, 61
Successive discrimination task, 129
Suicide, 220
Sulfone formation, 159
Sulfoxidation, see S-Oxidation
Sulfoxide formation, 159
Sulfur, 2
Sulfur trioxide, 137
Summation, 72
Supraspinal systems, 115
　control of, 84
Sural nerve, 141
Susceptibilities
　synapses, 113
　to HCH-CYCs, 121
　variations in, 99
Sustained movements, 82
Sustained synaptic action, 72
Sympathetic system, 62, 83, 86, 164
　compared to parasympathetic division, 86—87
Sympathoadrenal axis, 86
Symptoms, see also Signs, 101, 117, 118, 122,
　　123, 132, 164, 174—176
　acute mercurialism, 240
　acute poisoning, 132
　aryl phosphate ester poisoning, 174
　chronic mercurialism, 242
　DDT analogue poisoning, 115—116
　DDT poisoning, 106
　mercurial intoxication, 240
　mercurialism, 242, 243
　neurological, 106, 140
　poisoning, 99, 121—123, 132, 243, 248
　psychiatric, 170
　psychological, 178
　severity of, 122
Synapses, 58—59, 130
　axoaxonal, 72
　susceptibility and, 113
Synaptic action, 72
Synaptic afterdischarge, 124
Synaptic potentials, 71—72
Synthesis, 73
　of chlordecone, 137
　of norepinephrine, 75
　of proteins, 253
Synthetic processes, see Phase II detoxification

Systox, 7

T

Tabun, 6, 165
Taft's steric constant, 155
Target organs
    alkylmercurial, 243
    for DDT action, 133
    major, 138
    mercurial, 240, 242
    prime, 138
Taurine, 73, 114, 131
TDE, 94
Teased fiber preparation, 182
Technical DDT, 34
Technical dieldrin, 35
Tecoram, 224—225
Telecephalon, 61
Telodrin, 97, 98, 122, 123, 126, 132, 134, 136
    acute toxicity of, 92
Temik, see Aldicarb
Temporal development, 122
Temporal summation, 72
TEPP, 5, 152
Teratogenicity, 218
Testicular atrophy, 138
Tetrachlorodiphenylethane, 94
O,O,O,O-Tetraethyl phosphoric anhydride, see TEPP
Tetraethylpyrophosphate, see TEPP
Tetramethylthiuram disulfide, see Thiram
Thalamus, 61
Therapy, see Treatment
Thiocarbamates, 205
Thiodan, 97, 135
Thioether-containing organophosphorus, 159
Thioether oxidation, 159—160
Thioloisomers, 7
Thiolophosphoric anhydride bonds, 152
Thiol resin, 254
Thionoisomers, 7
Thionophosphoric anhydride bonds, 152
Thionophosphorus esters, 5, 7
Thiophosphorus esters, 5, 7
Thiram, 8, 225
Thorns, 56
Three-compartment model, 40, 46
Thyrotropin releasing hormone, 73
Tissues
    adipose, 39
    central nervous, 249
    chlordecone in, 138
    connective, 54
    DDT in, 34, 35
    enzyme changes in, 253
    esterases in, 210
    lipophilic chemicals in, 33
    pathological examination of, 176
    technical DDT in, 34
TOCP, 7, 172—174, 176, 177, 182—185, 188, 193
TOCP-active metabolite, 183
Toluidine, 27
Tonic neck reflexes, 81
Tonus of muscle, 82
Toxaphene, 98, 122
    acute toxicity of, 92
Toxicity, see also Poisoning, 139—140, 213—224, 243—247
    acute, see Acute toxicity
    carbaryl, 218
    CHIs, 98—99
    chlordecone, 137
    chronic, see Chronic toxicity
    convulsions as outstanding feature of, 122
    DDT, 101
    long-term, 11
    mammalian, 8
    mercurial fungicides, 240—249
    reproductive, 137
    selective, 209
    signs of, 164
    subacute, 218—219, 223—224
    to animals, 138—139
Toxic manifestations, 132
Toxicokinetics, 31
    parameters of, 27
Toxicology, 10—11
    behavioral, 178, 227
Toxic reactions, 53
Transection, 182
Transient current, 110
Transient phenomena, 72
Translocation of CHIs, 91
Transmitter inactivation, 73
Transmitter release, 73
Transplacental exposure, 248
Transport
    axoplasmic, 56, 182
    essential macromolecules, 182
    microtubular, 182
    rapid system of, 56
    slow, 56
Treatment, 132, 141
    of acute poisoning, 192
    of carbamate poisoning, 228—229
    of mercurial poisoning, 253—255
    organophosphorus esters and, 192—193
Tremor, 100, 115, 138
    causitive role in genesis of, 104
    intention, 82
    passive, 82
Triaryl esters of phosphoric acid, 7
Triaryl phosphates, 185
Triazine, 27
Trichlorfon, 7, 163, 174
    neuropathies and, 191
1,1,1-Trichloro-2,2-*bis*(p-chlorophenyl)ethane, 93
1,1,1-Trichloro-2,2-*bis*(p-methoxyphenyl)ethane, 95
Tricresyl phosphates, 7
1,2,2-Trimethylpropyl methylphosphonofluoridate, see Soman

Tri-orthocresyl phosphate, see TOCP
Trout, 35
Tumors, 138
Two-compartment model, 31, 40
Type I conjugations, 163
Type II reactions, 163

U

UDPGA, 212
Ultrastructural damage to ribosomes, 253
Unhydrolyzed acetylcholine, 164
Unicellular measurements, 63
Unit membrane, 54
Unmyelinated axons, 56
Urea, 27
Uridine diphospho-α-D-glucuronic acid, see UDPGA
Urinary mercury, 248
Uspulum, 235
Uterus, 129

V

Vacuolization of nerve cells, 224
Vaporizor, 136
Variations in susceptibility, 99
Vasoactive intestinal polypeptide, 73
Vasopressin, 73
Velsicol VCS-506, see Leptophos
Vertebrates, 104—106, 122
    DDT analogues and, 100—101
    peripheral nerve of, 104
Vesicles, 74
Vestibular sensory systems, 78
Virux X disease, 120
Visceral sensation, 77—78

Visual difficulty, 140
Visual discrimination task, 129
Visual nonspatial successive discrimination task, 129
Visual sensory systems, 78
Volatilization, 19
Volt, defined, 64
Voltage clamp analysis, 110
Volunteers, 39, 117

W

Waking EEGs, 179
Wallerian degeneration, 181
Water degeneration, 206
Weakness of muscle, 138, 191
Weight loss, 123
Wernicke's area, 89
Western grebes (*Aechmophorus occidentalis*), 21
West Pakistan, 245
White Leghorn cockerels, 35
White matter, 182
Whole-blood cholinesterase activities, 178
World pesticide markets, 16
World population, 91

X

"X" disease, 120
Xenobiotic compounds, see Exogenous compounds

Z

Zectran, 8, 206, 210, 211, 219
Zineb, 208, 225

DATE DUE

OCT 26 1991			

DEMCO 38-297

NO LONGER THE PROPERTY OF THE UNIVERSITY OF RI LIBRARY